Commentaries in Plant Science

Volume 2

OTHER TITLES OF INTEREST

Books

CHAPMAN, V. J.
Coastal Vegetation, 2nd Edition

FAEGRI, K. and VAN DER PIJL, L.
Principles of Pollination Ecology, 3rd Edition

FAHN, A.
Plant Anatomy, 2nd Edition

GOODWIN, T. W. and MERCER, E. I.
An Introduction to Plant Biochemistry

MAYER, A. M. and POLJAKOFF-MAYBER, A.
Germination of Seeds, 2nd Edition

REINHOLD, L. *et al.*
Progress in Phytochemistry, Volume 5

WAREING, P. F. and PHILLIPS, I. D. J.
The Control of Growth and Differentiation in Plants, 3rd Edition

Journals

Current Advances in Plant Science

Environmental and Experimental Botany

Phytochemistry

Commentaries in Plant Science

Volume 2

Edited by

HARRY SMITH

Department of Botany
The University of Leicester

PERGAMON PRESS

OXFORD · NEW YORK · TORONTO · SYDNEY · PARIS · FRANKFURT

U.K.	Pergamon Press Ltd., Headington Hill Hall, Oxford OX3 0BW, England
U.S.A.	Pergamon Press Inc., Maxwell House, Fairview Park, Elmsford, New York 10523, U.S.A.
CANADA	Pergamon Press Canada Ltd., Suite 104, 150 Consumers Rd., Willowdale, Ontario M2J 1P9, Canada
AUSTRALIA	Pergamon Press (Aust.) Pty. Ltd., P.O. Box 544, Potts Point, N.S.W. 2011, Australia
FRANCE	Pergamon Press SARL, 24 rue des Ecoles, 75240 Paris, Cedex 05, France
FEDERAL REPUBLIC OF GERMANY	Pergamon Press GmbH, 6242 Kronberg-Taunus, Hammerweg 6, Federal Republic of Germany

First edition 1981

British Library Cataloguing in Publication Data

Commentaries in plant science.
Vol. 2
1. Botany
581 QK45.2 80-41007

ISBN 0-08-025898-0

Printed in Great Britain by A. Wheaton & Co. Ltd., Exeter

PREFACE

Plant science is a rapidly developing subject of vital importance both for central, fundamental problems of modern biology and for the continued existence of mankind on this planet. These seventeen Commentaries in Plant Science have been chosen to present topical and provocative "mini-reviews" of some of the more exciting recent developments in pure and applied plant science, in a form which is easily digestible. The authors were encouraged to parade their own viewpoints, to be constructively critical of existing work in their fields, and to be very selective in their coverage of the literature. In eschewing the often uncritical, and always turgid, approach of the fully comprehensive review, we have tried to provide entertaining and stimulating, as well as instructive, reading. We hope that these articles will provoke discussion and experimentation amongst research workers in plant science; aid the university and college teacher in coming to grips with unfamiliar topics; and, above all, assist the hard-pressed student who requires topical information, but is overwhelmed by the mass of primary literature being published.

The articles were published originally in the monthly issues of the current awareness journal *Current Advances in Plant Science* between 1976 and 1980, on the recommendation of a distinguished editorial board. For this compilation the Commentaries have been brought up to date by their authors. The first volume of these Commentaries, published in 1977, was received with enthusiasm and it is planned to produce further volumes of Commentaries, in due course.

CONTENTS

LIST OF CONTRIBUTORS

ALLEN, J.F., Botany School, South Parks Road, Oxford OX1 3RA, U.K.

ASHMORE, M.R., Department of Plant Science, University of Leeds, Leeds LS2 9JT, U.K.

BARNETT, J.R., Plant Science Laboratories, University of Reading, Reading, Berks. RG6 2AS, U.K.

BURKE, M.J., Department of Horticulture, Colorado State University, Fort Collins, Colorado 80521, U.S.A.

CHOLLET, R., University of Nebraska, Laboratory of Agricultural Biochemistry, Lincoln, NE 68583, U.S.A.

DIGBY, J., Department of Biology, University of York, Heslington, York YO1 5DD, U.K.

DREW, M.C., Agricultural Research Council, Letcombe Laboratory, Wantage, OX12 9JT, U.K.

EVANS, A.M., Department of Applied Biology, University of Cambridge, Cambridge, U.K.

FIRN, R.D., Department of Biology, University of York, Heslington, York YO1 5DD, U.K.

GEORGE, M.F., Department of Horticulture, Colorado State University, Fort Collins, Colorado 80521, U.S.A.

GRIDLEY, H.E., Department of Applied Biology, University of Cambridge, Cambridge, U.K.

HART, J.W., Botany Department, University of Aberdeen, Aberdeen AB9 2UD, Scotland.

HIGGINS, T.J.V., CSIRO, Division of Plant Industry, Canberra 2601, Australia.

HOBSON, G.E., Glasshouse Crops Research Institute, Rustington, Littlehampton, West Sussex, BN16 3PU, U.K.

HOPWOOD, D.A., John Innes Institute, Norwich, NR4 7UH, U.K.

HOWLETT, B.J., School of Botany, University of Melbourne, Parkville, Victoria 3052, Australia.

INCOLL, L.D., Department of Plant Science, University of Leeds, Leeds LS2 9JT, U.K.

KNOX, R.B., School of Botany, University of Melbourne, Parkville, Victoria 3052, Australia.

LONG, S.P., Department of Plant Science, University of Leeds, Leeds LS2 9JT, U.K.

PATHAK, M.D., The International Rice Research Inst., Los Banos, Laguna, Philippines.

RATHNAM-SHAGUTURU, Dow Chemical U.S.A., Central Research, New England Laboratory, PO Box 400, Wayland, MA 01778, U.S.A.

RAVEN, J.A., Department of Biological Sciences, University of Dundee, Dundee DD1 4HN, Scotland.

SABNIS, D.D., Botany Department, University of Aberdeen, Aberdeen AB9 2UD, Scotland.

SAXENA, R.C., The International Rice Research Inst., Los Banos, Laguna, Philippines.

SMITH, F.A., Department of Biological Sciences, University of Dundee, Dundee DD1 4HN, Scotland.

SPENCER, D., CSIRO, Division of Plant Industry, Canberra 2601, Australia.

STACE, C.A., Botany Department, The University, Leicester LE1 7RH, U.K.

VASANTHE VITHANAGE, H.I.M., School of Botany, University of Melbourne, Parkville, Victoria 3052, Australia.

THE OCCURRENCE OF DEEP SUPERCOOLING IN COLD HARDY PLANTS[1]

MILON F. GEORGE[2] and MICHAEL J. BURKE[3]

Laboratory of Plant Hardiness, Department of Horticultural Science,
University of Minnesota, St. Paul, Minnesota 55108, U.S.A.

INTRODUCTION

It has long been accepted that cold hardy plants generally survive subfreezing temperatures by tolerating extracellular freezing and associated cellular dehydration and that intracellular freezing of supercooled cellular water is abnormal and does not occur in nature. Scarth and Levitt (33) demonstrated that ice formation occurs first on cell walls outside the protoplasm and vacuole. Ice growth under slow freezing conditions proliferates in intercellular spaces as water diffuses from the cell in response to decreasing tissue temperature. Soluble concentration or melting point depression of the cell sap keeps pace with ambient temperature so the cellular water does not freeze. Earlier microscopic work by Weigand (39) in 1906, conducted in the field at temperatures near –20°C, had shown that in buds and twigs of many woody species ice crystals always were found in the intercellular spaces of the tissue and that the cells were in a "more or less state of collapse". Even as early as 1860 the locus of ice formation had been observed to be intercellular (30). More recent work of Burke *et al.* (3) and Gusta *et al.* (16) using pulsed nuclear magnetic resonance spectroscopy suggests that freezing curves of plants which tolerate extracellular ice are very similar to those of ordinary salt solutions and that the primary difference between hardy and non-hardy varieties is their ability to tolerate cellular dehydration. Olien (24) has demonstrated that mechanical stress from ice formation is a freezing vector as well as cellular dehydration.

Depending on the particular species, cold hardy plant tissues can tolerate extracellular ice formation and cellular contraction to varying limits of low temperature. Siminovitch and Scarth (34) found hardy cabbage epidermis survives only to approximately –10°C while Weiser (40) notes that fully cold acclimated red osier dogwood bark and xylem can survive liquid nitrogen temperatures (–196°C) with no apparent injury. The reader is referred to Levitt's (21) treatise on environmental stress for detailed reviews of various proposed mechanisms dealing with low temperature injury in plants which exhibit extracellular ice formation.

Although the overwhelming evidence of early hardiness research supported the concept of extracellular freezing, a few examples were noted concerning lack of ice formation in plant parts at

[1] These studies were supported in part by Grants to M.J.B. from the National Science Foundation (BMS 74-23137), the Nerken Foundation and Minnesota Agricultural Experiment Station. Miscellaneous Journal Series Paper 1611 of the Minnesota Agricultural Experiment Station.

[2] Present address: Department of Horticulture, Virginia Polytechnic Institute and State University, Blacksburg, VA 24061.

[3] Present address: Department of Horticulture, Colorado State University, Fort Collins, Colorado 80521.

low temperature. Weigand (39) as stated above observed ice in twigs and buds during winter, but also showed that certain living vegetative buds could be cooled to -26.5°C before ice crystals were observed. Dorsey and Strausbaugh (6) found ice in the bud scales and axis of dormant plum buds, but not in the primordial region at -29.5°C. In neither case, however, did the researchers propose that the tissues were in fact supercooled. Among the first to propose deep supercooling as a freezing injury avoidance mechanism in plants were Tumanov and Krasavtsev (37) who observed low temperature freezing points or exotherms in oak, birch, fir, and pine branches. Tumanov and coworkers (38) have also observed low temperature freezing points in cherry flower buds. Krasavtsev (18) suggests that prolonged retention of supercooled water can lead to frost injury in marginally hardy trees such as apple, the injury occurring from intracellular freezing of a trapped fraction of supercooled cellular water at low temperature. Mild supercooling, less than 10°C has been proposed as an injury avoidance mechanism in olive leaves (20) and some fruit buds (17, 23). Despite these few exceptions to extracellular ice formation as the vehicle to avoidance of low temperature injury, Levitt (21) was justified to conclude in 1972 that pronounced supercooling plays no major role in frost resistance of most plants.

Recent evidence, however, suggests that deep supercooling may indeed play a significant part in frost resistance of many plants. Work on overwintering floral primordia (9, 12, 15, 27, 36) and hardwood xylem (4, 10, 11, 13, 25, 26) of many woody species, to be discussed below, suggests that these plant parts can only attain maximum cold hardiness near -40°C. Temperatures near -40°C are of physical significance since they are in the range of homogeneous ice nucleation temperatures for supercooled aqueous solutions (28). The implications for the northern distribution of plants utilizing supercooling as a frost avoidance mechanism are clear and are discussed further in the following sections.

SUPERCOOLING IN AQUEOUS SOLUTIONS

A supercooled system is in a "metastable equilibrium" as described by Glasstone (14). Actually most of the living systems to be described are in a transient non-equilibrium state where the rate of decay of the state is exceedingly slow. Glasstone uses the term "metastable equilibrium" to define an equilibrium state which is not the most stable state at a particular temperature. A metastable system will undergo a spontaneous transition on the addition of the stable phase. When water is cooled below its freezing point without ice formation it is in a metastable state. There is a theoretical as well as empirical limit on the stability of supercooled water. This limit is referred to as the homogeneous ice nucleation temperature where spontaneous change from the metastable state (liquid) to the stable state (ice) occurs. Many investigators have shown that very pure water droplets can be supercooled to low temperatures, but never below approximately -40°C (2, 7, 19, 28). Frenkel's (8) heterophase fluctuation theory for water, in which a distribution of constantly forming and reforming molecular clusters (some ice-like) is proposed to exist in the liquid state, has been used as a model for predicting the homogeneous nucleation temperature by nucleation theory. Above the homogeneous nucleation temperature the distribution of clusters both in terms of mole fraction and size is such that the rate of formation of critical size ice-like clusters capable of growing into macroscopic ice is extremely slow. Water, in principle, may remain supercooled within 2 or 3°C of its homogeneous nucleation point almost indefinitely. As noted by Rasmussen and MacKenzie (29), nucleation theory indicates the primary reasons why the cluster distribution reaches an unstable point at the homogeneous nucleation temperature. The relative number in terms of mole fraction and cluster size increases rapidly at the nucleation point along with a continuing exponential decrease in the critical cluster size, leading to rapid formation of a critical size ice nucleus. Although nucleation theory predicts the fundamental character of homogeneous ice formation, it does not predict precisely the experimental temperature at which ice nucleation occurs. This primarily results from the difficult task of assigning an accurate value to the surface tension between the solid and liquid phases. Fletcher's (7) theoretical derivation fits to a reasonable degree most experimental observations on supercooled water where the homogeneous nucleation temperature is $-38° \pm 1°C$.

Homogeneous ice nucleation is predicated on the concept that the water contains no heterogeneous nucleators. Foreign surfaces and suspended insoluble particles promote hetero-

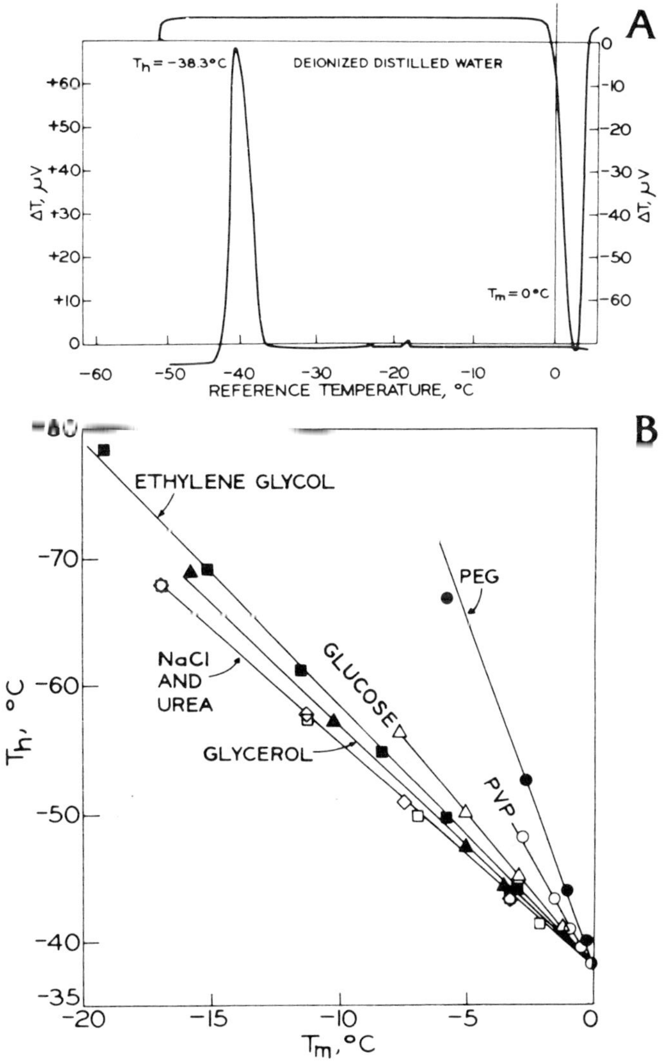

Figure 1. A. Thermogram for emulsified deionized distilled water. B. Homogeneous nucleation temperature (T_h) versus melting temperature of aqueous solutions (T_m). Rasmussen and MacKenzie (28).

geneous nucleation at higher temperatures. Increase in the nucleation temperature depends on the size and surface properties of the impurities (7) Solutes, however, generally increase the supercooling necessary for crystallization (28). It might be asked how water can be studied at temperatures near its homogeneous nucleation temperature (–38°C) since heterogeneous nucleators, even in very small amounts, will nucleate supercooled water at higher temperatures. One of the most successful methods for supercooling water to the homogeneous nucleation temperature is by homogenizing water into droplets in hydrocarbon solvents (28). Some of the water droplets freeze due to ice nucleation from heterogeneous impurities or from spontaneous crystallization but this is an almost undetectably small fraction of the droplets at temperatures above –38°C (Figure 1). A tristearate surfactant included in the emulsion isolates any heterogeneous ice nucleation above –38°C so that ice propagation does not occur By using this method, Rasmussen and MacKenzie (28) have shown that solutes including glucose, urea, polyvinylpyrrolidone, NaCl, glycerol, ethylene, ethylene glycol, and NH_4F depress the nucleation temperature proportionately more than the melting point depression (Figure 1).

Although it would be reasonable to expect the living cell to provide many sites for heterogeneous nucleations, MacKenzie ***et al.*** (22) indicate that ice-like structures do not generally exist in the cell and that the cytoplasm behaves much like a dilute aqueous solution. Their suggestion is based on experiments where individual yeast cells survived supercooling to –37°C.

DEEP SUPERCOOLING IN PLANT REPRODUCTIVE PARTS

Slight supercooling in fruit buds, especially at the first stages of anthesis, has been suggested as an injury avoidance mechanism in peach, but the extent of supercooling is generally less than –10°C (17, 23). Tumanov ***et al.*** (38) were the first to assign low temperature injury in reproductive parts to freezing of supercooled water. In calorimetric experiments conducted on cherry flower buds they found rapid freezing events between –20° and –30°C which were associated with injury to the tissues. In 1971, Graham (15) made an extensive study of low temperature freezing points in ***Rhododendron*** flower buds. He found that cold hardy ***Rhododendron*** floral primordia did not freeze at temperatures near –2°C as did the bud scales and stem axis to which they are attached, but froze at temperatures as low as –43°C. Primordia invariably survived low temperatures if they did not freeze, but death occurred to a primordium if it froze. Low temperature freezing points have since been observed in floral primordia of blueberry (36) and several ***Prunus*** species. (27)

Differential thermal analysis (d.t.a.), differential scanning calorimetry (d.s.c.) and nuclear magnetic resonance spectroscopy (n.m.r.) have been used to study water in floral primordia of ***Rhododendron*** during the dormant winter months (9, 12). A d.t.a. recording of freezing a whole flower bud (Figure 2) clearly displays the major freezing events or exotherms associated with ice formation. The large first exotherm results from freezing of water in the bud scales while succeeding exotherms are a consequence of ice nucleation in each individual primordium. Bud scale exotherms remain constant over the winter season but primordia freeze at lower average temperatures in midwinter than in fall or spring. Plants which exhibit extracellular ice formation follow the general rule that freezing and thawing occur at the same temperature. Cooling and thawing analysis of an excised primordium by d.s.c., d.t.a. and n.m.r. indicate the presence of only liquid or supercooled water and no ice from 0°C to the freezing point. Upon subsequent warming the primordium thaws gradually with a melting point near –2°C in a manner analogous to a solution containing approximately one mole of an osmotically active solution. This is not a characteristic of a melting point depression or eutectic freezing phenomenon which are well defined solid-liquid equilibrium processes that have the same

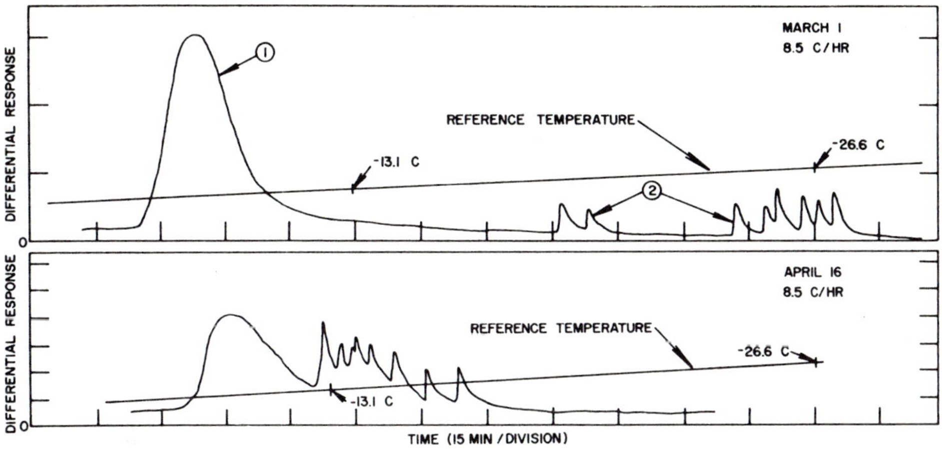

Figure 2. D.t.a. recordings of freezing profiles of whole azalea buds cooled at 8.5°C/hr. On 1 March, the bud is a typical midwinter one. On 16 April the convergence of exotherms indicates loss of hardiness in early spring. Exotherm 1 corresponds to freezing of water in the bud scales and stem axis upon which the primordia are attached. Exotherms of type 2 indicate freezing of an individual primordium. The number of primordia exotherms generally equals the number of primordia in the bud at a cooling rate of 8.5°C/hr. Each division on the differential response scale equals approximately 1.4°C. George *et al.* (9)

crystallization and melting temperatures. Heat of fusion measurements find that heat released when a primordium freezes is close to the value expected for supercooled water (12). From a physical standpoint the fraction of water freezing at subzero temperatures in floral primordia appears to be supercooled water.

Microscopic observation (12) of primordial leaves mounted in fluid fluorocarbon and cooled to subfreezing temperatures show that crystallization takes place randomly at a point in the leaf and then rapidly grows throughout the entire leaf. This sudden freezing of supercooled tissue involves intracellular freezing as observed by the sudden darkening of the cells. Siminovitch and Scarth (34) have shown that intracellular freezing in cortical cells of ***Catalpa*** and ***Cornus*** trees, brought on artificially by rapid cooling, causes visible mechanical disintegration and death to the protoplasm. In almost all cases reported, intracellular freezing has been found to be fatal in plant cells. If the primordial leaf is mounted in water and nucleated near 0°C, the tissues freeze at higher temperatures than if mounted and cooled in fluorocarbon. Ice growth proceeds gradually throughout the leaf and although the small cell size and experimental conditions prevented close observation, the mode of freezing was probably at least in part extracellular. No barrier to ice growth seems to exist between cells. Since in the whole bud primordia are surrounded by stem and bud scale tissues containing ice, there must exist a physical barrier to ice propagation between this ice and the supercooled primordial tissues. Freezing and associated injury to the floral primordia in winter cereals has been shown to depend somewhat on the ability of the vascular tissue to stop the advance of the freezing boundary (35). A study of whole ***Rhododendron*** buds cooled at different rates has found that primordia freeze at higher temperatures when cooled faster than approximately 8°C/hr supporting the concept of a physical barrier to ice growth between primordia and the scales and stem axis (9). Some underlying physical property of the primordia must also exist to prevent sublimation of the supercooled liquid water to the ice in the nearby tissues through the vapor phase. Regardless of the morphological details allowing supercooling in ***Rhododendron*** floral primordia, it can be stated that the freezing-thawing behavior, physical properties, and low temperature hardiness limit (–43°C) in the range of homogeneous nucleation temperatures for aqueous solutions all point to the existence of a "metastable state" or supercooled condition. The finding that the same phenomenon occurs in blueberry and a number of ***Prunus*** flower buds may indicate that supercooling is more widespread in plant reproductive parts than originally thought.

DEEP SUPERCOOLING IN XYLEM

Xylem of some woody species is the second plant tissue to exhibit pronounced low temperature freezing points. Krasavtsev (18) found that up to 40 percent of water in cold-hardy apple twigs remains unfrozen to –30°C at cooling rates as slow as 10°C/day. Quamme ***et al.*** (25) found that low temperature freezing points or exotherms near –40°C in mid-winter apple twigs caused injury to the xylem ray parenchyma. They also noted that maximum cold hardiness in apple xylem tissues throughout the season in Minnesota was about –41°C. Although Krasavtsev suggested this freezing was due to a supercooled water fraction, Weiser (40) interpreted the freezing event as a "pulling away" of "vital water" from protoplasmic constituents to the extracellular ice in xylem elements, bark tissues, etc.

Recent work by Burke ***et al.*** (4) and George and Burke (11) on killing low temperature exotherms in the xylem of shagbark hickory supports the concept of simple supercooling of cellular water in the ray parenchyma to its homogeneous nucleation temperature. Figure 3 shows d.t.a. and pulsed n.m.r. recordings of freezing hickory twigs (with bark removed) at four times during the winter and spring in Minnesota. Hickory xylem when cooled below 0°C exhibits essentially one major freezing peak which decreases to a limiting value near –45°C at mid-winter. At –15°C the time constant for freezing in fully acclimated hickory xylem is greater than 1 year, indicating the high stability of the fraction of water which freezes at low temperature. Rewarming of a frozen sample always finds a gradual thawing of the tissues with melting point near –2°C similar to a 1 molar osmotic solution. N.m.r. data indicates that 30 to 50 percent of the total tissue water freezes between –30°C and –70°C at midwinter. The temperatures observed for freezing in n.m.r. experiments are identical to the exotherm temperatures observed in d.t.a. experiments and indicate that the freezing of water is the origin of the exotherms (Figure 3). Heat of fusion measurements by d.s.c. find a heat release of 51 ± 3 cal/g. Heat of fusion for pure water at –38°C is 61.1 cal/g (29). D.s.c. experiments also demonstrate that the low temperature exotherm in hickory is the sum of many small ice nucleations in the wood. This information coupled with microscopic observation which shows that the predominant observable freezing occurs in the xylem ray parenchyma leads to the conclusion that the exotherm results from crystallization of supercooled cellular water in the parenchyma. It should be mentioned, however, that the microscopic observations are somewhat clouded by the fact that sectioning twigs thinner than about 1 mm causes the freezing to occur at higher temperature. This finding is not unexpected since xylem rays in hickory are large and multiseriate with heights often greater than 0.5 mm. Sectioning inevitably destroys the morphology and intactness of the complete ray.

Again as in the case with floral primordia the prevailing physical evidence points to supercooling as the freezing avoidance mechanism, but exactly what anatomical and morphological features allow supercooled water to exist in one plant part (xylem) adjacent to another with ice (bark) are not clear at the present time. George and Burke (11) present a hypothetical explanation in terms of the cohesion-tension properties of water which form the basis for explaining water transport in xylem vessels.

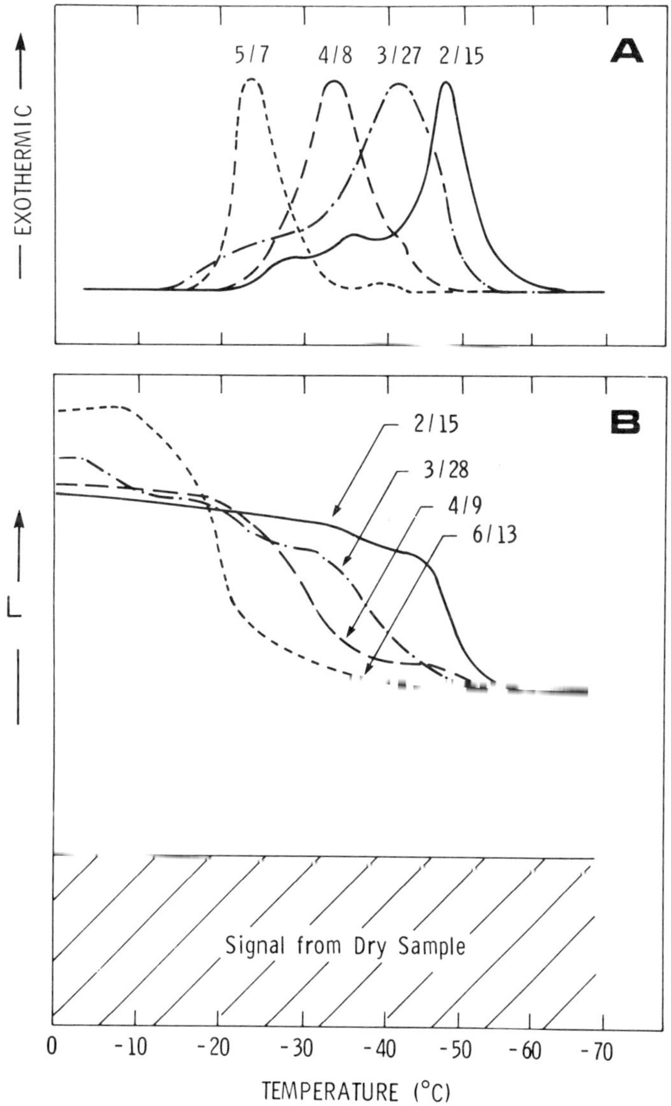

Figure 3. Changes in d.t.a. (A) and n.m.r. freezing curves (B) of shagbark hickory during deacclimation from winter. L represents unfrozen water. Burke *et al.* (4).

To examine the possibility of widespread existence of low temperature exotherms in xylem, d.t.a. experiments were performed on the wood of 49 woody species native to North America, collected in January in Minnesota (10). Of the 49 species 25 had low temperature exotherms near –40°C. These 25 were all indigenous to the eastern deciduous forest and represent 78 percent of the 32 angiosperm species studied from that region. No exotherms were found in the gymnosperms tested. Table 1 presents an extensive list of trees from Europe and Asia prepared by Dr. Akira Sakai

TABLE 1. (A) Sakai's (32) record of fully cold hardy tree species native to Asia and Europe which display xylem injury above –50°C. (B) Cold acclimated tree species of North America observed to have killing exotherms near –40°C in the xylem.

A. Asian and European tree species which display xylem injury above –50°C.		**B.** North American tree species with exotherms near –40°C.
Asia	*Europe*	
Acanthopanax senticosis	*Acer campestre*	*Acer rubrum*
Acer mandshuricum	*Acer platanoides*	*Acer saccharum*
Acer mono	*Acer pseudoplatanus*	*Betula lutea*
Acer okamotoanum	*Acer tatarium*	*Betula nira*
Acer triflorum	*Alnus glutinosa*	*Carpinus caroliniana*
Alnus hirsuta	*Alnus icana*	*Carya cordiformis*
Carpinus coreana (C. Turczaninovii)	*Carpinus betulas*	*Carya ovata*
Castanea bungeana (C. Mollissima)	*Castanea sativa*	*Celtis occidentalis*
Castanea dunfer	*Cornus mas*	*Cercis canadensis*
Cornus controversa	*Cornus sunguineana*	*Cornus alternifolia*
Cornus kousa	*Corylus avellana*	*Cornus racemosa*
Corylus heterophylla	*Corylus maxima*	*Fagus grandifolia*
Corylus sieboldiana	*Fagus orientalis*	*Fraxinus americana*
Fraxinus chinensis var. rhynchophylla	*Fagus sylvatica*	*Fraxinus nigra*
Fraxinus densata	*Fraxinus excelsior*	*Fraxinus pennsylvanica*
Fraxinus mandshurica	*Fraxinus ornus*	*Gleditsia triacanthos*
Hydrangea paniculata	*Juglans regina*	*Junglans nigra*
Juglans mandshurica	*Quercus pe traceae*	*Ostrya virginiana*
Kalopanax septemlobus	*Quercus pyrenaica*	*Platanus occidentalis*
Malus sieboldi	*Quercus robur*	*Prunus serotina*
Phellodendron amurense	*Ulmus carpiniforia*	*Quercus alba*
Pterocarya stenoptera	*Ulmus glabra*	*Quercus macrocarpa*
Pyrus ussuriensis		*Quercus rubra*
Quercus dentata		*Ulmus americana*
Quercus mongolica var. grosseserrata		*Ulmus thomasii*
Rosa pendulina		
Rosa rugosa		
Sorbus alnifolia		
Ulmus davidiana var. japonica		
Ulmus laciniata		

(32) which are observed to have xylem injury in mid-winter above –50°C. North American trees studied by George *et al.* (10) having exotherms are also given. The widespread occurrence of low temperature exotherms in deciduous hardwoods of North America leads one to suspect that exotherms also exist in many of the European and Asian species. Occurrence of exotherms throughout the natural range in a single woody species is discussed in the next section. In any case, supercooling as a mechanism of frost resistance in xylem is seen to be a common phenomenon throughout deciduous tree species of North America.

DEEP SUPERCOOLING AND PLANT DISTRIBUTION

Most higher plant cells have an osmotic concentration less than 2 molar (21). The combination of this information with homogeneous nucleation temperatures determined for dilute aqueous solutions (28) places a limit of approximately –38°C to –47°C on supercooling for plant cells. Killing exotherm temperatures in fully cold hardy trees of the deciduous forest generally fall in this range (10). Utilization of supercooling as a frost avoidance mechanism will then have a definite consequence on the northern distribution of those plants using it. In North America, angiosperm species with exotherms in the xylem generally have northern limits near regions where the probability of –40°C minimum environmental temperatures occur at least once per year (10). Angiosperm species native to the northern ranges of the boreal forest do not have low temperature exotherms in the xylem. Figure 4 illustrates this finding for red oak (exotherm) and trembling aspen (no exotherm).

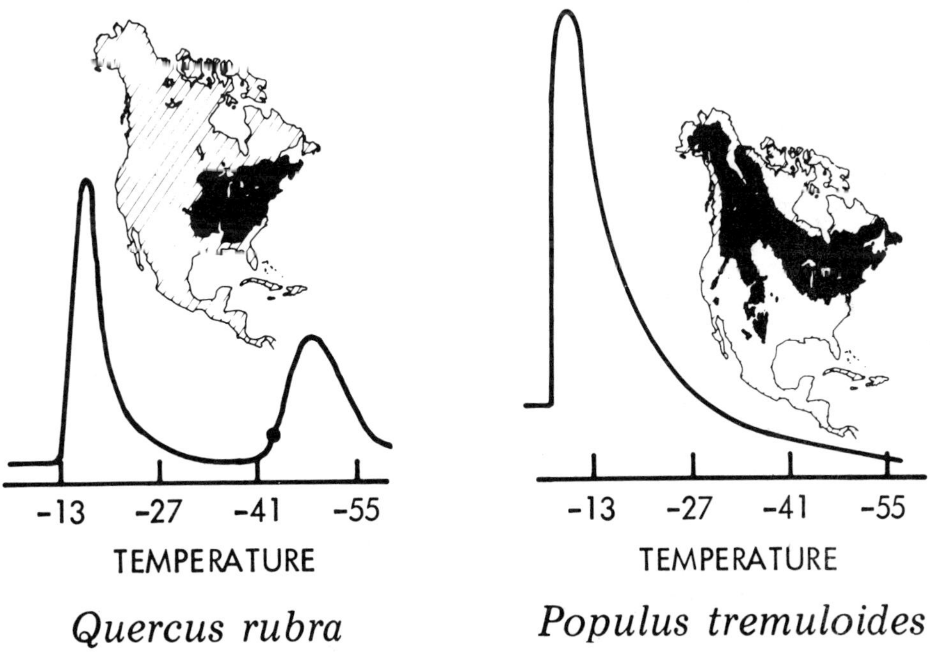

Figure 4. D.t.a.'s of the wood and the natural geographic ranges of red oak (***Quercus rubra*** L.) and trembling aspen (***Populus tremuloides*** Michx.). The exotherm near –40°C in red oak is from intracellular freezing in the xylem ray parenchyma and causes injury. Exotherms near –10°C are from extracellular freezing and do not injure the tissues. George *et al.* (10).

An interesting question arises when a particular species exhibits a low temperature exotherm; do exotherms exist throughout the natural range of the species? To determine if supercooling does exist throughout the range of a species exhibiting it, George *et al.* (13) performed d.t.a. experiments on twigs collected in mid-winter from trees over the natural range of red oak, yellow birch, black walnut, and wild black cherry. The trees sampled were from seeds collected from native stands and grown on single sites. For the four species tested exotherms were found in all trees. Exotherm temperatures were generally close to –40°C and the lowest exotherm temperature recorded was –46.4°C. The results suggest that woody species with exotherms are not evolving away from supercooling as a freezing resistance mechanism and are therefore physically limited in their migration northward.

CONCLUSION

Avoidance of freezing injury by supercooling has been implicated in the survival of many overwintering insects (1, 31) and even in Antarctic fish (5). However, only in the recent studies cited above has deep supercooling been shown to occur to a wide degree in higher plants. The finding that cold hardy floral primordia of some ornamental species and xylem ray parenchyma of a large number of deciduous forest hardwoods can only attain maximum hardiness near the homogeneous ice nucleation temperatures for dilute aqueous solutions (–40°C) has practical as well as fundamental importance. It is practical in that a limit is imposed on breeding for cold hardiness in some species and fundamental for promoting a further understanding of the physiological restrictions imposed on plant distributions in nature. Although further research is required to understand the anatomical and morphological details of supercooling, it is safe to say that supercooling does play an important role in the winter survival and ecological distribution of many plant communities.

ACKNOWLEDGEMENT

The authors would like to express their appreciation to Dr. Akira Sakia, Institute of Low Temperature Science, Hokkaido University, Sapporo, Japan for the cold hardiness data of Asian and European woody species. This data reflects in part the extensive nature of Dr. Sakai's contributions to the understanding of winter hardiness in many plant species.

REFERENCES

1. ASAHINA, E., AOKI, K., and SHINOZAKI, J. The freezing process of frost hardy caterpillars. ***Bull. ent. Res.*** **45** (1954):329-339.

2. BIGG, E. K. The supercooling of water. ***Proc. Phys. Soc. B.*** **66** (1953):688-694.

3. BURKE, M. J., BRYANT, R. G., and WEISER, C. J. Nuclear magnetic resonance of water in cold acclimating red osier dogwood stem. ***Pl. Physiol.*** **54** (1974):392-398.

4. BURKE, M. J., GEORGE, M. F., and BRYANT, R. G. Water in plant tissues and frost hardiness. ***In*** R. B. DUCKWORTH (Ed.), ***Water Relations of Foods.*** (Academic Press; New York) (1975) pp. 111-135.

5. DEVRIES, A. L. and WOHLSCHLAG, D. E. Freezing resistance in some Antarctic Fishes. ***Science*** **163** (1969):1073-1075.

6. DORSEY, M. J., and STRAUSBAUGH, P. D. Winter injury to plum during dormancy. ***Bot. Gaz.*** **76** (1923):113-142.

7. FLETCHER, N. H. The Chemical Physics of Ice. (Cambridge University Press; Cambridge, England) (1970).

8. FRENKEL, J. Kinetic Theory of Liquids. (Oxford U.P., Oxford; England) (1946).

9. GEORGE, M. F., BURKE, M. J., and WEISER, C. J. Supercooling in overwintering azalea flower buds. ***Pl. Physiol.*** **54** (1974):29-35.

10. GEORGE, M. F., BURKE, M. J., PELLETT, H. M., and JOHNSON, A. G. Low temperature exotherms and woody plant distribution. ***Hort-Science* 9** (1974):519-522.

11. GEORGE, M. F., and BURKE, M. J. Cold hardiness and deep supercooling in xylem of shagbark hickory. ***Pl. Physiol.*** (1975) Submitted.

12. GEORGE, M. F., and BURKE, M. J. Supercooling in overwintering azalea flower buds: additional freezing parameters. ***Pl. Physiol.*** (1975) Submitted.

13. GEORGE, M. F., HONG, S. G., and BURKE, M. J. Cold hardiness and deep supercooling of hardwoods. Its occurrence in provenance collections of red oak, yellow birch, black walnut and black cherry. ***Ecology*** (1975) Submitted.

14. GLASSTONE, S. Textbook of Physical Chemistry, Edition 2. (D. Van Nostrand Company, Inc.; New York) (1946).

15. GRAHAM, R. P. Cold injury and its determination in selected ***Rhododendron*** species. Master's Thesis. University of Minnesota, St. Paul (1971).

16. GUSTA, L. V., BURKE, M. J., and KAPOOR, A. Determination of unfrozen water in winter cereals at sub-freezing temperatures. ***Pl. Physiol.* 56** (1975):707-9.

17. JOHNSTON, E. S. Undercooling of peach buds. ***Am. J. Bot.* 9** (1922):93-98.

18. KRASAVTSEV, O. A. Rate of water outflow from cells of frost resistant plants at negative temperatures. ***Soviet Pl. Physiol.* 17** (1970):417-422.

19. LANGHAM, E. J., and MASON, B. J. The heterogeneous and homogeneous nucleation of supercooled water. ***Proc. Roy. Soc. Lond. A.* 247** (1958):493-504.

20. LARCHER, W. Das Assimilationsuermögen von ***Quercus ilex*** und ***Olea europea*** im Winter. ***Ber. dt. bot. Ges.* 72** (1959).

21. LEVITT, J. Responses of Plants to Environmental Stresses. (Academic Press; New York) (1972).

22. MACKENZIE, A. P., RASMUSSEN, D. H., and MACAULAY, M. N. Deep supercooling of ***Saccharomyces cerevisiae. Society for Cryobiology, 10th. Ann. Meeting*** (1973), Key Biscayne, Fla.

23. MODLIBOWSKA, I. Some factors affecting supercooling of fruit blossoms. ***J. hort. Sci.* 87** (1962): 249-261.

24. OLIEN, C. R. A comparison of desiccation and freezing as stress vectors. ***Cryobiology* 8** (1971): 244-248.

25. QUAMME, H., STUSHNOFF, C., and WEISER, C. J. The relationship of exotherms to cold injury in apple stem tissues. ***J. Am. Soc. Hort. Sci.* 97** (1972):608-613.

26. QUAMME, H., STUSHNOFF, C., and WEISER, C. J. Winter hardiness of several blueberry species and cultivars in Minnesota. ***HortScience* 7** (1972):500-501.

27. QUAMME, H. A. An exothermic process involved in freezing injury to flower buds of several ***Prunus*** species. ***J. Am. Soc. Hort. Sci.* 99** (1974):315-317.

28. RASMUSSEN, D. H., and MACKENZIE, A. P. Effect of solute on ice-solution interfacial free energy; calculation from measured homogeneous nucleation temperatures. ***In*** H. H. G. JELLINEK (Ed.), ***Water Structure at the Water-Polymer Interface.*** (Plenum Publishing Corporation; New York) (1972) pp. 126-145.

29. RASMUSSEN, D. H. and MACKENZIE, A. P. Clustering in supercooled water. ***J. chem. Phys.* 59** (1973):5003-5013.

30. SÄCHS, J. Krystallbildungen bei dem Gefrieren und Veränderung der Zellhäute bei dem Aufthauen saftiger Pflanzentheile, mitgetheilt von W. Hofmeister. ***Ber. über die Verhand. der König. Sächs. Ges. der Wiss. zu Leipzig. math.-phys. Classe,* 12** (1860):1-50.

31. SALT, R. W. Principles of insect cold-hardiness. ***Annu. Rev. Entomol.* 6** (1961):55-74.

32. SAKAI, A. (1975) Personal Communication.

33. SCARTH, G. W., and LEVITT, J. The frost-hardening mechanism of plant cells. ***Pl. Physiol.* 12** (1937):51-78.

34. SIMINOVITCH, D., and SCARTH, G. W. A study of the mechanism of frost injury to plants. ***Can. J. Res. C.* 16** (1938):467-481.

35. SINGLE, W. V. Studies on frost injury to wheat. II. Ice formation within the plant. ***Aust. J. Agric. Res.*** **15** (1964):869-875.

36. STUSHNOFF, C., and BIERMAN, J. (1974) Personal Communication.

37. TUMANOV, I. I., and KRASAVTSEV, O. A. Hardening of northern woody plants by temperatures below zero. ***Soviet Pl. Physiol.*** **6** (1959): 663-673.

38. TUMANOV, I. I., KRASAVTSEV, O. A., and TRUNOVA, T. I. Investigation of the ice formation process in plants by measuring heat evolution. ***Soviet Pl. Physiol.*** **17** (1969):754-760.

39. WEIGAND, K. M. Some studies regarding the biology of buds and twigs in winter. ***Bot. Gaz.*** **103** (1906):373-424.

40. WEISER, C. J. Cold resistance and injury in woody plants. ***Science*** **169** (1970):1269-1278.

ADDENDUM

INTRODUCTION

Since the first publication of this manuscript, freezing avoidance by supercooling of tissue water has been found in many additional overwintering plant species (4, 5, 8, 10, 12, 13, 14). Of particular importance is that supercooling has now been implicated as a survival mechanism in tissues of certain temperate zone conifers and in tissues of timberline flora of the western United States. The following discussion briefly summarizes some of the data from recent work in these areas.

DEEP SUPERCOOLING IN TEMPERATE ZONE CONIFERS

Sakai (11) has reported the existence of low temperature freezing points associated with injury between −20 and −35°C in winter buds of a number of hardy conifers (Addendum Table 1). Anatomical and morphological characteristics of the overwintering bud are suggested to play a role in isolating the primordial shoot from extracellular ice in the budscales and the stem axis to which it is attached. Similarly, recent evidence (1, 3, 6, 7) shows that freezing avoidance by supercooling of tissue water to its homogeneous nucleation temperature (approximately −40°C) occurs in reproductive and vegetative structures of particular conifers native to North America (Addendum Table 1). Supercooling, therefore, seems to be an important mechanism of freezing resistance in conifers as well as in angiosperms. As noted above for angiosperms (see section "Deep Supercooling and Plant Distribution", p. 9) this observation may help to explain why many North American conifers are limited to regions where −40°C temperatures do not occur.

DEEP SUPERCOOLING IN TIMBERLINE FLORA

Becwar ***et al.*** (1, 2, 3) have proposed that the winter hardiness limit of supercooling is a cause of timberline in the Rocky Mountains of Colorado, U.S.A. They predict that the average annual minimum temperature at timberline in the continental mountain regions of the western United States is approximately −40°C. This prediction correlates with measurement of lethal low temperature freezing points near −40°C in specific tissues of flora which do not grow above timberline in Colorado (Addendum Table 1). It is pointed out, however, that this explanation does not apply to the Cascade Mountains of Washington and Oregon, to timberline in polar regions, or to timberline in equatorial regions.

In concluding this short addendum, it is reasonable to say that freezing avoidance by supercooling is more widespread than originally assumed. Despite this fact and although the biophysics of supercooling has been fairly well established, only a small amount of evidence is available (9, 11) on the precise anatomical, morphological, and physiological features which allow significant fractions of tissue water to remain unfrozen at low subfreezing temperatures while ice exists in nearby tissues.

ADDENDUM REFERENCES

1. BECWAR, M.R., RASASHEKAR, C. and BURKE, J.J. Deep undercooling of water in conifers and western hardwoods: Its −40°C winter hardiness limit and implications regarding timberline. Am. Soc. of Pl. Physiol., Annual Meeting, Ohio State University, Columbus, Ohio (August 1979).

2. BECWAR, M.R., RAJASHEKAR, C., HANSEN, K.J. and BURKE, M.J. Deep undercooling of tissue water and winter hardiness limitations in timberline flora. ***Pl. Physiol.*** (1980). In press.

3. BECWAR, M.R. Deep undercooling and winter hardiness limitations in conifers and conifer mistletoe parasites. Ph.D. Thesis. Colorado State University, Fort Collins (1980).

4. BIERMANN, J., STUSHNOFF, C. and BURKE, M.J. Differential thermal analysis and freezing injury in cold hardy blueberry flower buds. ***J. Am. Soc. Hort. Sci.*** **104** (1979): 444–449.

5. BURKE, M.J. and STUSHNOFF, C. Frost hardiness: A discussion of possible molecular causes with particular reference to deep supercooling of water. ***In*** H. MUSSELL and R.C. STAPLES (Eds.), ***Stress Physiology in Crop Plants.*** (John Wiley & Sons, New York) (1979) pp. 197–227.

6. GEORGE, M.F., SMITH, D.E. and GUYETTE, R.P. A preliminary report on freezing avoidance in ***Juniperus virginiana.*** Missouri Academy of Science, Annual Meeting, Jefferson City, Missouri (April, 1979).

7. GEORGE, M.F. and CARRASQUILLA, M.L. Freezing avoidance in male cones of ***Juniperus virginiana.*** Am. Soc. of Pl. Physiol., Annual Meeting, Ohio State University, Columbus, Ohio (August 1979).

8. PIERQUET, P., STUSHNOFF, C. and BURKE, M.J. Low temperature exotherms in stem and bud tissues of ***Vitus riparia*** Michx. ***J. Am. Soc. Hort. Sci.*** **102** (1977): 54–55.

9. QUAMME, H.A. Mechanism of supercooling in overwintering peach flower buds. ***J. Am. Soc. Hort. Sci.*** **103** (1978): 57–61.

10. RAJASHEKAR, C. and BURKE, M.J. The occurrence of deep undercooling in the genera ***Pyrus, Prunus***, and ***Rosa***: A preliminary report. ***In*** P.H. LI and A. SAKAI (Eds.), ***Plant Cold Hardiness and Freezing Stress: Mechanisms and Crop Implications.*** (Academic Press, New York) (1978) pp. 213–225.

11. SAKAI, A. Low temperature exotherms of winter buds of hardy conifers. ***Plant and Cell Physiol.*** **19** (1978): 1439–1446.

12. SAKAI, A. Deep supercooling of winter flower buds of ***Cornus florida*** L. ***HortScience*** **14** (1979) 69–70.

13. SAKAI, A. and MIWA, S. Frost hardiness of Ericoideae. ***J. Am. Soc. Hort. Sci.*** **104** (1979): 26–28.

14. SAKAI, A. and HAKODA, N. Cold hardiness of the genus ***Camellia. J. Am. Soc. Hort. Sci.*** **104** (1979): 53–57.

ADDENDUM TABLE 1. Temperate zone conifers and timberline flora of the western United States that exhibit low temperature freezing points in winter

Temperate Zone Conifers:

- *Abies balsamea* (needles, winter buds)
- *A. firma* (winter buds)
- *A. homolepis* (winter buds)
- *A. veitchii* (winter buds)
- *A. sachalinensis* (winter buds)
- *Juniperus virginiana* (male cones, needles, twigs)
- *Larix leptolepis* (winter buds)
- *Pinus resinosa* (needles)
- *P. strobus* (needles)
- *Pseudotsuga menziesii* (winter buds)
- *Tsuga canadensis* (needles)

Timberline Conifers:

- *Abies concolor* (twigs)
- *A. lasiocarpa* (needles, twigs)
- *Juniperus communis* (twigs)
- *J. scopulorum* (twigs)
- *Picea engelmanii* (needles, twigs)
- *Pinus flexis* (needles)
- *Pseudotsuga menziesii* (needles, twigs)

Timberline Angiosperms:

- *Acer glabrum* (twigs)
- *Arceuthobium vaginatum* (seeds)
- *A. americanum* (seeds)
- *A. cyanocarpum* (seeds)
- *Arctostaphylos uva-ursi* (twigs)
- *Artemisia tridentata* (twigs)
- *Cercocarpus montanus* (twigs)
- *Gutierrezia sarothrae* (twigs)
- *Physocarpus monogynus* (twigs)
- *Purshia tridentata* (twigs)
- *Shepherdia canadensis* (twigs)
- *Symphoricarpos sps.* (twigs)
- *Rhus trilobata* (twigs)
- *Ribes cereum* (twigs)
- *Rubus sps.* (twigs)
- *Vitis riparia* (twigs)

THE STUDY OF INFRASPECIFIC VARIATION

C. A. STACE

Botanical Laboratories, School of Biology, University of Leicester, Leicester, U.K.

INTRODUCTION

In recent decades the systematic study and cataloguing of minor infraspecific variation in plants has become one of the casualties of changing fashion. The emphasis has moved towards the investigation of interspecific or even higher relationships or, at the infraspecific level, the comparison of populations showing rather wide differences which might be expected to present the best opportunities for the study of speciation. Those botanists, notably genecologists, who have studied minor variation have mostly not been taxonomists and have mostly not expressed their discoveries in taxonomic terms.

These trends are reflected in the classification of infraspecific variation. The last half-century has seen a gradual decline in the use of the lower levels of the hierarchy for which provision is made in the International Code of Botanical Nomenclature (variety, subvariety, form, subform), together with the rise to prominence of the highest infraspecific rank (subspecies). Thus, whereas critical Floras written 50 or more years ago[3] sought to describe infraspecific variation in some detail, using the whole spectrum of available categories, modern Floras rarely do this. Most of them use only the subspecific category and, although some[15, 62] convincingly argue that they cannot attempt a critical treatment with the time, manpower and finances available, they nevertheless set the style which others follow. Indeed, there are some[49] who would argue that botanists should emulate zoologists, who generally recognise only the subspecies, and a number belittle the over-ambitious attempts in the past of those such as Ascherson & Graebner and Jordan by disrespectful comparison with stamp-collecting (as opposed to philately) or bird-watching (as opposed to ornithology), which they suppose to be similarly unfruitful pursuits.

But the inadequacies of the systems proposed by workers such as these, and adopted in this country by Druce and others, is largely a measure of the inadequacy of their means of investigation and presentation, and should not be used as an excuse for not attempting better ones with the greatly improved resources now available. There are also more positive reasons for undertaking investigations of infraspecific variation, and for giving it appropriate formal taxonomic status. A more intensive study of minor variation in European vascular plants is vital for the immediate future. Such variation should be recognised taxonomically, and particular attention needs to be paid to areas of genetic diversity in southern Europe which have hitherto been relatively neglected

CENTRES OF GENETIC DIVERSITY

From the detailed anlyses of Vavilov and others the notion has developed that certain restricted areas of the world have been particularly important as origins of plant taxa. Vavilov[66], who was interested in the origins of cultivated plants, found the greatest diversity in south-western Asia, which he therefore assumed to have been the origin of the greatest number of crop plants. To what extent the present centres of diversity can be considered the ultimate centres of origin has been very much debated[70]. In many instances they represent secondary centres of origin, particularly in the case of cultivated plants whose domestication has been so closely linked to the development of civilizations and their migrations and requirements.

Whatever the case there is now remarkable agreement over great numbers of taxa, both wild and cultivated, which has led to the recognition of a relatively small number of centres of diversity[66, 69]. In temperate Eurasia the most important centre lies to the south of the Black and Caspian Seas, in a belt stretching roughly from western Iran to Turkey and Syria, which has given rise to names such as the 'fertile crescent', 'nuclear zone', 'Middle East arc', and 'gene centre belt', because of the origin there of so many important crops, especially cereals and legumes.

The Mediterranean zone, which is a continuation westwards of the above is of less such importance (although *Beta vulgaris, Vicia faba, Lactuca sativa, Raphanus sativus* and *Brassica oleracea* are among the exceptions), but is nevertheless a major centre of diversity of non-cultivated plants. Moreover a high proportion of central and northern European species and genera show their greatest variation there, and it was from there that species migrated north after the last glaciation. Therefore the student of the European flora frequently needs to study the variation of his taxa in the Mediterranean zone, and failure to do this often leads to a one-sided and ill-balanced view of the plants concerned.

The Mediterranean zone itself is extremely diverse, with great variations in altitude, climate and soils, and its landmass is greatly dissected into islands and peninsulas of varying degrees of isolation. Such a pattern of long-established landmasses, with their varied topography and environmental conditions, the latter often very favourable to plant growth, is strongly conducive to population differentiation and is one cause of the far greater range of variation found in the Mediterranean zone than further north. The other major factor is the phenomenon of biotype depletion, which is generally shown by taxa migrating away from their centre of origin and which is the principle underlying the general coincidence of centres of diversity and centres of origin.

THE STUDY OF GENETIC DIVERSITY

The investigation of regional, ecological and other infraspecific variation was at first carried out for its intrinsic interest for example by taxonomists, who wished to classify it, or by agriculturists, who selected the most promising variants for growing and breeding. In the last half century, dating largely from the pioneer work of Turesson[61], minor genetic variation has been studied by genecologists, who have sought to use it in their micro-evolutionary studies as an indicator of adaptation, selection, speciation and migration.

Most genecological work has been based on morphological characters, and the degree to which these vary from population to population. The morphological divergence shown by related ecotypes ranges from vast differences which have been used by taxonomists to recognise distinct species, e.g. in *Geum*[26], to differences only detectable with certainty by statistical methods, e.g. in *Dactylorhiza*[31] and *Gentianella*[48], where it is much debated whether or not they should receive any taxonomic recognition at all. Indeed, in the last 25 years or so it has become increasingly realised[32] that differentiation between populations may be well-marked physiologically (and, therefore, ecologically), but hardly or not at all morphologically.

Examples of such physiological ecotypes showing adaptation to differing soil conditions (e.g. salinity[29, 60], pH[36, 53], heavy metals[11, 20, 39, 68], nutrients[18, 27, 54, 55], moisture[40]), climate (e.g. day-length[44], temperature[41], length of seasons[16], shade[36]), and biotic influences (e.g. grazing[57], parasitism[4], cultivation[52]) have been investigated, and they show all grades of associated morphological differentiation (cryptic, semi-cryptic and non-cryptic ecotypes). This is not surprising, since there is no reason to assume that physiological and morphological characters should evolve at similar rates, any more than do morphological characters and breeding barriers, but it is important to bear in mind that the extent of ecotypification may be underestimated by the measurement of morphological divergence alone.

Much variation, both structural and cryptic, is not, however, ecotypic or ecoclinal, but represents sporadic mutations varying in frequency and in the extent to which they have been investigated, e.g. pubescent/glabrous variants in many species, flower colour in various species, heteromorphic/homomorphic *Primula vulgaris*[17], chemical races in *Lotus corniculatus*[38], *Trifolium repens*[19], *Pinus* species[42] and *Eucalyptus* species[45], host-specific races of many parasites, and isozyme races in a number of plants.

Unfortunately, it is in many cases not even known whether the variation recorded is genetically fixed or merely ecophenetic. Transplanting does not always provide a clear-cut answer and even where genetical experiments have been undertaken the evidence may be equivocal. Moreover, there is still argument as to whether ecophenetic variation should be taxonomically recognised; in general it is not thus recognised in vascular plants but it frequently is in the bryophytes and lichens.

But nearly all the variation discussed above, being based upon studies of species over small parts of their total range, and often parts well removed from their centres of diversity, represents only the tip of the iceberg. This is well known to even the most casual observer. The British field botanist who studies alien species, or who spends some time in Mediterranean countries, encounters 'British' species showing extremes of structural variation well beyond the ranges exhibited in Britain. Similarly, he often finds that species with well-marked ecological preferences in Britain are far more catholic in southern Europe, and that his concept of the habitat of a species breaks down. It is important that many more studies are directed to the investigation of variation in southern Europe as well as in the north. Without such studies only a small part of their total range of variation is uncovered, and valuable data of evolutionary significance are lost. By studying the variation pattern (i.e. the distribution of morphs) over the whole range of a species its pathways of migration will be much more easily traced, and this perhaps represents the only feasible direct method of discovering centres of origin. Moreover, the more detailed and diverse the methods of investigation the more data will accumulate and the more easily the pattern will be understood. Thus, if the methods include those such as chemical analysis, cytological studies, experiments on physiological tolerance and isozyme studies as well as the description of structural characteristics, the results will be more significant, but it must be remembered that if the former are exploited at the expense of the gathering of morphological data much important information will be lost.

Breeding programmes are needed to ascertain more precisely than is possible by the use of conventions such as A, A′, B the degrees of genomic homology exhibited by related taxa. Such programmes should involve a high proportion of inter-racial intra-specific crosses as well as inter-specific hybridisations, and might be expected to throw more light on the general problem of genome homology as expressed by de Wet & Harlan.[21]

The geographical pattern of research in Europe today is, however, not likely to produce significant advances in the investigation of infraspecific variation in the near future. Such an opinion, which has been clearly expressed previously[34, 35], is not subjective. There is in general an inverse relationship not only between the number of taxonomists and the number of taxa present in any country, but also between the number of taxa in a given area and the number of investigations being carried out on them. Valentine[63] has compiled a list of biosystematic research in progress in Europe (excluding the British Isles). Of 241 entries only 28 refer to work being carried out in Mediterranean countries (excluding France), and this is increased only to about 44 if all the projects biased towards Mediterranean floras are totalled. His comparable list for the British Isles[64] shows that only about 9 of 93 entries are primarily involved with Mediterranean taxa.

In the case of cultivated plants the importance of studying (and preserving) minor genetic variation both in crop species and in their wild relatives, especially in the native areas of genetic diversity, has been realised for some time. Important contributions include those by Zohary and collaborators[6, 47, 71, 72, 73] in the Near and Middle East on cereals and legumes, and those by Borrill and collaborators[7, 8, 9, 10, 14] on herbage grasses in the western Mediterranean. In the last few years work of this nature has received an important boost in the shape of joint sponsorship by the Food and Agriculture Organization of the United Nations Organization and by the International Biological Programme of the International Council of Scientific Unions, which has led to a number of important conferences, consultative committees, expeditions and more permanent organizations and their corresponding publications[23, 24, 25].

The need for the selection and conservation of native variation in economically important groups (in other words the preservation of potentially desirable new crop characteristics) is very clear, and it is therefore not surprising that the same sort of attention has not been paid to non-economic species. Of the very few programmes of intensive study of wild Mediterranean floras the most outstanding examples are those of Runemark, Snogerup, Strid and collaborators[50, 56, 58] in the Aegean Sea area, where they have investigated several genera at the population level. Moreover there are a few signs that more studies of this sort might be initiated in the future. For example in 1971 the *Société pour l'échange des plantes vasculaires d'Europe occidentale et du bassin Méditerranéen* arose, from a similar but more restricted organization, in order to exchange, study and comment upon "plantes rare, intéressantes ou critiques, en vue d'améliorer la connaissance de la flora des régions susmentionnées et de constituer des riches collections de référence", and in 1974 the *Organization for the phyto-taxonomic investigation of the Mediterranean area* was founded with the purpose of stimulating and co-ordinating research, promoting the exchange of information, and encouraging the training of Mediterranean plant taxonomists. The effectiveness of such bodies in terms of the promotion of basic research remains to be proved.

The task of amassing a reasonable knowledge of the variation of the Mediterranean flora is daunting. There are perhaps in the order of 14,000 species (excluding agamospecies) in the Flora Europaea region, and the majority of these are confined to or are most variable in the Mediterranean zone. For example the Iberian peninsula alone contains about 6000 species[34], of which approximately 20% are endemics, and this last figure does not include species confined to the Iberian peninsula together with adjacent parts of Mediterranean Europe and/or North Africa. In addition there are some thousands of species present in the Asian and African Mediterranean countries which do not reach Europe. Clearly it is not possible to study all of these in detail, and some sort of list of priorities is required.

Just as ecologists might survey a large area by sampling representative habitats, taxonomists often work on selected genera to exemplify general situations. But with a knowledge of the nature of centres of genetic diversity (whereby it is possible to predict areas in which taxa might exhibit maximum variation) it is equally logical for taxonomists also to survey floras by sampling areas rather than taxa, and this carries the advantage of ease of collection of material and it focusses attention upon the centres of diversity. It is therefore suggested that programmes of research such as the Lund School's study of the Aegean region are likely to prove most fruitful, and that the Iberian peninsula, or parts of it, is particularly well suited for similar treatment. Indeed, it is the lack of detailed biosystematic research in the Mediterranean basin which has led to a situation where so many aspects of modern biosystematic theory are based on North American studies. The broad similarities between the Iberian peninsula and California in terms of geography and biology have been pointed out by previous authors[51], and it is clear that the vast range of situations in Spain, encompassing cool temperate to semi-desert conditions, Atlantic and Mediterranean coasts, deciduous, coniferous and broad-leaved evergreen forests, mountain ranges exceeding 3000 m altitude with varying degrees of isolation, coastal and inland saline areas, gypsaceous deposits, etc., provide opportunities for evolutionary studies equal to those in North America.

But it is not to be assumed that the discoveries awaiting the biosystematist in the Mediterranean region will exactly mirror those made in the New World. New investigations in fresh fields will reveal novel situations, and it is reasonable to suspect that new theories of evolution, or at least modified theories with a different bias, might result. There are indeed important differences between southern Europe and southern North America, the most significant of which is the very long history of gradually increasing cultivation and other human interference in the former compared with the relatively very recent and sudden exploitation of the latter. It has already been suggested[12, 30] that introgression (as opposed to hybridisation in general), shown by Anderson[1, 2] and others to be so important an evolutionary phenomenon in parts of North America, might in fact be less significant generally, being largely dependent on human interference with the natural vegetation and therefore less prevalent in more stable communities.

THE CONSERVATION OF GENETIC DIVERSITY

Although the ideals of conservation have existed for a very long time it was not until the 1950s that they became widely held and entered the political scene. From that time conservation has been generally considered a practical application of ecological study, and in the minds of the majority of the public at large the terms 'conservation' and 'ecology' have become intimately associated. This is not entirely a good thing. It has been pointed out by others[34, 67] that principles of conservation should not be decided entirely by ecologists, but that taxonomists have an equally important role to play. In practical terms, the ecologically most attractive areas are not always those most in need of conservation; indeed, areas possessing the greatest genetic diversity are often marginal habitats which can appear, especially in Mediterranean countries, very unattractive and which are therefore in greatest danger.

Since the Mediterranean zone possesses some of the oldest of man's cultures the conservation of its flora is intimately connected with agricultural practice. Fortunately the realization by crop breeders has now fully arrived that the conservation of areas of genetic diversity is a high priority in terms of world food supply[24, 25, 33]. The deleterious effects on the native vegetation of the increasing cultivation of uniform monocultures, the replacement of local cultivars by supposedly superior pan-cultivars, the management of marginal areas, and the increasing use of herbicidal, weeding and screening processes have been described by several authors[5, 22, 43] The result, genetic erosion, is the loss of genetic variability and with it the best source of new, desirable crop characteristics, and is already leading to the use of the term 'centres of former genetic diversity'[5]. It can only be halted by a concerted effort by taxonomists, ecologists and plant breeders to pinpoint the centres of genetic diversity and to devise methods of conserving them.

Broadly, it is in the short term easier to conserve plants in cultivation than in the wild, particularly since the most important wild areas are relatively remote. Hence the most significant advances have been made in various botanic gardens ('seed-banks' or 'gene-banks'), where stocks of wild-collected material are stored and regularly screened. Many such schemes exist at present, such as that at Kew, England[59], the 'germ plasm collection of crucifers' in Madrid[28], the gene-bank of the National Institute of Agrobotany, Hungary[37], and others[23]. Nevertheless it might be the case that the only effective way to prevent genetic erosion is to prevent niche erosion in the native regions themselves[22]. Certainly gene-banks alone cannot conserve more than a small fraction of wild variation, and in any major conservation programme the preservation of wild populations should be the main objective.

There are thus many difficulties in the way of successful conservation practice, but the greatest and most basic of them all is that one cannot even begin the task until the material, or at least representative parts of it, has been carefully investigated. Only when the variation of wild plants has been documented, and its distribution has been mapped, can one have a clear idea of exactly what must take priority. It is primarily the task of the taxonomist to provide the data upon which worthwhile programmes of conservation of genetic diversity can be based.

THE NAMING OF GENETIC DIVERSITY

As stated in the Introduction, it has become unfashionable to name infraspecific variants other than subspecies. Many modern authors recognise as subspecies variants which are of interest and deserving of naming but which do not correspond with that rank as usually defined 20 or 30 years ago, i.e. a geographical race. Thus the subspecific rank today embraces cryptic or semi-cryptic polyploids, maritime or alpine ecotypes, physiological races, and other morphs differing in relatively minor characteristics, as well as the traditional geographical races. In other words the subspecies has become the dumping ground for many sorts of situations, much as the variety was 100 years ago.

Although some botanists have difficulty in remembering plant names, and try to keep the number of names to a minimum, the naming of plant taxa contributes greatly to their investigation. The provision of a name attracts attention and facilitates reference to a plant in many different ways. It is logical, therefore, to advocate the naming of infraspecific variants and their incorporation into modern, critical Floras, since this will lead to the accumulation of more information on them. There can be no doubt that the lack of generally known or accepted names for varieties, etc., has led to their neglect. The Critical Supplement to the Atlas of the British Flora[46], for example, mapped only two varieties of British vascular plants (*Veronica scutellata* var. *villosa* and *Aster tripolium* var. *discoideus*), whereas many others would have revealed equally or more interesting distributions (e.g. in *Primula vulgaris, Spergula arvensis, Bidens cernua, Geum rivale*, etc.). Valentine[65] has recently stressed the value of naming variants of a polymorphic series.

It is not easy to reverse these trends. Some of the problem lies in the provisions of the International Code of Botanical Nomenclature. Whereas the Code is flexible in certain directions, it lays down that the system of taxonomic ranks must be hierarchical. Thus forms must be grouped into varieties, and varieties into subspecies, and so on. Moreover it allows the use of a name only once within one species, unless the taxa have the same type specimen.

Thus if one recognises a white-flowered variant occurring sporadically in a species as f. *albus*, but later it is found that this species consists of two subspecies, then the name f. *albus* can only refer to white-flowered plants within one of the subspecies (whichever is represented by the type specimen of f. *albus*), and a new name is needed for white-flowered plants within the other subspecies. Secondly the case of reticulate variation is not easily treated. If a species exists as white- and coloured-flowered variants, and glabrous- and hairy-leaved variants, how does one decide which character is the higher in the hierarchy?; or, if one recognises both non-typical conditions at the same rank, what does one call plants bearing both non-typical characters? Thirdly, there is much uncertainty concerning the ranks which should be given to plants showing the same variation but different geographical distributions. White-flowered plants of *Primula vulgaris*, for example, may be sporadic variants in Britain, part of a polymorphic situation in an area by the Caspian Sea, or a distinct subspecies in the Balearics[65]. Similarly white-flowered *Dactylorhiza fuchsii* is a sporadic variant in England, but in Ireland is recognised as subsp. *okellyi*.

One is therefore forced to the conclusion that the Code is unsatisfactory in its provisions for infraspecific variation; in particular it is failing to express it adequately and thus failing to focus attention on to it. There is no good reason to believe that a system suitable for supraspecific classification should necessarily be so for infraspecific categories, and there seems to be no alternative to a radical reappraisal and recasting of the relevant Articles of the Code. Such a revised Code needs to take account of the following requirements:

1. It should reflect field variation as far as possible.
2. It should be readily applicable in practice.
3. It should be easily understandable, with an easily recognisable set of categories.
4. It should be flexible but unambiguous.
5. It should initially be acceptable to the great majority of botanists.

In order to satisfy all these demands any new set of Articles would need careful formulation over an extended period, with the advice of taxonomists with varied interests (i.e. not solely nomenclaturists) sought, and the compilers should not shirk from radical changes where they are needed.

It is not feasible to make detailed proposals concerning any new Articles here, but it seems that many of the problems outlined above would be overcome (although some would remain and others, less important, appear) if the hierarchical arrangement of infraspecific categories were discontinued. Burtt[13] has already suggested such action, which could be carried out in two ways: the abandonment of any named infraspecific categories or the use of only one named category, resulting in a trinomial system as practised in zoology at present and to which an approach was made by Raven's proposals[49]; or the use of several categories for different situations but all of them at the same rank, i.e. the same level in the hierarchy, as suggested in the forthcoming new bacteriological code of nomenclature. Of these two approaches the latter seems infinitely preferable for, despite Raven's claims to the contrary, many taxonomists do wish to recognise more than one sort of variation within a species, since it certainly exists, but are at present dissuaded by the Code. For such a system to become operable certain obvious problems, particularly the question of the rules of priority and typification, would need to be overcome, but, to quote Burtt[13], "So far I have merely attempted to sow the germs of the idea that a basic re-appraisal of the current hierarchical use of subspecies and variety is more to be desired than a continuation of the argument on precisely how they shall be defined." To this one might add that the codes of nomenclature should be subservient to, not dictate, taxonomic practice.

CONCLUSIONS

1. Infraspecific variation should be studied critically by all available means.

2. Taxa should be studied over the whole of their range, with particular emphasis on their centres of diversity, to produce an inventory of genetic resources.

3. The variation discovered should be categorized using precisely applicable names, and consideration should be given to amending the International Code of Botanical Nomenclature to provide a suitable 'legal' taxonomic framework.

4. The results should be incorporated in a new style of Critical Flora which integrates and assesses data from the whole range of available sources. Such Floras should be compiled for selected well-studied areas in parallel with the continued preparation of basic Floras of less investigated regions, not in competition with them.

5. The information gathered should be used to investigate evolutionary patterns, and especially to discover whether the major evolutionary mechanisms are the same in all parts of the world.

6. The taxonomic data accumulated should be used to direct programmes of conservation of genetic diversity in the same way that ecological data are used to determine programmes of conservation of habitat diversity.

7. The knowledge obtained on the range and distribution of wild genetic diversity should be used both to improve present crops and to extend the range of useful crop species.

REFERENCES

1. ANDERSON, E. Introgressive hybridisation. *Biol. Rev.* **28** (1953) 280-307.

2. ANDERSON, E. and HUBRICHT, L. Hybridisation in *Tradescantia*, 3. The evidence for introgressive hybridisation. *Am. J. Bot.* **25** (1938) 396-402.

3. ASCHERSON, P. F. A. and GRAEBNER, K. O. P. P. Synopsis der Mitteleuropäischen Flora. Gebrüder Borntraeger, Leipzig (1896-1938).

4. BALL, P. W. *Viscum* L. In T. G. TUTIN *et al.* *Flora Europaea*, Vol. 1. Cambridge University Press (1964) pp 72-73.

5. BENNETT, E. The origin and importance of agroecotypes in south-west Asia. In *Plant Life of South-West Asia*, P. H. DAVIS, P. C. HARPER and I. C. HEDGE (eds.) Botanical Society of Edinburgh (1971) 219-234.

6. BEN-ZE'EV, N. and ZOHARY, D. Species relationships in the genus *Pisum* L. *Israel J. Bot.* **22** (1973) 73-91.

7. BORRILL, M. *Dactylis marina*, sp. nov., a natural group of related tetraploid forms. *J. Linn. Soc. (Bot.)* **56** (1960) 431-439.

8. BORRILL, M. The pattern of morphological variation in diploid and tetraploid *Dactylis*. *J. Linn. Soc. (Bot.)* **56** (1960B) 441-452.

9. BORRILL, M. Studies in *Festuca*, 3. The contribution of *F. scariosa* to the evolution of polyploids in sections *Bovinae* and *Scariosae*. *New Phytol.* **71** (1972) 523-532.

10. BORRILL, M., TYLER, B. and LLOYD-JONES, M. Studies in *Festuca*, 1. A chromosome atlas of *Bovinae* and *Scariosae*. *Cytologia* **36** (1971) 1-14.

11. BRADSHAW, A. D. Populations of *Agrostis tenuis* resistant to lead and zinc poisoning. *Nature* **169** (1952) 1098.

12. BRIGGS, D. and WALTERS, S. M. *Plant variation and evolution*. Weidenfeld and Nicholson, London (1969).

13. BURTT, B. L. Infraspecific categories in flowering plants. *Biol. J. Linn. Soc.* **2** (1970) 233-238.

14. CHANDRASEKHARAN, P. and THOMAS, H. Studies in *Festuca*, 6. Chromosome relationships between *Bovinae* and *Scariosae*. *Z. Pflanzenzüchtg.* **66** (1971) 76-86.

15. CLAPHAM, A. R., TUTIN, T. G. and WARBURG, E. F. *Flora of the British Isles*. Cambridge University Press (1952).

16. CLAUSEN, J., KECK, D. D. and HIESEY, W. M. Experimental studies on the nature of species, 1. The effect of varied environments on western North American plants. *Carnegie Inst. Washington Publ.* **520** (1940).

17. CROSBY, J. L. Selection of an unfavourable gene-complex. *Evolution, Lancaster, Pa* **3** (1949) 212-230.

18. CROSSLEY, G. K. and BRADSHAW, A.D. Differences in response to mineral nutrients of populations of ryegrass, *Lolium perenne* L., and orchard grass, *Dactylis gomerata* L. *Crop. Sci.* **8** (1968) 383-387.

19. DADAY, H. Gene frequencies in wild populations of *Trifolium repens*, 1. Distribution by latitude. *Heredity, Lond.* **8** (1954) 61-78.

20. DAVIES, M. S. and SNAYDON, R. W. Physiological differences among populations of *Anthoxanthum odoratum* L. collected from the Park Grass Experiment, Rothamsted, 2. Response to Aluminium. *J. appl. Ecol* **10** (1973) 47-55.

21. DE WET, J. M. J. and HARLAN, J. R. Chromosome pairing and phylogenetic affinities. *Taxon* **21** (1972) 67-70.

22. FRANKEL, O. H. Genetic conservation in perspective. In O. H. FRANKEL and E. BENNETT (eds.) *Genetic Resources in Plants—Their Exploration and Conservation*. Blackwell Scientific Publications, Oxford (1970) pp. 469-489.

23. FRANKEL, O. H. (ed.) Survey of crop genetic resources in their centres od diversity. Food and Agriculture organisation of the United Nations (1973)

24. FRANKEL, O. H. and BENNETT, E. (eds.) *Genetic Resources in Plants—Their Exploration and Conservation*. Blackwell Scientific Publications, Oxford (1970).

25. FRANKEL, O. H. and HAWKES, J. G. (eds.) *Crop Genetic Resources for Today and Tomorrow.* Cambridge University Press (1975).

26. GAJEWSKI, W. A cytogenetic study of the genus *Geum* L. *Monographiae bot.* **4** (1957).

27. GIGSON, A. and RORISON, I. H. The response of some ecologically distinct plant species to nitrate- and ammonium-nitrogen. *J. Ecol.* **60** (1972) 93-102.

28. GÓMEZ-CAMPO, C. Preservation of West Mediterranean members of the Cruciferous tribe *Brassiceae*. *Biol. Conservation* **4** (1972) 355-360.

29. HANNON, N. and BRADSHAW, A. D. Evolution of salt tolerance in two co-existing species of grass. *Nature, Lond.* **220** (1968) 1342-1343.

30. HEISER, C. B. Introgression re-examined. *Bot. Rev.* **39** (1973) 347-366.

31. HESLOP-HARRISON, J. A synopsis of the dactylorchids of the British Isles. *Ber. geobot. Forsch. Inst. Rübel* **1953** (1954) 53-82.

32. HESLOP-HARRISON, J. Forty years of genecology. *Adv. ecol. Res.* **2** (1964) 159-247.

33. HESLOP-HARRISON, J. Genetic resource conservation: the end and the means. *J. Roy. Soc. Arts* (1974) 157-169.

34. HEYWOOD, V. H. The preservation of the Euro pean flora—the taxonomist's role. *Bull. Jard. bot. nat. belg.* **41** (1971) 153-166.

35. HEYWOOD, V. H. Taxonomy in crisis? or Taxonomy is the digestive system of biology. *Act. bot. Acad. scient. hungar.* **19** (1973) 139-146.

36. HUTCHINSON, T. C. Ecotype differentiation in *Teucrium scorodonia* with respect to susceptibility to lime-induced chlorosis and to shade factors. *New Phytol.* **66** (1967) 439-453.

37. JÁNOSSY, A. Gene-bank work of the National Institute of Agrobotany. *Agrobotanika* **15** (1973) 7-9.

38. JONES, D. A. On the polymorphism of cyanogenesis in *Lotus corniculatus* L., 3. Some aspects of selection. *Heredity, Lond.* **25** (1973) 633-641.

39. JOWETT, D. Population studies on lead tolerant *Agrostis tenuis. Evolution, Lancaster, Pa* **18** (1964) 17-80.

40. LEE, J. A. and STEWART, G. R. Desiccation injury in mosses, 1. Intra-specific differences in the effect of moisture stress on photosynthesis. *New Phytol.* **70** (1971) 1061-1068.

41. LEWIS, M. C. Genecological differentiation of leaf morphology in *Geranium sanguineum* L. *New Phytol.* **68** (1969) 481-503.

42. MIROV, N. T. *The genus* Pinus. The Ronald Press, New York (1967).

43. OCHOA, C. Native potatoes in Bolivia, Chile and Peru. In *Survey of Crop Genetic Resources in Their Centres of Diversity*, O. H. FRANKEL (ed.) Food and Agriculture Organization of the United Nations (1973) 116-124.

44. OLMSTED, C. E. Growth and development in range grasses, 4. Photoperiodic responses in twelve geographic strains of wide-oats grama. *Bot. Gaz.* **106** (1944) 46-74.

45. PENFOLD, A. R. and MORRISON, F. R. The occurrence of a number of varieties of *Eucalyptus dives* as determined by chemical analyses of the essential oils, 1. *J. Proc. Roy. Soc. N.S.W.* **61** (1927) 54-67.

46. PERRING, F. H. (ed.) *Critical Supplement to the Atlas of the British Flora.* Thomas Nelson, London (1968).

47. PLITMANN, U. Biological Flora of Israel, 4. *Vicia sativa* subsp. *amphicarpa* (Dorth.) Aschers. & Graebn. *Israel J. Bot.* **22** (1973) 178-194.

48. PRITCHARD, N. M. *Gentianella* in Britain, 1. *G. amarella, G. anglica* and *G. uliginosa. Watsonia* **4** (1959) 169-193.

49. RAVEN, P. H. Proposals for the simplification of infraspecific terminology. *Taxon* **23** (1974) 828-831.

50. RUNEMARK, H. Studies in the Aegean flora, 13. *Tordylium* L. (Umbelliferae). *Bot. Notiser* **121** (1968) 233-258.

51. SANDWITH, N. Y. Review of A California Flora by P. A. MUNZ. *Watsonia* **5** (1961) 110-111.

52. SINSKAIA, E. N. and BEZTUZHEVA, A. A. The forms of *Camelina sativa* in connection with climate, flax and man. *Bull. sppl. Bot. Genet. pl. Breed.* **25** (1931) 98-200.

53. SNAYDON, R. W. The growth and competitive ability of contrasting natural populations of *Trifolium repens* L. on calcareous and acid soils. *J. Ecol.* **50** (1962) 439-447.

54. SNAYDON, R. W. and BRADSHAW, A. D. Differences between natural popuiations of *Trifolium repens* in response to mineral nutrients, 1. Phosphates. *J. exp. Bot.* **13** (1962) 422-434.

55. SNAYDON, R. W. and BRADSHAW, A. D. Differences between natural populations of *Trifolium repens* in response to mineral nutrients, 2. Calcium, magnesium and potassium. *J. appl. Ecol.* **6** (1969) 185-202.

56. SNOGERUP, S. Studies in the Aegean flora, 9. *Erysimum* sect. *Cheiranthus,* B. Variation and evolution in the small-population system. *Opera Bot.* **14** (1967) 1-85.

57. STAPLEDON, R. G. Cocksfoot grass (*Dactylis glomerata* L.): ecotypes in relation to the biotic factor. *J. Ecol.* **16** (1928) 71-104.

58. STRID, A. Studies in the Aegean flora, 16. Biosystematics of the *Nigella arvensis* complex with special reference to the problem of non-adaptive radiation. *Opera Bot* **28** (1970) 1-169.

59. THOMPSON, P. A. and BROWN, G. E. The seed unit at the Royal Botanic Gardens, Kew. *Kew Bull.* **26** (1972) 445-456.

60. TIKU, B. L. and SNAYDON, R. W. Salinity tolerance with the grass species *Agrostis stolonifera. Plant & Soil* **35** (1971) 421-431.

61. TURESSON, G. The genotypical response of the plant species to the habitat. *Hereditas* **3** (1922) 211-350.

62. TUTIN, T. G. *et al. Flora Europaea,* Vol. 1. Cambridge University Press (1964).

63. VALENTINE, D. H. Biosystematic research in progress in Europe, February 1969. Private issued, Manchester.

64. VALENTINE, D. H. Experimental taxonomy in Great Britain—Research in progress—March 1974. Private issued, Manchester.

65. VALENTINE, D. H. The taxonomic treatment of polymorphic variation. *Watsonia* **10** (1975) 385-390.

66. VAVILOV, N. I. The origin, variation, immunity and breeding of cultivated crops. Chronica Botanica, Waltham, Massachusetts (1949-1950).

67. WALTERS, S. M. Taxonomic and floristic aspects of plant conservation. In *Plant life of South-West Asia,* P. H. DAVIS, P. C. HARPER and I. C. HEDGE (eds.) Botanical Society of Edinburgh (1971) 293-296.

68. WILKINS, D. A. A technique for the measurement of lead tolerance in plants. *Nature* **180** (1957) 37-38.

69. ZEVEN, A. C. and ZHUKOVSKY, P. M. Dictionary of cultivated plants and their centres of diversity. Centre for agricultural publishing and documentation, Wageningen (1975).

70. ZOHARY, D. Centers of diversity and centers of origin. In *Genetic Resources in Plants—Their Exploration and Conservation,* O. H. FRANKEL and E. BENNETT (eds.) Blackwell Scientific Publications, Oxford (1970A) 33-42.

71. ZOHARY, D. Wild wheats. In *Genetic Resources in Plants—Their Exploration and Conservation,* O. H. FRANKEL and E. BENNETT (eds.) Blackwell Scientific Publications, Oxford (1970B) 239-247.

72. ZOHARY, D. Origin of South-West Asiatic cereals: wheats, barley, oats and rye. In *Plant life of South-West Asia,* P. H. DAVIS, P. C. HARPER and I. C. HEDGE (eds.) Botanical Society of Edinburgh (1971) 235-263.

73. ZOHARY, D. The wild progenitor and the place of origin of the cultivated lentil: *Lens culinaris. Economic Bot.* **26** (1972) 326-332.

CYTOPLASMIC pH REGULATION AND ELECTROGENIC H^+ EXTRUSION

J. A. RAVEN and F. A. SMITH*

Department of Biological Sciences, University of Dundee, Dundee DD1 4HN, Scotland

INTRODUCTION

The extreme sensitivity of enzyme-catalysed reactions to pH changes suggests that regulation of cytoplasmic pH is a fundamental feature of cell homeostasis [15, 16, 72–75, 91, 92]. One process which could alter cytoplasmic pH is the passive movement of H^+ across the plasmalemma. At external pH values below pH 9, the passive driving forces of electrical potential difference and chemical concentration difference across the plasmalemma, tend to move H^+ inwards, i.e. cause cytoplasmic acidification. The other major sources of cytoplasmic pH perturbation are essential biosyntheses required for cell growth. Growth with CO_2 or carbohydrate as the C source leads to excess H^+ production in the cytoplasm when NH_4^+ is the N source, and excess OH^- production when NO_3^- is the N-source[72–74].

Thus both passive H^+ transport and metabolism of H^+ could produce large changes of cytoplasmic pH if there is no pH regulation[74]. Measurements of cytoplasmic pH[18, 37, 72, 73, 102] suggest that it *is* regulated to a value close to pH 7.0. Possible mechanisms for cytoplasmic pH regulation include (1) pre-existing cytoplasmic buffers, (2) biochemical production of H^+ or OH^-, or (3) active transport of H^+ or OH^-.

(1) Pre-existing cytoplasmic buffers: this mechanism is only effective in the short term, as the buffers are themselves produced by pH-perturbing reactions. The Appendix gives estimates of the cytoplasmic buffer capacity in the physiological range (pH 6–8), and shows that a pH change of 0.3 units can occur in a short time in response to the H^+ or OH^- changes occasioned by passive H^+ flux or by N assimilation.

(2) Biochemical synthesis of H^+ or OH^- has been analysed by Davies[15, 16]. Here excess OH^- production (e.g. during NO_3^- assimilation) is viewed as increasing the cytoplasmic pH, which in turn increases the rate of a pH sensitive reaction (PEP carboxylase) in the synthesis of a strong acid (malic acid) from neutral and weakly acid substrates (carbohydrate and CO_2). Conversely, excess H^+ production (e.g. during NH_4^+ assimilation or passive H^+ entry) lowers cytoplasmic pH. This inhibits malic acid synthesis, and activates malic enzyme which breaks down $malate^{2-}$ to $pyruvate^-$, CO_2 and OH^-. These reactions have sufficient capacity to deal with most naturally-occurring pH perturbations, and have been shown to occur *in vivo*[15, 16, 72–75, 101]. The pH responses of the carboxylation and

*Permanent address: Department of Botany, University of Adelaide, South Australia 5001, Australia.

decarboxylation reactions are such that they would regulate cytoplasmic pH (to within a small fraction of a pH unit) just on the alkaline side of neutrality.

It is clear that a cell using CO_2 or carbohydrate as its C source can use this 'biochemical pH stat' to counter long-term OH^- synthesis: the malic acid is synthesised from exogenous neutral materials. Generation of OH^- to counter excess cytoplasmic H^+ by malate^{2-} breakdown in the 'biochemical pH stat' can only occur to the extent that the cell has a supply of malate^{2-}. For a cell using CO_2 or carbohydrate as its C source, this malate^{2-} can only have arisen from malic acid synthesis and disposal of excess H^+ [see (3) below]. Thus, over a period of a quarter of a cell generation time or longer, disposal of excess H^+ cannot occur via the 'biochemical pH stat' alone, and H^+ extrusion from the cell is required

(3) H^+ or OH^- transport out of the cytoplasm into the surrounding aqueous medium is a common response to excess cytoplasmic H^+ or OH^- in plants[72–74] and animals[1, 99]. At external pH values below 9, application of the Nernst equation, using measured values of E_m and of cytoplasmic and external pH, shows that passive driving forces on H^+ are directed inwards at the plasmalemma[72, 73, 92, 102]. Thus the net H^+ efflux found during NH_4^+ assimilation or the accumulation of salts of organic acids is an active process, while the net OH^- efflux associated with NO_3^- assimilation is passive, in solutions with a pH below 9. In solutions of higher pH, it is likely that net H^+ efflux is passive, while net OH^- efflux is active. The regulatory properties which would be predicted if these fluxes constitute a 'biophysical pH stat' are discussed below.

The central vacuole of plant cells has a lower pH than the cytoplasm[73, 102]. Consideration of the free energy gradient for H^+ across the tonoplast shows that production and maintenance of this low vacuolar pH requires active H^+ transport from the cytoplasm to the vacuole. This is consistent with a lysosomal origin of plant vacuoles[19, 21, 60]. The limited capacity of the vacuole as a sink for excess cytoplasmic H^+ is discussed by Raven and Smith[74].

The relative importance of the 'biochemical pH stat' and the biophysical pH stat' in the regulation of the pH of plant cell cytoplasm is discussed at the end of this paper.

NATURE OF THE H^+ FLUXES IN PLANT CELLS

General considerations on ion transport

Ion transport through plant cell membranes can either be passive (in response to the driving forces of transmembrane differences in electrical potential, and in chemical concentration of the transported ion), or active (in a direction not accounted for by the passive driving forces, and hence requiring coupling to some exergonic process: a biochemical reaction in primary active transport, or some other solute flux in secondary active transport). All active movements, and many passive ones, require some membrane components additional to the basic bimolecular leaflet[69]. Such mediated transport can be by uniport, i.e. the membrane carrier transports a single solute species. When the uniported solute is an ion, transport moves charge across the membrane and can contribute to the transmembrane potential difference, and to the membrane conductance: such transport is electrogenic. When the membrane carrier transports two solute species, the transport is said to be antiport if the two solute species move in opposite directions, or symport if the two species move in the same direction. Antiport and symport can be either electroneutral or electrogenic, depending on the species transported and the stoichiometry: and, unlike uniport, can incorporate secondary active transport.

The electrogenic H^+ efflux pump

Early attempts to explain the resting potential in plant (mainly giant algal) cells were in terms of passive electrogenic ion fluxes; active ion transport was thought of as electroneutral[13]. The equation used was of the constant field (Goldman) type, i.e.:

$$E_m = \frac{RT}{F} \ln \frac{P_K.K_o + P_{Na}.Na_o + P_{Cl}.Cl_i}{P_K.K_i + P_{Na}.Na_i + P_{Cl}.Cl_o}$$

where E_m is the electrical potential of the cell interior taking the external potential as zero; P_K, etc., the membrane permeabilities to K, etc; the subscripts *o* and *i* refer to the external and internal concentrations respectively of the various ions; and *R, T* and *F* have their usual meanings.

This equation was unsatisfactory for a number of reasons[34, 69, 83, 95].

(1) At least in the presence of Ca^{2+}_o and at low K^+_o, this equation does not account quantitatively for the resting potential, or its lack of variation with changes in K_o, Na_o or Cl_o.

(2) Although most of the measured values of the resting potential are within the range permitted by the Goldman equation, it is sometimes found that E_m is more negative than E_K, the most negative potential which can be produced by passive ion diffusion.

(3) The plasmalemma conductance of freshwater algal coenocytes is greater than that calculated from tracer passive fluxes of K^+, Na^+ and Cl^-

(4) Passive ion fluxes commonly show larger responses to changes in the electrical potential component of the passive driving force, and smaller responses to changes in the chemical concentration component of the passive driving force, than expected.

Kitasato[41] suggested, on the basis of his work on *Nitella clavata*, that H^+ fluxes could account for some of these discrepancies. The large response of E_m to changes in external pH in the range pH 4–6 was consistent with a high P_H; this could account in part for the anomalously high conductance, and necessitates terms for H^+ in the Goldman equation. The values of E_m greater than E_K, and the depolarisation caused by DNP, can be explained in terms of electrogenic active H^+ efflux. This again requires an additional term in the Goldman equation, and could also contribute to membrane conductance (see below). These additional terms for electrogenic active and passive H^+ transport in the Goldman equation enable it to be fitted more readily to real membrane behaviour.

Evidence has recently accumulated for values of E_m greater than E_K in plant cells. These very negative potentials are rapidly and reversibly decreased by inhibition of energy metabolism. This is consistent with the operation of a hyperpolarising electrogenic pump[27, 28]. This could involve *either* cation efflux *or* anion influx. Evidence as to which ion is transported electrogenically can be obtained by removing the suspected ion from the side of the membrane from which its active transport occurs. This is not possible in the case of cytoplasmic H^+; *faute de mieux* electrogenic H^+ efflux is invoked if removal of anions from the bathing medium does *not* cause a *rapid* decrease in E_m. In aquatic and terrestrial glycophytes the removal of such rapidly transported anions as Cl^- and NO_3^- from the bathing medium produces effects on E_m which do not favour the view that the active influx of these anions involves net negative charge influx[11, 58, 97]. However, in the marine alga

Acetabularia the metabolically maintained ($E_m - E_K$) of some −80 mV is reduced by the removal of Cl_o[24, 25, 78]. In the marine alga *Halicystis* and the very closely related[69] *Bryopsis*, removal of Cl_o causes a rapid and reversible decrease in E_m, which in these algae is more positive than E_m[8, 54]. In *Limonium*, active Cl^- extrusion from the salt glands appears to electrogenic[33].

Thus, at least in glycophytes, values of E_m more negative than E_K is best explained in terms of active electrogenic cation efflux. Quantitatively, Na^+ efflux is inadequate; the only other cation which is actively extruded is H^+. Evidence that the hyperpolarisation is indeed due to electrogenic H^+ extrusion comes from experiments in which there is a correlation between negative values of E_m and net H^+ efflux (ideally, the tracer H^+ flux should be used, but this cannot be measured). Such a correlation has been shown for *Neurospora* hyphae[83, 84] *Beta* storage tissue discs[61–63] in $KHCO_3$ or K_2SO_4 solutions, and for a number of higher plant tissues in which H^+ extrusion and hyperpolarisation can be manipulated by the addition of indoleacetic acid or fusicoccin[49, 50, 59, 104].

The common feature of these experiments in which net H^+ efflux is correlated with hyperpolarisation is that there is a large net H^+ production in the cytoplasm due to organic acid synthesis, and an absence of major 'short-circuiting' of either the ΔE or the Δ pH generated by the H^+ pump. Patently there must be some passive electrogenic ion movement balancing the net H^+ efflux. This appears to be largely a net passive influx of monovalent cations in the experiments cited above, with a low P_{cation} requiring that a large value of ($E_m - E_K$) be produced in order that net passive cation influx shall equal net H^+ efflux.

However, such a correlation of hyperpolarisation with net H^+ efflux is not always found. It will be seen in the following section that this is not necessarily evidence against the occurrence of an electrogenic H^+ extrusion, although it is clear that net H^+ active efflux in the absence of a hyperpolarisation could be accounted for by electroneutral active H^+ efflux, while hyperpolarisation in the absence of net H^+ efflux could be explained by the occurrence of some other electrogenic active flux.

Short-circuiting the electrogenic H^+ pump

(1) *Passive cation influx*. When the product of external cation concentration and P_{cation} is high, a net passive cation influx could equal net active H^+ efflux at relatively low values of E_m. Provided the major passively moving cation is not H^+ [see (2) below], this results in a short-circuiting of the electrical, but not the chemical, component of the proton-motive force. A possible example of this is a charophyte cell at high external K^+ concentration. Here there is an apparent increase in P_K at external K concentrations above about 1mM[79, 94]. Under these conditions E_m is responsive to changes in K^+ concentration, K^+ tracer influx is increased, and E_m is never higher than E_K.

Another possible example is NH_4^+ assimilation. Here there is a considerable H^+ generation within, and active H^+ efflux from, the cell[74], yet NH_4^+ depolarises E_m more effectively than does an equal, low concentration of K^+[29]. This is consistent with the larger net influx of NH_4^+ than of K^+ during cell growth with NH_4^+ as N-source[69]. Further measurements are needed, especially of the effect of NH_4^+ on plasmalemma conductance.

(2) *Passive H^+ uniport.* A high P_H short-circuits not only the electrical (ΔE) component of the proton-motive force as does (1) above, but also the chemical (Δ pH) component. Evidence for a high P_H comes from the finding that decreasing pH 6 gives large depolarisations in many plant cells (40–60mV per pH unit):[4, 24, 36, 39–41, 77, 79, 81, 83, 88, 95, 96]. In a number of green tissues the

pH dependency is much greater in the light than in the dark; this is correlated with a maximal light-induced hyperpolarisation at pH 6–8 (e.g.[7, 95]). This suggests that there may be an increased passive electrogenic H^+ flux paralleling light stimulation of the electrogenic H^+ efflux[7], such as was found for active and passive electrogenic Cl^- transport in *Acetabularia*[25]. The occurrence of a high P_H, at least at low external pH values, is consistent with the large net H^+ influx observed at low external pH in *Nitella clavata*[76].

The large value for passive H^+ uniport suggested by these results helps to account for the 'conductance anomaly' mentioned above, but means that both pH regulation via active H^+ efflux, and any associated co-transport phenomena, are energetically inefficient. It is, however, possible that a portion of the apparent H^+ uniport is in fact H^+ symport in a complex bearing a net positive charge [see (3) below]; this reduces the apparent energetic inefficiency.

(3) *Passive electrogenic H^+ symport.* This has similar consequences for ΔE and ΔpH as does (2), but coupling to a neutral solute can lead to secondary active transport of this solute (e.g. sugars or uncharged amino-acids). The symport is detected as a net H^+ influx associated with the neutral solute influx[22, 42, 82, 86], a glucose-induced decrease in E_m[23, 37, 85, 86], or as active influx of neutral solute in the absence of biochemical energy supplies but in the presence of a proton-motive force[22, 82]. Anion influx can be coupled in a similar fashion, provided the stoichiometry of H^+ and anion binding to the carrier is such as give net positive charge influx[10, 11, 58], see (4) below.

Passive electrogenic H^+ symport can thus employ both the electrical and chemical components of the proton-motive force generated by the electrogenic H^+ pump to energise secondary active solute transport.

(4) *Passive electroneutral H^+ symport (or OH– antiport).* This resembles (3), but involves 1:1 stoichiometry of anion influx with H^+ influx (or OH^- efflux). This results in a short-circuit of the ΔpH component of the proton-motive force. The hyperpolarisation resulting from electrogenic H^+ extrusion will tend to be short-circuited by passive cation influx, and the result could be an accumulation of salts such as KCl[88, 89], where Cl^- is transported by secondary active transport (see above). Evidence for such a system has been produced for freshwater giant algal cells, in which net Cl^- influx is found when an inside-alkaline pH gradient is imposed across the plasmalemma, but with decreased biochemical energy sources for transport[67, 70, 89]. However, these experiments do not prove that no biochemical energy input is involved. Indeed, such an input is required to allow net Cl^- influx when Cl_i^-/Cl_o^- is greaterthan OH_i^-/OH_o^+, if coupling of Cl^- and OH^- is 1:1[45, 47, 48, 93].

This system is analogous to K^+ organate$^-$ accumulation, except that in the latter intracellular OH^- resulting from H^+ extrusion is neutralised by organic acid synthesis rather than Cl^-/OH^- exchange. Effects of E_m will be more complex, however, due to additional limitations on H^+ extrusion by Cl^-/OH^- exchange. The system can explain the observations of Higinbotham and Pierce[32] that E_m values more negative than E_K occur during net KCl accumulation in the absence of net organic acid changes, although this is not always found[61–63], possibly due to variable values of P_K.

Electroneutral symport, or antiport dependent on H^+ extrusion, can also be invoked to explain the high values of E_m often found in the presence of exogenous NO_3^- (e.g.[30, 31]), whose assimilation generally leads to a net OH^- efflux[74]. In higher plants net OH^- efflux is usually less than net NO_3^- influx and assimilation; thus if there were an electroneutral NO_3^-/OH^- antiport, there would also have to be an active H^+ efflux to maintain cytoplasmic pH in the face of an OH^- efflux (in exchange for NO_3^-) in excess of the OH^- generated in the cytoplasm during NO_3^- assimilation. This combination of an electrogenic H^+ efflux and a larger electroneutral OH^- efflux can explain the hyperpolarisation as well as the net OH^- efflux observed during NO_3^- assimilation. Under these conditions only a portion of the driving force for net passive OH^- efflux is generated by active H^+ extrusion; the rest is generated by net OH^- production in the cytoplasm due to NO_3^- assimilation. Thus the metabolic coupling of any secondary active transport linked to this OH^- efflux is via the reductive assimilation of NO_3^-.

(5) *Conclusions.* It is clear that coupling of the influx of a very permeant cation to the electrical potential difference generated by active electrogenic H^+ efflux can, in the presence of net H^+ production within the cell, lead to a large net H^+ efflux without E_m exceeding E_K (see ref.[64] for discussion of analogous phenomena related to the K^+/Na^+ pump in animal tissue). Conversely, in the absence of net H^+ generation within the cell, recycling of extruded H^+ via an electroneutral anion symport (anion-OH^- antiport) can allow a hyperpolarisation due to the electrogenic H^+ efflux even when there is no *net* H^+ efflux. In the case of nitrate assimilation there may even be a net OH^- efflux by an electroneutral mechanism in the presence of a high E_m due to electrogenic H^+ efflux.

While it would be extravagant to claim that every ion transport phenomenon in plants is associated with electrogenic H^+ transport, the cases considered above *do* show that the absence of a correlation between hyperpolarisation and net H^+ efflux need not rule out the occurrence of an electrogenic H^+ extrusion.

Energetics of active H^+ transport

In various eukaryote tissues, evidence consistent with coupling of active H^+ efflux at the plasmalemma to metabolism via ATP[87, 97, 98], a high-energy-precursor of ATP[43] and redox reactions of photosynthesis[46] has been presented. With respect to active H^+ transport at the tonoplast, Lambert[44] has shown that H^+ transport into the luteoids of *Hevea* latex uses ATP. These luteoids probably share a common, lysosomal, origin with plant vacuoles (see Introduction, and D'Auzac and Lioret[14]). ATP appears to be involved in H^+ transport in animal lysosomes[51].

Regulation of H^+ transport

If cytoplasmic pH is to be regulated by H^+ and OH^- fluxes, it would be predicted that H^+ and OH^- efflux should respond to cytoplasmic pH like the OH^--generating (malic enzyme) and H^+-generating (PEP carboxylase) components of the 'biochemical pH stat' respectively. Thus H^+ efflux from the cytoplasm (normally active, but passive at high values of external pH) should be stimulated by small decreases in cytoplasmic pH, while OH^- efflux should be stimulated by increased cytoplasmic pH. Whether such pH changes can be measured by the currently available techniques for determining cytoplasmic pH depends on the sensitivity of the fluxes to cytoplasmic pH changes.

Cram[12] discusses regulation of transport processes in plant cells. In most cases it appears that net active transport is not close to thermodynamic equilibrium with their energy source, i.e. at 'static head' in the terminology of Caplan[9]. While operations near 'static head' gives good energy conversion for the (small) fluxes which occur, it does not give the optimal balance of rate and efficiency[55], and is relatively inflexible as regards regulation.

Nevertheless, Walker and Smith[102] have presented evidence which suggests that the H^+ pump in *Chara corallina* is running close to static head over a range of values of pH_0. The notion that the H^+ pump is close to thermodynamic equilibrium with its biochemical energy source is supported by evidence that the pump responds to applied electrical currents (i.e. it possesses a conductance: see refs.[81, 95, 105]; cf. also ref.[3]) as well as external pH changes. However, at 'static head' there are restrictions on regulation of pH *vis à vis* regulation of membrane potential[5, 26, 103]. In particular, no increase in proton motive force is possible without an increase in the free energy available from the biochemical energy source (e.g. an increased [ATP] / [ADP] + [P_i]). Hence the chemical concentration component (i.e. the pH difference across the plasmalemma) can only be increased at the expense of the electrical potential difference, and vice versa. This has obvious implications for situations in which large net H^+ fluxes are known to occur. For example, the suggestion[73, 92] that H^+ pump activity can be stimulated by metabolic triggers other than energy supply (e.g. by hormones) would not be valid at static head.

RELATIVE IMPORTANCE OF THE BIOCHEMICAL AND THE BIOPHYSICAL pH-STATS

The biophysical pH-stat, mediating net H^+ or OH^- efflux from the cytoplasm to the medium can deal, in the short *or* the long term, with both excess H^+ and excess OH^- in the cytoplasm. However, it can only operate if there is a large extracellular aqueous sink for the excreted H^+ or OH^-. In submerged algae H^+ or OH^- excretion is the major pH-stat mechanism[74]. The biophysical pH stat can also operate in submerged vascular plants, and in the less bulky of the underground organs of terrestrial vascular plants.

However, in aerial shoots and in bulky underground organs of terrestrial vascular plants, there is a very limited aqueous extracellular volume (i.e. the cell walls) within diffusive range of the H^+ or OH^- excreted from cells[71, 74]. Unregulated H^+ or OH^- extrusion from such cells into the cell walls would very soon produce damaging changes in *extracellular* pH.

The solution to this problem, which terrestrial vascular plants have adopted, is to carry out the *biophysical* pH stat reactions involving net H^+ or OH^- excretion, in the roots. Combined operation of the biochemical and biophysical pH stat in the roots occurs in relation to the pH stress occasioned by NH_4^+ or NO_3^- assimilation. The roots thus supply the shoot with a mixture of nutrients, which can be incorporated during shoot growth without resulting in pH changes, that cannot be dealt with by H^+ production in a biochemical pH stat[71, 74].

This does not, however, completely exclude active H^+ extrusion at the plasmalemma from the normal metabolism of shoot cells. One possible exception is the involvement of IAA-induced H^+ efflux in cell expansion, related to an increased cell wall plasticity at low pH values[17, 104]. The small extracellular aqueous volume in intact shoots is compatible with a small H^+ extrusion from the expanding shoot cells (balanced by cation influx), leading to a sufficient decrease in cell wall pH to increase the wall plasticity. The large increase in vacuolar solutes (largely K^+ organate$^-$) would then come about *not* from *de novo* organic acid synthesis and K^+/H^+ exchange, such as can occur when tissue is immersed in a large volume of solution[20], but from K^+ organate$^-$ import from seedling reserves[53], or from roots and rhizomes[2, 74].

A case in which turgor generation in shoots does arise from organic acid synthesis and K^+/H^+ exchange is stomatal opening in many plant species[52, 56, 57, 65]. Here the H^+ lost from the guard cells probably exchanges with K^+ in the Donnan free space of the leaf. The capacity of the Donnan free space to supply K^+ and take up H^+ appears to be adequate in many plant species[35, 66, 90].

In *Vicia faba* the H^+ content of the Donnan free space of the epidermis would be doubled during stomatal opening, if it took up all of the excess H^+ generated in the guard cells. The possibility of abnormal patterns of ion exchange must once again be borne in mind, when epidermal strips or leaf slices are bathed in large volumes of solution. It may be concluded that the biochemical pH stat has a particular significance in vascular plants, but that excess H^+ production can still only be dealt with in the long term by excretion from the roots. As we have suggested previously[74], much of the complexity of the location of N metabolism and the long distance transport of organates and organic N may be related to pH regulation in cells isolated from large extracellular sinks for H^+ or OH^-. The extent to which an explanation for stress-induced changes in metabolism may be sought in the need for pH regulation[92] remains to be seen.

APPENDIX

Cytoplasmic buffer capacity in plant cells

In most cells the major buffering substances operative in the pH range pH 6–8 are phosphate compounds with a pK_a of 6.8–7.0, and certain amino acids (histidine, pK_a = 6.0, –SH of cysteine, pK_a = 8.3, and an $-NH_2$ of cystine, pK_a = 7.85) both free and in proteins. For the microalga *Chlorella*, the content of free and protein amino-acids and of phosphates with a pK_a of 6.8–7.0 (refs.[38, 68, 69]) suggests that the cytoplasmic buffer capacity is about 20 mMol H^+ $litre^{-1}$ pH $unit^{-1}$ in the physiological range pH 6–8. Similar computations for the cytoplasm of *Hydrodictyon* and charophytes (based on data in Raven[68, 69]) suggest that the values are similar. This agrees with estimates for *Nitella clavata*[6, 76] and for animal cell cytoplasm[1, 99, 100].

In the absence of significant vacuolar buffer capacity in these freshwater giant algal cells, a net H^+ efflux of 20 pMol cm^{-2} s^{-1} (see refs.[41, 81, 98]) would change the pH of the cytoplasm (1 μl cm^{-2}) by 0.3 pH units in 5 min. For comparison, the amount of charge transfer needed to hyperpolarise the cell by 50 mV is only 0.5 p^{-mole} H^+ cm^{-2}, assuming a membrane capacitance of 1 μF cm^{-2}; such a hyperpolarisation will only take a fraction of a second if the net H^+ flux of 20 pMol cm^{-2} s^{-1} were not accompanied by other charge-transfer processes.

REFERENCES

1. AICKIN, C. C. and THOMAS, R. C. Microelectrode measurements of the internal pH of crab muscle fibre. *J. Physiol.* **252**, (1975) 803–816.

2. ALLSOP, A. Seasonal changes in the organic acids of rhubarb (*Rheum hybridum*). *Biochem. J.* **81**, (1937) 1820–1829.

3. ANDERSON, W. P., HENDRIX, D. L. and HIGINBOTHAM, N. The effect of cyanide and carbon monoxide on the electrical potential and resistance of cell membranes. *Pl. Physiol.* **54**, (1974) 712–716.

4. ANDRIANOV, V. K., VOROBEYEVA, I. A. and KURELLA, G. A. Study of the nature of resting potential in *Nitella* cells – I. Medium pH effect on resting potential of *Nitella* cells. *Biofizika* **13** (1968) 335–336.

5. BARR, C. E. and RYAN, T. E. Hypothesis: the trifunctional H^+ pump in *Nitella*. *Pl. Physiol.* **50** (1974) s-50.

6. BARR, C. E., KOH, M. S. and RYAN, T. E. NH_3 efflux as a means for measuring H^+ extrusion in *Nitella*. In *Membrane Transport in Plants*, ZIMMERMANN, U. and DAINTY, J. (eds.), Springer, Berlin (1974) 180–185.

7. BENTRUP, F. W., GRATZ, H. J. and UMBEHAUEN, H. The membrane potential of *Vallisneria* leaf cells: evidence for light-dependent proton permeability changes. In *Ion Transport in Plants*, ANDERSON, W. P. (ed.), Academic Press, London (1973) 171–182.

8. BLINKS, L. R. The relation of bioelectric phenomena to ionic permeability and to metabolism in large plant cells. *Cold Spring Harbor Symp. quant. Biol.* **8** (1940) 204–215.

9. CAPLAN, S. R. Non-equilibrium thermodynamics and its application to bioenergetics. In *Current Topics in Bioenergetics*, Vol. 4, SANADI, D. R. (ed.), Academic Press, N Y (1971) 1–79.

10. COCKBURN, M., EARNSHAW, P. and EDDY, A. A. The stoicheiometry of the absorption of protons with phosphate and *L*-glutamate by yeast cells of the genus *Saccharomyces*. *Biochem. J.* **146** (1975) 705–712.

11. CRAM, W. J. Relationships between chloride transport and electrical potential differences in carrot root cells. *Aust. J. biol. Sci.* **2** (1975) 301–310.

12. CRAM, W. J. Negative feed back regulation of transport in cells. The maintenance of turgor, volume and nutrient supply. In *Encyclopedia of Plant Physiology – Short Distance Transport.* LÜTTGE, U. and PITMAN, M. G. (eds.), Springer, Berlin (1976) 284–316.

13. DAINTY, J. Ion transport and electrical potentials. *Ann. Rev. Pl. Physiol.* **13** (1962) 379–402.

14. D'AUZAC, J. and LIORET, C. Mise en evidence d'un mecanisme d'accumulation du citrate dans les luteoides de latex d'*Hevea brasiliensis*. *Physiolog. Veg.* **12** (1975) 617–635.

15. DAVIES, D. D. Metabolic control in higher plants. In *Biosynthesis and its Control in Plants*, MILLBORROW, B. V. (ed.), Academic Press, London (1973*a*), 1–20.

16. DAVIES, D. D. Control of and by pH. *Symp. Soc. exp. Biol.* **27** (1973*b*) 513–529.

17. DAVIES, P. J. Current theories on the mode of action of auxin. *Bot. Rev.* **39** (1973) 140–171.

18. DAVIS, R. F. Photoinduced changes in electrical potentials and H^+ activities of the chloroplast, cytoplasm and vacuole of *Phaeoceros laevis*. In *Membrane Transport in Plant Cells*, ZIMMERMAN, U. and DAINTY, J. (eds.), Springer, Berlin (1974) 197–201.

19. DE DUVE, C., DE BARSEY, Y., POOLE, B. TROUET, A., TULKERS, P. and VAN HOOF, F. Lysosomotropic agents. *Biochem. Pharmacol.* **23** (1974) 2495–2531.

20. DHINDSA, R. A., BEASLEY, C. A. and TING, I. P. Osmoregulation in cotton fiber. Accumulation of potassium and malate during growth. *Pl. Physiol.* **56** (1975) 394–398.

21. DOI, E., OHTSURU, C. and MATOBA, T. Lysosomal nature of plant vacuoles – II. Acid hydrolases in the central vacuole of Charophyta. *Pl. Cell Physiol., Tokyo* **16** (1975) 581–588.

22. EDDY, A. A. and NOWACKI, J. A. Stoicheiometric proton and potassium ion movements accompanying the absorption of amino-acids by the yeast *Saccaromyces carlsbergiensis*. *Biochem. J.* **122** (1971) 701–711.

23. ETHERTON, B. and NUOVO, C. J. Rapid changes in membrane potential of oat coleoptile cells induced by amino acids and carbohydrates. *Pl. Physiol.* **53** (1974) s-49.

24. GRADMANN, D. Einfluss von Licht, Temperatur und Aussenmedium auf das elektrische Verhalten von *Acetabularia mediterranea*. *Planta* **93** (1970) 323–353.

25. GRADMANN, D., WAGNER, G. and GLASEL, R. H. Chloride efflux during light-triggered action potentials in *Acetabularia mediterranea*. *Biochim. biophys. Acta* **323** (1973) 151–155

26. HASTINGS, D. F. Evolution of the cell potential as an energy conserving mechanism. *Biophys. J.* **15** (1975) 114a.

27. HIGINBOTHAM, N. Electropotentials of plant cells. *Ann. Rev. Pl. Physiol.* **24** (1973) 25–46.

28. HIGINBOTHAM, N. and ANDERSON, W. P. Electrogenic pumps in higher plant cells. *Can. J. Bot.* **52** (1974) 1011–1121.

29. HIGINBOTHAM, N., ETHERTON, B. and FOSTER, R. J. The effects of external K, NH_4, Na, Ca, Mg and H ions on the cell transmembrane potential of *Avena* coleoptiles. *Pl. Physiol.* **39** (1964) 196–203.

30. HIGINBOTHAM, N. ETHERTON, B. and FOSTER, R. J. Mineral ion contents and cell transmembrane electropotentials of pea and oat seedling tissue. *Pl. Physiol.* **42** (1967) 37–46.

31. HIGINBOTHAM, N., GRAVES, J. S. and DAVIS, R. F. Evidence for an electrogenic ion transport pump in cells of higher plants. *J. membrane Biol.* **3** (1970) 210–222.

32. HIGINBOTHAM, N. and PIERCE, W. S. Potassium uptake with respect to cation-anion balance in pea epicotyl segments. In *Membrane Transport in Plants,* ZIMMERMANN, U. and DAINTY, J. (eds.), Springer, Berlin (1974) 406–411.

33. HILL, A. E. and HILL, B. S. The electrogenic chloride pump of the *Limonium* salt gland. *J. membrane Biol.* **12** (1973) 129–144.

34. HOPE, A. B. and WALKER, N. A. *The Physiology of Giant Algal Cells.* Cambridge University Press (1975) 201–13.

35. JACOBY, B. Light sensitivity of ^{22}Na, ^{86}Rb and ^{42}K absorption by different tissues of bean leaves. *Pl. Physiol.* **55** (1975) 978–981.

36. JESCHKE, W. D. Lichtabhängige Veränderungen des Membranpotentials bie Blattzellen von *Elodea densa. Z. Pflanzenphysiol.* **62** (1970) 158–172.

37. JONES, M. G. K., NOVACKY, A. and DROPKIN, V. H. Transmembrane potentials of parenchyma cells and nematode-induced transfer cells. *Protoplasma* **85** (1975) 15–38.

38. KANAZAWA, T. Changes of amino acid composition of *Chlorella* cells during their life cycle. *Pl. Cell Physiol., Tokyo* **5** (1964) 333–354.

39. KISHIMOTO, U. Electrical characteristics of *Chara corallina* — I. The membrane potential. *A. Rep. Sci. Works, Fac. Sci. Osaka* **7** (1959) 115–129.

40. KISHIMOTO, U. and AKABORI, H. Protoplasmic streaming of an internodal cell of *Nitella flexilis. J. gen. Physiol.* **42** (1959) 1167–1183.

41. KITASATO, H. The influence of H^+ on the membrane potential and ion fluxes in *Nitella. J. gen. Physiol.* **52** (1968) 60–87.

42. KOMOR, E. and TANNER, W. The hexose-proton symport system of *Chlorella vulgaris.* Specificity, stoichiometry and energetics of sugar-induced proton uptake. *Eur. J. Biochem.* **44** (1974*a*) 219–223.

43. KOMOR, W. and TANNER, W. The nature of the energy metabolite responsible for sugar accumulation in *Chlorella vulgaris. Z. Pflanzenphysiol.* **71** (1974*b*) 115–128.

44. LAMBERT, C. Influence de l'ATP sur le pH intraluteoidique et sur la penetration du citrate dans les luteoides du latex d'*Hevea brasiliensis. C.r. hebd. Séanc. Acad. Sci. Paris* Series D **281** (1975) 1705–1708.

45. LUCAS, W. J. and SMITH, F. A. Influence of light intensity on H^+ efflux and Cl^- transport models. *Aust. J. Plant Physiol.* **3** (1976) **443–456.**

46. LÜTTGE, U. PALLAGHY, C. K. and OSMOND, C. B. Coupling of ion transport in green cells of *Atriplex spongiosa* leaves to energy sources in the light and the dark. *J. membrane Biol.* **2** (1970) 17–30.

47. MACROBBIE, E. A. C. Ion transport. In *Algal Physiology and Biochemistry,* STEWART, W. D. P. (ed.) Blackwell Scientific Publications, Oxford, (1974) 678–713.

48. MACROBBIE, E. A. C. Ion transport in plant cells. In *Current Topics in Membranes and Transport,* Vol. 7, BONNER, F. and KLEINZELLER, A. (eds.), Academic Press, New York (1975) 1–48.

49. MARRÈ, E., COLOMBO, R., LODO, P. and RASI-CALDOGNO, F. Correlation between proton extrusion and stimulation of cell enlargement. Effects of fusicoccin and of cytokinins on leaf fragments and isolated cotyledons. *Pl. Sci. Lett.* **2** (1974) 139–150.

50. MARRÈ, E., LADO, P., FERRONI, A. and BALLARIN DENTI, A. Transmembrane potential increase induced by auxin, benzyladenine and fusicoccin. Correlation with proton extrusion and cell enlargement. *Pl. Sci. Lett.* **2** (1974) 257–265.

51. MEGO, J. L. Further evidence for a proton pump in a mouse kidney phagolysosome: effect of nigericin and DNP on the stimulation of intralysosomal proteolysis by ATP. *Biochem. biophys. Res. Commun.* **67** (1975) 571–575.

52. MEIDNER, H. and WILLMER, C. M. Mechanics and metabolism of guard cells. *Current Adv. Pl. Sci.* **7** (1975) 1–15.

53. MEISS, A. N. The formation of asparagine in etiolated seedlings of *Lupinus albus.* L. *Conn. AES New Haven Bull.* **553** (1952) 1–74.

54. MUNDAY, J. C. Jr. Membrane potentials in *Bryopsis plumosa. Bot. Mar.* **15** (1972) 61–63.

55. ODUM, H. T. and PINKERTON, R. C. Time's speed regulator: the optimum efficiency for maximum power output in physical and biological systems. *Am. Sci.* **43** (1955) 331–343.

56. PEARSON, C. J. Fluxes of potassium and changes in malate within epidermis of *Commelina cyanea* and their relations with stomatal aperture. *Aust. J. Plant Physiol.* **2** (1975) 85–90.

57. PEARSON, C. J. and MILTHORPE, F. L. Structure, carbon dioxide fixation and metabolism of stomata. *Aust. J. Plant Physiol.* **1** (1974) 221–236.

58. PICKARD, W. F. Does the resting potential of *Chara braunii* have an electrogenic component? *Can. J. Bot.* **51** (1973) 715–724.

59. PITMAN, M. G., SCHAEFER, N. and WILDES, R. A. Relation between permeability to potassium and sodium ions and fusicoccin-stimulated hydrogen-ion efflux in barley roots. *Planta* **126** (1975) 61–74.

60. PITT, D. Lysosomes and cell function. Longmans, London (1975)

61. POOLE, R. J. The influence of intracellular potential on potassium uptake by beetroot tissue. *J. gen. Physiol.* **49** (1966) 551–563.

62. POOLE, R. J. The H^+ pump in red beet. In *Ion Transport in Plants*, ANDERSON, W. P. (ed.), Academic Press, London (1973) 129–134.

63. POOLE, R. J. Ion transport and electrogenic pumps in storage tissue cells. *Can. J. Bot.* **52** (1974) 1038–1088.

64. RARAPORT, S. I. The sodium–potassium exchange pump: relation of metabolism to electrical properties of the cell – I. Theory. *Biophys. J.* **10** (1970) 246–259.

65. RASCHKE, K. Stomatal action. *Ann. Rev. Pl. Physiol.* **26** (1975) 309–340.

66. RASCHKE, K. and HUMBLE, G. D. No uptake of anions required by opening stomata of *Vicia faba:* guard cells release hydrogen ions. *Planta* **115** (1973) 47–58.

67. RAVEN, J. A. Time course of chloride fluxes in *Hydrodictyon africanum* in alternating light and darkness. In *Membrane Transport in Plants*, ZIMMERMANN, U. and DAINTY, J. (eds.), Springer, Berlin (1974*a*) 167–172.

68. RAVEN, J. A. Phosphate transport in *Hydrodictyon africanum. New Phytol.* **73** (1974*b*) 421–432.

69. RAVEN, J. A. Transport in algal cells. In *Encyclopedia of Plant Physiology* – Short Distance Transport, LUTTGE, U. and PITMAN, M. G. (eds.), Springer, Berlin (1976*a*), 129–188.

70. RAVEN, J. A. Transport at algal membranes. In *Perspectives in Experimental Botany*, Vol. 2 – Botany, SUNDERLAND, N. (ed.), Pergamon Press, Oxford (1976*b*) 381–389.

71. RAVEN, J. A. The evolution of vascular land plants in relation to supracellular transport processes. *Adv. Bot. Res.* **5** (1977a) 153–219.

72. RAVEN, J. A. and SMITH, F. A. The regulation of intracellular pH as a fundamental biological process. In *Ion Transport in Plant Cells,* ANDERSON, W. P. (ed.), Academic Press, London (1973) 271–278.

73. RAVEN, J. A. and SMITH, F. A. Significance of hydrogen ion transport in plant cells. *Can. J. Bot.* **52** (1974) 1035–1048.

74. RAVEN, J. A. and SMITH, F. A. Nitrogen assimilation and transport in vascular land plants in relation to intracellular pH regulation. *New Phytol.* **76** (1976*a*) 415–431.

75. RAVEN, J. A. and SMITH, F. A. The evolution of chemiosmotic energy coupling. *J. Theor. Biol.* **57** (1976*b*) 301–312.

76. RENT, R. K., JOHNSON, R. A. and BARR, C. E. Net H^+ transport in *Nitella clavata. J. membrane Biol.* **7** (1972) 231–244.

77. RICHARDS, J. L. and HOPE, A. B. The role of protons in determining membrane electrical characteristics in *Chara corallina. J. membrane Biol.* **16** (1974) 121–144.

78. SADDLER, H. D. W. The membrane potential of *Acetabularia mediterranea. J. gen Physiol.* **55** (1970) 802–821.

79. SAITO, K. and SENDA, M. The light-dependent effect of external pH on the membrane potential of *Nitella. Pl. Cell. Physiol., Tokyo* **14** (1973*a*) 147–156.

80. SAITO, K. and SENDA, M. The effect of external pH on the membrane potential of *Nitella* and its linkage to metabolism. *Pl. Cell Physiol., Tokyo* **14** (1973*b*) 1045–1052.

81. SAITO, K. and SENDA, M. The electrogenic ion pump revealed by the external pH effect on the membrane potential of *Nitella.* Influence of external ions and electric current on the pH effect. *Pl. Cell Physiol., Tokyo* **15** (1974) 1007–1016.

82. SEASTON, A., INKSON, C. and EDDY, A. A. The absorption of protons with specific amino acids and carbohydrates by yeast. *Biochem. J.* **134** (1973) 1031–1043.

83. SLAYMAN, C. L. Movement of ions and electrogenesis in microorganisms. *Am. Zool.* **10** (1970) 377–392.

84. SLAYMAN, C. L. Proton pumping and generalised energetics of transport: a review. In *Membrane Transport in Plants,* ZIMMERMANN, U. and DAINTY, J. (eds.), Springer, Berlin (1974) 107–119.

85. SLAYMAN, C. L. and GRADMANN, D. Electrogenic proton transport in the plasma membrane of *Neurospora. Biophys. J.* **15** (1975) 968–971.

86. SLAYMAN, C. L. and SLAYMAN, C. W. Depolarisation of the plasma membrane of *Neurospora* during active transport of glucose: evidence for a proton-dependent transport system. *Proc. natn. Acad. Sci. U.S.A.* **71** (1974) 1935–1939.

87. SLAYMAN, C. L., LONG, W. S. and LU, C. Y-H. The relationship between ATP and an electrogenic pump in the plasma membrane of *Neurospora crassa. J. membrane Biol.* **14** (1973) 305–338.

88. SMITH, F. A. The mechanism of chloride transport in Characean cells. *New Phytol.* **69** (1970) 903–917.

89. SMITH, F. A. Stimulation of chloride transport in *Chara* by external pH changes. *New Phytol.* **71** (1972) 595–601.

90. SMITH, F. A. and FOX, A. L. The free space of citrus leaf slices. *Aust. J. biol. Sci.* **2** (1975) 441–446.

91. SMITH, F. A. and RAVEN, J. A. H^+ fluxes, cytoplasmic pH and the control of salt accumulation in plants. In *Membrane Transport in Plants,* ZIMMERMANN, U. and DAINTY, J. (eds.), Springer, Berlin (1974) 380–385.

92. SMITH, F. A. and RAVEN, J. A. Hydrogen ion transport and role in the cell. In *Encyclopedia of Plant Physiology* – Short Distance Transport, LÜTTGE, U. and PITMAN, M. G. (eds.), Springer, Berlin (1976) 317–346.

93. SMITH, F. A. and WALKER, N. A. Chloride transport in *Chara corallina* and the electrochemical potential difference for hydrogen ions. *J. exp. Bot.* **27 (1976) 451–459.**

94. SPANSWICK, R. M. The potassium absorption isotherm in *Nitella translucens. Abst. 11th Int. Bot. Congr., Seattle* (1969) 206.

95. SPANSWICK, R. M. Evidence for an electrogenic pump in *Nitella translucens* – I. The effects of pH, K^+, Na^+, light and temperature on the membrane potential and resistance. *Biochim. biophys. Acta* **288** (1972) 73–89.

96. SPANSWICK, R. M. Electrogenesis in photosynthetic tissues. In *Ion Transport in Plants,* ANDERSON, W. P. (ed.), Academic Press, London (1973) 113–128.

97. SPANSWICK, R. M. Evidence for an electrogenic ion pump in *Nitella translucens* – II. Control of the light-stimulated component of the membrane potential. *Biochim. biophys. Acta* **332** (1974*a*) 387–398.

98. SPANSWICK, R. M. Hydrogen ion transport in giant algal cells. *Can. J. Bot.* **52** (1974*b*) 1029–1034.

99. THOMAS, R. C. Intracellular pH of snail neurones measured with a new pH-sensitive glass microelectrode. *J. Physiol.* **238** (1974*a*) 159–180.

100. THOMAS, R. C. The effect of bicarbonate ion on the intracellular buffering power of snail neurones. *J. Physiol.* **241** (1974*b*) 103–104.

101. WAGER, H. G. The effect of subjecting peas to air enriched with carbon dioxide – II. Respiration and the metabolism of the major acids. *J. exp. Bot.* **25** (1974) 338–351.

102. WALKER, N. A. and SMITH, F. A. Intracellular pH in *Chara corallina* measured by DMO distribution. *Pl. Sci. Lett.* **4** (1975*a*) 125–132.

103. WALKER, N. A. and SMITH, F. A. The cell membrane of *Chara* – a two-proton ATPase fixes the P.D.? *Abst. 12th Int. Bot. Congr., Leningrad* (1975*b*), 568.

104. YAMAGATA, Y. and MASUDA, Y. Comparative studies on auxin and fusicoccin action on plant growth. *Pl. Cell Physiol., Tokyo* **16** (1975) 41–52.

105. ZIMMERMANN, U. and DAINTY, J. (eds.), Membrane transport in plants, Springer, Berlin, (1975), 162–163.

ADDENDUM

More recent reviews of the subject matter of this commentary may be found in references 106–110.

REFERENCES

106. RAVEN, J. A. and SMITH, F. A. The chemi-osmotic approach. In SPANSWICK, R. M. and LUCAS, W. J. (eds.) *Membrane Transport in Plants*, North-Holland, Amsterdam (1979) in press.

107. RAVEN, J. A., SMITH, F. A. and SMITH, S. E. Ions in osmoregulation. In RAINS, D. W. and VALENTINE, R. C. (eds.) *Genetic Engineering of Osmoregulation: Impact on Plant Productivity for Food, Chemicals and Energy*, Plenum Publishing Corporation, New York (1980) in press.

108. SMITH, F. A. and RAVEN, J. A. The evolution of H^+ transport and its role in photosynthetic energy transduction. In DEAMER, D. W. (ed.) *Light Transducing Membranes: Structure, Function and Evolution*, Academic Press, New York (1978) pp. 233–251.

109. SMITH, F. A. and RAVEN, J. A. Intracellular pH and its regulation. *Ann. Rev. Pl. Physiol.* **30** (1979) 289–311.

110. WALKER, N. A. Ion transport in giant algal cells. In SPANSWICK, R. M. and LUCAS, W. J. (eds.) *Membrane Transport in Plants*, North-Holland, Amsterdam (1979) in press.

A CRITICAL ASSESSMENT OF THE CHOLODNY-WENT THEORY OF SHOOT GEOTROPISM

J. DIGBY and R. D. FIRN

Department of Biology, University of York, Heslington, York YO1 5DD, U.K.

"By doubting we come to questioning, and by questioning we perceive the truth" Abelard 1122A.D

The Cholodny-Went theory of shoot geotropism, which was first put forward nearly 50 years ago, is still widely taught in school and university plant physiology courses. Most authors of plant physiology texts treat the theory in a somewhat dogmatic manner, indeed many texts offer no alternative explanation of shoot geotropism. The theory states that when a shoot is placed horizontally, gravity induces the movement of auxin from the upper to the lower side of the shoot. The resulting auxin concentration gradient causes the differential growth which results in geotropic curvature[22]. A number of reviews of shoot geotropism have been written which adequately summarize the evidence which is consistent with the Cholodny-Went hypothesis[1,3,23] but evidence which does not support the hypothesis has received less attention. The almost universal acceptance of the Cholodny-Went hypothesis as the only explanation of shoot geotropism has led to a situation where investigators have tended to design experiments to prove the hypothesis. This is indeed unfortunate because a hypothesis can never be proved simply by accumulating results consistent with it. A hypothesis can, however, be disproved by the establishment of evidence which is inconsistent with it. The aim of this review is to bring together evidence from a number of sources; evidence which seems inconsistent with the Cholodny-Went hypothesis. The actual experimental results of many authors have been examined and in many cases new interpretations of the data have been attempted.

In judging the validity of the Cholodny-Went hypothesis one should apply the same criteria as one applies when assessing the involvement of a hormone in any process. Thus if auxin causes the differential growth evident in geotropic curvature one should be able to establish that the appropriate amount of the hormone is in the correct place at the correct time.

1. IS THE MAGNITUDE OF GRAVITY-INDUCED AUXIN REDISTRIBUTION SUFFICIENT TO ACCOUNT FOR THE OBSERVED GROWTH DIFFERENTIAL?

Although many studies have demonstrated a gravity-induced lateral transport of endogenous auxin[1,3,23] the magnitude of this redistribution is never very large. One recent detailed study failed to find any significant gravity induced redistribution of applied auxin in *Helianthus*[17] The amount of auxin diffusing out of the lower half of a horizontally placed shoot rarely exceeds that diffusing out of the upper half by more than a factor of two[23]. Thus it may be inferred that in a horizontal shoot the concentration of diffusible auxin in the lower half may be up to twice that in the upper half. Could such a relatively small difference in auxin levels between the upper and lower sides of the shoot produce the growth differentials which are known to occur during geotropic curvature? To answer this crucial question one needs to know:-

(a) The growth rates of the upper and lower sides of the shoot during geotropic curvature.
(b) The effect of auxin at different concentrations on the short term growth rates of such tissues.

Whilst such information exists, unfortunately it was not obtained in a single study of tissue grown under identical conditions. The kinetics of geotropic curvature in ***Avena*** coleoptiles was studied by Navez & Robinson[15] and it is possible to calculate from their data (reproduced in Fig. 1) that at the

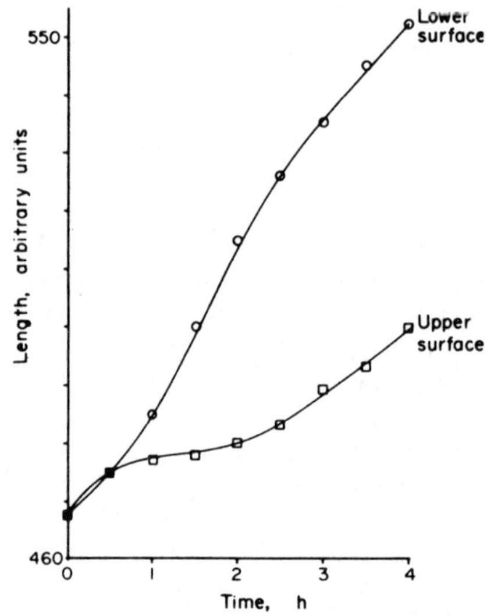

Fig. 1. The kinetics of ***Avena*** coleoptile geotropic curvature. Data redrawn from Navez and Robinson[15]. Rates of bending were calculated from the slopes of the 2 curves between 30 and 90 min.

onset of negative geotropic curvature the lower side of the coleoptile is growing 12 times faster than the upper side. Using the dose-response curve published by Cleland[6] for short term promotion of ***Avena*** coleoptile elongation by auxin (reproduced as Fig. 2) it is evident that the growth differential observed by Navez & Robinson would require that endogenous auxin concentrations in the upper and lower halves should differ by a factor in excess of 100. Indeed it is apparent that auxin cannot produce a twelve-fold increase in growth rate of the coleoptiles used by Cleland and we have not found evidence in the literature for an auxin stimulation of this magnitude. The discrepancy is not confined to ***Avena***, indeed such a discrepancy would exist for any tissue where geotropic curvature was relatively rapid and where the dose-response curve for auxin action on tissue elongation was log/linear. Thus in pea internode segments geotropic curvature can proceed at a rate of 17°/hr[14] and using the method of Ganot & Reinhold[12] one can estimate that the rate of tissue elongation of the lower half of such segments must be at least double that of the upper half. The dose-response curve for the effect of auxin on pea internode segment elongation (Yamamoto ***et al.***[24] – reproduced in Fig. 3) would require that the endogenous auxin concentration in upper and lower halves of such sections would have to differ by a factor of at least 10 if auxin were causing the geotropic curvature. Our own recent studies on ***Helianthus*** hypocotyl geotropic curvature[11] revealed a similar situation.

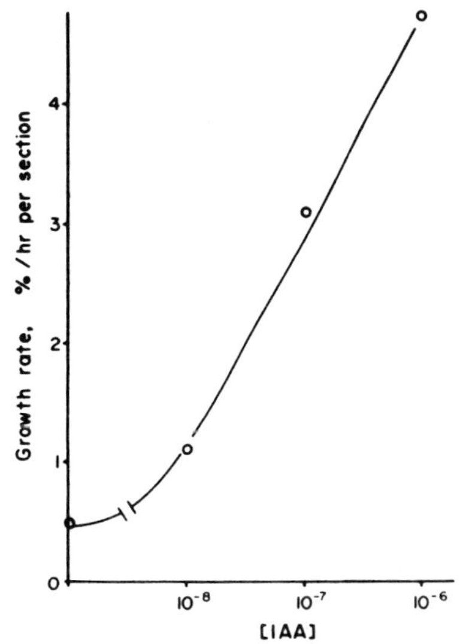

Fig. 2. The effect of auxin concentration on the short term growth of ***Avena*** coleoptile sections. Data redrawn from Cleland.[4] The dose–response curve was constructed from the initial steady state growth rates.

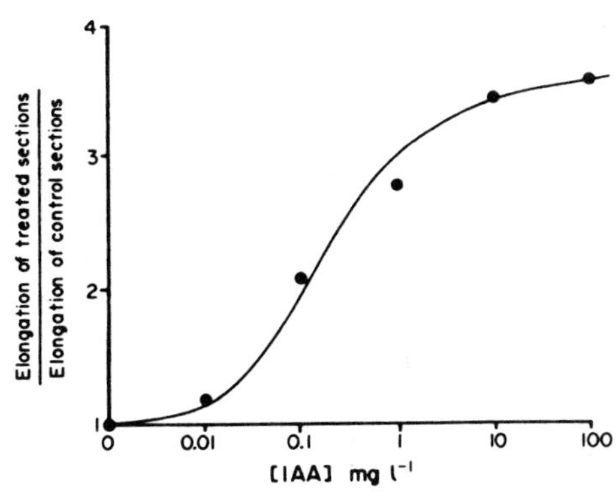

Fig. 3. A typical auxin dose–response curve. The effect of auxin concentration on the growth of light-grown pea epicotyl segments. Data redrawn from Yamamoto ***et al.***[24]

The lower halves of horizontal sections grew 2-4 times faster than the upper halves during the period of geotropic curvature. Once again the observed auxin sensitivity of the tissue was such that the normal gravity-induced auxin redistribution could not account for the observed differential growth rates. This type of discrepancy was apparently not evident when Cholodny-Went hypothesis was first proposed because at that time it was generally believed that the response of a tissue to an applied hormone was directly proportional to the dose applied. This belief came mainly from studies on the ***Avena*** curvature bioassay where, under the special conditions employed, a linear relationship existed between the applied auxin concentration and the curvature induced[22]. However it was subsequently shown that the action of auxin on cell elongation generally showed a log/linear relationship[16]. As cell elongation causes the curvature in the ***Avena*** curvature bioassay it seems that the most frequently quoted[22] linear/linear dose-response curve for auxin action on ***Avena*** curvature is anomolous. A study of the literature pertaining to the ***Avena*** curvature bioassay reveals that the conditions employed were carefully manipulated so that the curvature induced by auxin was proportional to the applied dose. It was known for instance that this relationship applied only over a narrow range of auxin concentrations, it did not apply for auxin concentrations which induced a curvature of more than 20° in the coleoptile[5].

The results of Söding[21], however, showed that conditions existed where coleoptiles responded to a wider range of auxin concentrations and under these conditions the relationship between the logarithm of the auxin concentration and the induced curvature was indeed linear. The atypical linear/linear dose-response curves sometimes reported for the effect of auxin on ***Avena*** curvature were thus unfortunately reponsible for encouraging the belief that small changes in hormone concentration ($< 5X$) were physiologically significant. We are not inclined to this view, in the light of the overwhelming evidence that dose-response curves for all plant hormones are log/linear[4] Thus we conclude that the gravity-induced auxin redistribution in horizontal stems is too small to cause the observed growth differential.

2. CAN EXOGENOUSLY APPLIED AUXIN INFLUENCE GEOTROPIC CURVATURE IN A MANNER PREDICTED BY THE CHOLODNY-WENT HYPOTHESIS?

The Cholodny-Went hypothesis dictates that it should be possible to influence the geotropic curvature of shoots in a predictable manner by the strategic application of auxin. Thus, by abolishing the gravity-induced concentration gradient of endogenous auxin by applying exogenous auxin it should be possible to abolish geotropic curvature. If the gravity-induced gradient of endogenous auxin is reversed it should be possible to cause plants to grow towards gravity. There is in fact data in the literature which shows that applied auxins do not influence the geotropic curvature of ***Zea, Avena*** and ***Helianthus*** tissue in a manner which would be predicted by the Cholodny-Went hypothesis. The significance of this data as it relates to the Cholodny-Went hypothesis has usually not been commented upon by the investigators. Thus some results recently published by Phillips & Hartung[17] show that in light-grown sunflower internode sections the application of auxin to horizontal sections, such that the upper half contained 4 times the amount of auxin found in the lower half actually increased negative geotropic curvature dramatically (see Fig. 4). Similarly the results of Shaw ***et al.,***[20] showed that the auxin concentration gradient between the upper and lower halves of ***Avena*** and ***Zea*** coleoptiles could be greatly modified by the application of auxin with a micropipette, yet such auxin applications did not greatly disturb the observed geotropic curvature (see Fig. 4). These results are clearly not consistent with the Cholodny-Went hypothesis.

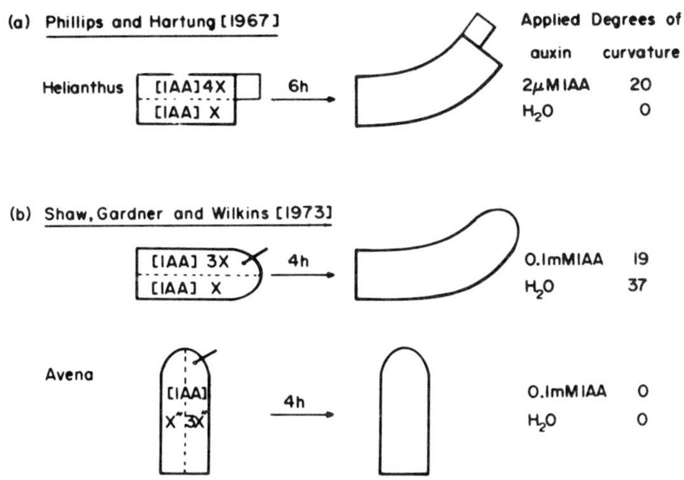

Fig. 4. Modification of auxin ratios in horizontal and vertical plant material and its effect on curvature. (a) ***Helianthus*** Data from Phillips ***et al.***[17] Sections (3 cm) were taken from light grown ***Helianthus*** first internodes and agar blocks were placed on the upper half of the cut end ± 2μM IAA. In separate experiments curvature and the distribution of radioactive IAA in upper and lower half sections was determined after 6 hr. (b) ***Avena*** Data from Shaw ***et al.***[20] Coleoptile apices 15 mm in length were used. Between 0.0003 and 0.03 μ1 of a 10^{-4}M solution of IAA was injected as shown. Curvature was measured after 4 hr. Distribution of radioactive IAA between upper and lower sides was determined after 30 and 60 min.

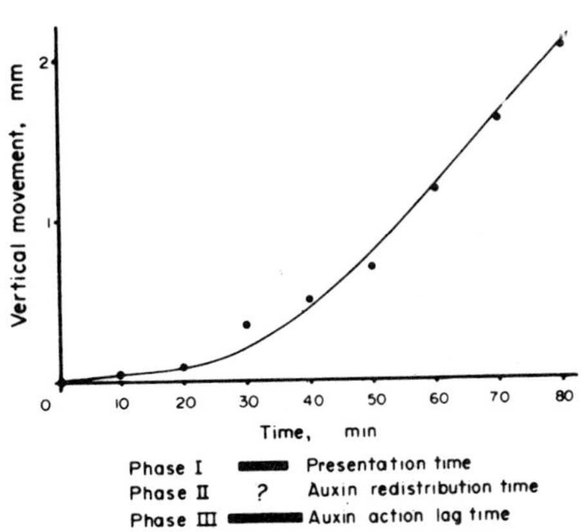

Fig. 5. Kinetics of geotropic curvatures of etiolated ***Helianthus*** hypocotyl sections. Data Firn & Digby (unpublished). 2 cm segments were located horizontally at 25°C. The vertical movement of an ink spot placed near the apical end of the section was followed with a travelling microscope. The data shown is the average of 3 sections.

3. DOES THE GRAVITY-INDUCED AUXIN REDISTRIBUTION PRECEDE GEOTROPIC CURVATURE?

The Cholodny–Went hypothesis not only demands that gravity-induced auxin redistribution must precede geotropic curvature but also demands that the kinetics of geotropic bending must be related to the kinetics of auxin redistribution.

Time course studies of the development of geotropic curvature have indicated that geotropic curvature can begin soon after the shoot is placed horizontally. In ***Avena*** negative geotropism is clearly evident after 30 min[2] but there is one report of negative geotropic curvature in ***Avena*** after only 8 min[18]. Our own studies on the geotropic curvature of ***Helianthus*** have shown that negative geotropic curvature is always evident 20–30 min after hypocotyls are placed horizontally but some hypocotyls begin to curve upwards after only 10 min of geostimulus. If the Cholodny–Went hypothesis is valid, these reaction times must include in sequence:-

Phase I, a period for the tissue to sense the geostimulus – the presentation time.
Phase II, a period for auxin redistribution to occur.
Phase III, a period for auxin to influence the growth rate of the cells.

As the lag time for auxin action (Phase III) in most tissues[9], including ***Helianthus***[11] is approx 10 min it is evident that Phase II must have begun at least 10 min before geotropic curvature is measurable. Is auxin redistribution evident 10 min before bending begins? Unfortunately the studies which have been carried out of the short term changes in auxin redistribution due to gravity have not included an adequate study of the kinetics of geotropic curvature. Filner ***et al.***[10], claim to show that a downward movement of IAA begins 10–20 min after ***Zea*** coleoptile segments are placed horizontally (the timing is variable from plant to plant) and state that negative geotropic curvature begins in their coleoptiles after 15 min. Claiming that auxin acts "without any significant lag period"

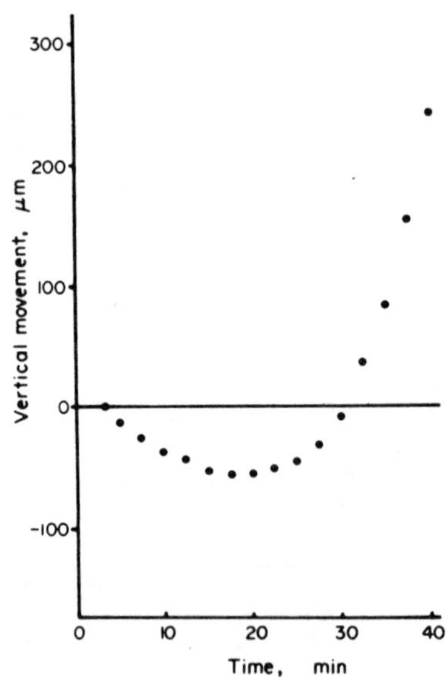

Fig. 6. Geotropic curvature in ***Zea*** coleoptiles. Data from Hild & Hertel.[13] Coleoptile apical sections (3 cm) were located horizontally and the vertical displacement of the tip measured with a travelling microscope. The data shown is the average of 26 coleoptiles.

the authors state that their results "provide further confirmation of the Cholodny-Went theory". As auxin is known to act after a lag of 12-15 min in ***Zea*** tissue[8] it is evident that these results do not support the Cholodny-Went hypothesis.

In another study of ***Zea*** coleoptile curvature Hild & Hertel[13] produced detailed kinetics of geotropic curvature and claimed to show that auxin redistribution preceded the observed curvature. The initial period of positive (downward) geotropic curvature which sometimes occurs in ***Zea*** (see Fig. 6) is said to be due to an initial upward movement of auxin. Examination of the data however casts grave doubt on their interpretation as this initial downward movement is evident 5 min after the coleoptile is placed horizontally. Gravity perception, auxin redistribution and auxin action in less than the lag time for auxin action is unrealistic. Likewise the transition from positive to negative curvature occurs within a 10 min period, and this would require the co-ordinated movement of the auxin from the upper to the lower side of the coleoptile in order to produce the accelerated growth rate on the lower side which will cause negative geotropic curvature to begin.

Thus we believe that once again there is no unequivocal evidence that gravity-induced auxin redistribution precedes geotropic curvature by a period at least equal to the lag time for auxin action.

CONCLUSION

The validity of the Cholodny-Went hypothesis of shoot geotropism seems extremely suspect in the light of the evidence reviewed in this paper. There is simply no unequivocal evidence that auxin redistribution occurs rapidly enough to cause geotropic curvature and there are good reasons for believing that the magnitude of auxin redistribution is much too small to be the cause of the observed differential growth. If auxin redistribution does not cause geotropic curvature it is legitimate to ask whether geotropic curvature causes auxin redistribution. Most investigations of lateral transport of exogenously applied radioactive auxin have used isolated tissue sections and the applied hormone must inevitably have been taken up into the free space of the tissue. Geotropic curvature involves expansion of the cells in the lower half of the tissue and in a closed system, such as the isolated segment, the water taken up into those lower cells must have come from the upper half. This water flow from the upper half to the lower half would begin as soon as geotropic curvature was evident and would carry some radioactive auxin from the upper to the lower half. We have recently proposed a new theory of shoot geotropism which includes a feature which may also be relevant to the observed lateral transport of auxin[11]. It was suggested that a pH gradient may be established across a shoot which has been placed horizontally, this gradient possibly being a result of an increased respiration rate of cells in the lower epidermal layers. Should the pH in the cell walls in the lower half of the horizontal shoot be lower than that in the upper half of the segment one might predict a movement of auxin from the upper to the lower half of the shoot. This would be as a result of the p*K*a of auxin and the known effect of pH on auxin uptake into cells[7,19]. In the cell walls in the lower tissues, auxin would tend to become un-ionized hence would pass into the cell more readily, reducing the concentration of auxin in the cell wall. The greater uptake of auxin by cells in the lower half of the horizontal section would thus create an auxin accumulation system and could account for the observed concentration gradient of auxin across the shoot following geotropic stimulation.

REFERENCES

1. ANKER, L. Ortho-geotropism in shoots and coleoptiles In ***Encyclopedia of Plant Physiology,*** Vol. XVII, chap. 2. Springer, Berlin (1962) 103–152.

2. ANKER, L. The circumnutation of the ***Avena*** coleoptile, its autonomous nature and its interference with the geotropic reaction. ***Acta bot. neerl.*** **21** (1972) 71–76.

3. AUDUS, L.J. Geotropism. In ***The Physiology of Plant Growth and Development*** (edited by M.B. WILKINS) McGraw-Hill, London (1969) 203–242.

4. ANDUS, L.J. ***Plant Growth Substances.*** Leonard Hill, London (1972).

5. BOYSEN-JENSEN, P. ***Growth Hormones in Plants.*** (Translated by G.S. AVERY and P.R. BURKHOLDER) McGraw-Hill, New York (1936).

6. CLELAND, R. The dosage–response curve for auxin-induced cell elongation: a re-evaluation. ***Planta*** **104** (1972) 1–9.

7. COLLANDER, R. Cell membranes: their resistance to penetration and their capacity for transport. In ***Plant Physiology: A Treatise*** (edited by F.C. STEWARD) Vol. II, Academic Press, New York (1959) 3–102.

8. DELA FUENTE, R.K. and LEOPOLD, A.C. Time course of auxin stimulation of growth. ***Pl. Physiol.*** **46** (1970) 186–189.

9. EVANS, M.L. Rapid responses to plant hormones. ***Annu. Rev. Pl. Physiol.*** **25** (1974) 195–223.

10. FILNER, B., HERTEL, R., STEELE, C. and FAN, U. Some aspects of geotropism in coleoptiles. ***Planta*** **94** (1970) 333–354.

11. FIRN, R.D. and DIGBY, J. In press.

12. GANOT, D. and REINHOLD, L. The acid growth effect and geotropism. ***Planta*** **95** (1970) 62–71.

13. HILD, V. and HERTEL, R. Initial phases of gravity-induced lateral auxin transport and geotropic curvature in corn coleoptiles. ***Planta*** **108** (1972) 245–258.

14. KANG, B.G. and BURG, S.P. Relation of phytochrome-enhanced geotropic sensitivity to C_2H_4 production. ***Pl. Physiol.*** **50** (1972) 132–135.

15. NAVEZ, A.E. and ROBINSON, T.W. Geotropic curvature of ***Avena*** coleoptiles. ***J. gen Physiol.*** **16** (1933) 135.

16. NITSCH, J.P. Methods for the investigation of natural auxins and growth inhibitors. In ***The Chemistry and Mode of Action of Plant Growth Substances.*** (edited by R.L. WAIN & F. WIGHTMAN) Butterworths, London (1956) 3–31.

17. PHILLIPS, I.D.J. and HARTUNG, W. Longitudinal and lateral transport of [3,4-^3H] gibberellin A_1 and 3-indolyl (acetic acid-2-^{14}C) in upright and geotropically responding green internode segments from ***Helianthus annus. New Phytol.*** **76** (1976) 1–9.

18. PICKARD, B.G. Second positive phototropic response patterns of the oat coleoptile. ***Planta*** **88** (1969) 1–33.

19. RUBERY, P.H. and SHELDRAKE, A.R. Carrier-mediated auxin transport. ***Planta*** **118** (1974) 101–121.

20. SHAW, S., GARDNER, G. and WILKINS, M.B. The lateral transport of 1AA in intact coleoptiles of ***Avena sativa*** and ***Zea mays*** during geotropic stimulation. ***Planta*** **115** (1973) 97–111.

21. SÖDING, H. Wirkt der Wuchsstoff artspezifisch? ***Jahrb. Wiss. Bot.*** **82** (1936) 534–554.

22. WENT, F.W. and THIMANN, K.V. Phytohormones, MacMillan, N.Y. (1937).

23. WILKINS, M.B. Geotropism. ***Annu. Rev. Pl. Physiol.*** **17** (1966) 379–408.

24. YAMAMOTO, R., MAKI, K., YAMAGATA, Y. and MASUDA, Y. Auxin and hydrogen ion actions on light-grown pea epicotyl segments — I. Tissue specificity of auxin and hydrogen ion actions. ***Pl. cell Physiol.,*** Tokyo **15** (1974) 823–831.

ADDENDUM

Since this article was first published, more detailed studies of the growth rate changes occurring in horizontal hypocotyls and coleoptiles have been published[27,30,31]. The pattern of growth rate changes were similar to those shown by ***Avena*** coleoptiles.[15]

The assertion in the article that the "classical" dose-response curve[22] for the action of IAA on ***Avena*** curvature was atypical has been found to have more support in the literature. Bentley and Bickle[25] published a dose-response curve for the ***Avena*** curvature assay which showed a typical log/linear dose-response curve.

Further compelling evidence against a lateral movement of IAA from the top to the bottom half of a horizontal shoot as being the cause of geo-curvature comes from studies on the georesponse of semicylinders isolated from shoots. Such studies suggest that the lower half of a horizontal shoot is not dependent on the upper half of such a shoot during geocurvature.[28]

Some of the arguments advanced in the article have been challenged in a review.[32] However, it may be noted that the defence of the Cholodny-Went model referred only to coleoptiles and the author emphasized elsewhere in the review that auxin redistribution seemed unable to account for geocurvature in dicotyledons. With regards to auxin redistribution in coleoptiles, it still remains to be shown by an unambiguous technique that endogenous IAA moves in large amounts across a horizontal coleoptile during the lag period. Whilst studies of exogenous radiolabelled IAA do show a lateral movement of IAA in horizontal coleoptiles, the amount moving is very small. Two points are sometimes overlooked in studies of lateral auxin movement. Firstly, longitudinal transport continues in horizontal shoots and an auxin molecule in the upper half of a horizontal shoot has a much greater change of being moved longitudinally than laterally.[26] Secondly, applications to the upper half of a horizontal shoot always seem to result in less than 50% of the applied hormone moving laterally. These findings are hardly indicative of a well-developed lateral transport system.

Finally, evidence against gibberellins playing a role in geotropism has been advanced.[29]

REFERENCES

25. BENTLEY, J.A. and BICKLE, A.S. Studies on plant growth hormones, II. Further biological properties of IAN. ***J. exp. Bot.*** **3** (1952) 406–423.

26. CANE, A.R. and WILKINS, M.B. Independence of lateral and differential longitudinal movement of IAA in geotropically stimulated coleoptiles of ***Zea mays. Pl. Physiol.*** **44** (1969) 1481–1487.

27. DIGBY, J. and FIRN, R.D. An analysis of the changes in growth rate occurring during the initial stages of geocurvature in shoots. ***Plant Cell & Environment*** **2** (1979) 145–148.

28. FIRN, R.D. and DIGBY, J. The role of the peripheral cell layers in the geotropic curvature of sunflower hypocotyls: a new model of shoot geotropism. ***Aust. J. Plant Physiol.*** **4** (1977) 337–347.

29. FIRN, R.D., DIGBY, J. and PINSENT, C. Evidence against the involvement of gibberellins in the differential growth which causes geotropic curvature. ***Z. Pflanzenphysiol. Bd.*** **82** (1977) 179–185.

30. FIRN, R.D., DIGBY, J. and RILEY, H. Shoot geotropic curvature – the location, magnitude and kinetics of gravity induced differential growth in horizontal sunflower hypocotyls. ***Ann. Bot.*** **42** (1978) 465–468.

31. FIRN, R.D. and DIGBY, J. A study of the autotropic straightening reaction of a shoot previously curved during geocurvature. ***Plant, Cell and Environment*** **2** (1979) 149–154.

32. WILKINS, M.B. Gravity and light-sensing guidance systems in primary roots and shoots. In ***Integration of Activity in Higher Plants*** (edited by D.H. JENNINGS) ***Proc. Symp. Soc. Exp. Biol.*** **31** (1977) 275–335.

COLCHICINE AND PLANT MICROTUBULES: A CRITICAL EVALUATION

JAMES W. HART and DINKAR D. SABNIS

Botany Department, University of Aberdeen, Aberdeen AB9 2UD, Scotland

INTRODUCTION

Microtubules are ubiquitous components of the eukaryote cell. In recent years, the microtubules of animal cells have been subject to intense investigation and their constituent proteins have been isolated. The discovery that the antimitotic alkaloid, colchicine, interacts directly with the microtubule subunit was of prime importance in the original isolation of tubulin, the major protein of microtubules. The use of colchicine to disrupt labile microtubules *in vivo*, also permitted the identification of a variety of cellular activities dependent upon these organelles. Colchicine has similarly been applied to plant tissues in attempts to elucidate the interaction of microtubules in certain aspects of plant cell development. However, relatively high concentrations of the drug seem to be required to depolymerise plant microtubules, and such concentrations of colchicine can also affect cellular activity through mechanisms probably unrelated to a direct action on microtubules.

In this Commentary, the various actions of colchicine on cellular processes in general are briefly reviewed, preparatory to considering the effects of the drug on plant systems. Difficulties in interpreting many of these effects of colchicine on plant cells are discussed and procedures are suggested for distinguishing those processes which directly involve microtubules. Possible reasons for the seeming stability towards colchicine shown by the microtubules of many types of plant cell are also discussed.

COLCHICINE AND MICROTUBULES

Several recent reviews have dealt with microtubules[34, 50] and their responses to pharmacological agents.[43, 71] Differential responses to drugs and other depolymerising agents allow the arbitrary classification of microtubules into stable or labile types. Those of the latter category, which includes cytoplasmic microtubules of animal cells, are thought to exist in dynamic equilibrium with a pool of soluble subunits.[36, 66] It is generally believed that colchicine, by complexing with the subunits in this pool,[8, 72] changes the equilibrium to result in the depolymerisation of these microtubules. The stable microtubules of cilia and flagella are not depolymerised in the presence of colchicine, although the drug can inhibit the assembly of these structures.[3, 15] In addition, stable microtubules can be rendered soluble to provide tubulin subunits which show characteristic colchicine binding activity.[73]

The binding of colchicine to tubulin *in vitro* is irreversible and noncovalent, with a number of specific characteristics which are common to preparations from a variety of animal sources.[71, 70] Colchicine binding is an unstable property of tubulin in solution, with a half-life for decay of activity in the order of hours, even when conditions are optimised for maximum stability. Factors involved in stabilising colchicine binding activity are: source and concentration of tubulin itself;[70] pH, temperature and ionic strength of the medium;[70, 69] and the presence of the nucleotide, GTP,[68] or the alkaloid, vinblastine.[69] Colchicine is bound mole for mole to a tubulin dimer of mol. wt 115,000–120,000 daltons. The affinity constant for binding is high, of the order of $0.5–2.0 \times 10^6$ l/mole for tubulins derived from various sources.[61] It is important to note that lumicolchicine, an isomer produced by exposure of the drug to light,[22] neither itself binds to tubulin nor influences the binding of colchicine to tubulin.[12, 69, 72] The presence of colchicine binding activity, in conjunction with considerations of its lability and response to other drugs, can be used as a diagnostic feature for the initial identification of tubulin in biological samples.[71]

Colchicine can also be used at the physiological level to delineate processes dependent upon microtubules. Investigations of the effects of exogenously applied colchicine on a variety of cell types have indicated the involvement of microtubules in movement of whole cells, intracellular movement of various organelles and inclusions, maintenance of cell shape and processes of secretion.[1, 50, 34] It has been suggested that colchicine affects those processes which involve assembly and disassembly of microtubules.[43] The limitations of such ablation experiments have been discussed,[60] not least of which is the inability of colchicine to depolymerise stable-type microtubules, thus curtailing the usefulness of negative results. In addition, if concentrations greater than 10^{-6} M colchicine[60] are required to achieve a physiological effect, it should be ensured that: (a) general metabolism is unaffected; (b) the inhibition of a process is correlated with disappearance of microtubules; (c) removal of the drug, if resulting in release from physiological inhibition, is accompanied by reappearance of microtubules. The lack of effect of lumicolchicine on physiological processes involving microtubules,[2] presents an additional invaluable control treatment (see later).

OTHER EFFECTS OF COLCHICINE

Such cautionary comments are appropriate, since effects of colchicine on various aspects of metabolism have been described whose bases are probably not related to actions of the drug on microtubules. Early workers noted effects of high concentrations of colchicine (10^{-2}M) on cell respiration and on ATP levels.[17, 20] A voluminous later literature exists concerning inhibition of the synthesis of cellular macromolecules by colchicine (at 10^{-4}M or greater)[16, 38, 57] and there is evidence of complex formation between colchicine and DNA.[35] However, many of the studies on the effects of colchicine on biosynthetic processes involve the application of radiolabelled nucleotides or amino acids to whole cells or tissues, and distinction has not always been made between effects on uptake of tracer into the cell and effects on its incorporation into the macromolecule. Thus, whether colchicine does indeed bring about the observed changes in the rates of incorporation of the labelled substrates into the various types of macromolecules by directly inhibiting any of the mechanisms of DNA, RNA or protein, synthesis remains controversial.[46]

Colchicine has recently been found to bring about changes both in the ultrastructure of the cell membrane[75] and in membrane-mediated transport processes.[5, 46] In addition, earlier work showed that colchicine can affect the *in vivo* activities of various enzymes located in the cell membrane,[19, 24, 47] including phosphatases in plant cell membranes.[59] While certain of these effects may be due to metabolic products of the applied colchicine,[64] the drug itself (at 10^{-4}M) inhibits the activity *in vitro* of two purified glucose-utilising enzymes.[25]

Studies of the interaction *in vitro* of radiolabelled colchicine with cellular components also suggest that the drug can bind to moieties other than tubulin. The kinetics of colchicine binding to components from bovine anterior pituitary slices have been interpreted as showing the presence of high and low affinity sites.[62] Membrane fractions from liver cells also bind colchicine in a nonspecific, unsaturable mode.[65] A factor in serum binds colchicine[46] and both the soluble and particulate fractions from chick brain show nonspecific colchicine binding at high concentrations of the drug (10^{-4}M).[4]

Thus, various types of evidence suggest that colchicine can affect cellular processes by interaction with components other than microtubules, and many of these effects involve membrane-associated features. The effects are observed at concentrations of colchicine several orders of magnitude greater than those normally required for disruption of labile cytoplasmic microtubules. It is also noteworthy that, in contrast to its lack of effect on microtubules and on tubulin, lumicolchicine, in the relatively few cases where it has been tested, is as effective as colchicine on these other processes, both at the physiological level[5, 46, 75] and in binding assays.[4, 46, 65]

IN VIVO EFFECTS OF COLCHICINE ON PLANT CELLS

In ultrastructural investigations of plant cells, the presence of oriented microtubules has been correlated with the maintenance of cell shape, the plane of cell division, formation of the cell plate and with certain features of cell differentiation involving wall architecture.[34, 48, 53] Colchicine has been widely used in ablation-type experiments in attempts to confirm a role for microtubules in these phenomena. However, a fundamental difficulty in interpreting the significance of such experiments arises from the relatively high concentrations of colchicine required. The insensitivity of even plant mitosis to chemical agents, including colchicine, is well documented.[17, 18, 20] A colchicine concentration of 5mg/l (1.25 $\times$ 10^{-5}M), which inhibits cell division in *Phaseolus* cultures,[37] is among the lowest reported to be effective on plant tissues, yet even this is an order of magnitude greater than the levels routinely used on animal cells. More commonly, higher concentrations are necessary to disrupt cytoplasmic microtubules in plant cells. For example, in cells of wheat coleoptiles,[52] disappearance of certain cytoplasmic microtubules (notably not all) results from the use of colchicine solutions of 0.1–0.3% (2.5–7.5 $\times$ 10^{-3}M) and in root cells,[9] from exposure to 500 ppm of the drug. In callus cultures of *Nicotiana*,[74] colchicine at 0.1% has a variable effect on the microtubule population and 0.5–1.0% is required to depolymerise 70–80% of the microtubules. Microtubules could still be detected in hyphae of a water-mould grown in near-lethal concentrations of colchicine.[31]

A classical response of plant tissues to colchicine is seen in the formation of the morphologically-abnormal, isodiametric cells of C-tumours, an effect requiring long-term exposure to solutions of 500–2000 ppm drug in grass roots.[9] Other examples of changes in cell shape in response to colchicine also involve drug dosages of these orders of magnitude.[26, 48] In *Nitella*, various morphological abnormalities, including branched flagella, appear during spermatogenesis in solutions of 0.1% colchicine, a level of drug equivalent to half the lethal concentration for some strains.[67] Colchicine treatment of the naked, unicellular alga, *Ochromonas*, leads to loss of its characteristic pear-shape, accompanied by depolymerisation of the cytoplasmic microtubules. Again, high drug concentrations of 5–10mg/ml (1.25–2.5 $\times$ 10^{-2}M) are required.[10]

In interphase celks of higher plants, cytoplasmic microtubules lie close to the cell membrane and it has been suggested that they may be involved in the guidance and alignment of dictyosome-derived vesicles containing wall precursors. Correlations between the orientation of wall microfibrils and subjacent cytoplasmic microtubules have been drawn in many,[48] but not all,[49] ultrastructural studies. Such co-alignments are well-exemplified in the long xylem fibres of *Salix*, where wall microfibrils are oriented differently in the middle and at the extremities of each cell.[54] However, the degree of involvement of microtubules in wall microfibril orientation[48] is controversial,[49, 53] and colchicine has figured prominently as an experimental tool used to depolymerise microtubules and disrupt wall microfibril orientation. Other examples of an association between microtubules and wall architecture are afforded by cells undergoing irregular or sculptured deposition of secondary wall material, for example during xylogenesis where microtubules are often specifically grouped under the developing wall thickenings.[48, 55] Again, colchicine inhibits xylem vessel differentiation, but concentrations of 0.04–0.4% drug (10^{-3}–10^{-2}M) are required.[23, 33, 56]

Differentiation of xylem vessels in excised tissue segments is induced by auxin, a response inhibited by colchicine.[56] Various studies have attempted, through the use of colchicine, to implicate microtubule action in the specific responses of plant tissues to other hormones. For example, the drug, at 3 $\times$ 10^{-4}M, inhibits gibberellin-induced cell expansion in bean epicotyls and reverses the kinetin-induced inhibition of elongation.[63] Similar concentrations of colchicine inhibit the irradiation-induced elongation of wheat coleoptile segments.[39]

Thus, in studies of the interactions of colchicine with plant tissues, while the involvement of microtubules in the particular process is often conjectured, the drug concentrations of 10^{-4}M or greater are well above those levels at which, in animal cells, a specific effect on microtubules has been indicated. Indeed, examination of plant cells after such colchicine treatments can reveal changes in other ultrastructural features, such as in the arrangements of endoplasmic reticulum and of ribosomes.[52]

The use of colchicine alone, at relatively high concentrations, to distinguish a specific role for microtubules particularly in processes determining cell wall architecture may be of limited value. It is becoming more apparent that both the synthesis[30] and oriented assembly[11] of plant cell wall components involve active sites located within the cell membrane and, as previously described, such concentrations of colchicine can also affect many aspects of membrane function, including mobility of membrane components.[75] The effect of exogenously supplied proline in reversing the colchicine-induced inhibition of xylem differentiation[55] may in fact involve action on the source of such a 'nonspecific' effect of colchicine. Similarly, the use of high concentrations of colchicine to study hormonal effects on tissue segments *in vitro*, far from being a convenient means for implicating microtubules directly in hormone-induced processes, may be subject to complication by 'nonspecific' effects of colchicine, not only on metabolism, but also on nutrient or even hormone transport.

It will be recalled that the isomer lumicolchicine *is* as effective as colchicine itself on many aspects of cellular activity, but does *not* affect microtubule-dependent processes, the latter including plant mitosis.[42] Therefore, in investigations of plant processes where high concentrations of colchicine are necessary for effectiveness, it would seem advisable that caution be exercised in inferring a direct involvement of microtubules, at least until equivalent concentrations of lumi-colchicine have been shown to be *in*effective.

INTERACTIONS OF COLCHICINE WITH PLANT COMPONENTS *IN VITRO*

Recent examinations of plant preparations for a tubulin-like moiety with characteristic colchicine binding activity have met with limited success. Although protein electrophoretically similar to tubulin has been isolated from a variety of plants,[14, 21] colchicine binding activity in plant preparations is low.[27, 28, 29, 32] The apparent affinity constant for binding is at least an order of magnitude lower than the colchicine-tubulin interaction.[27] The nature of the binding component is uncertain, and there is evidence that not all drug binding is attributable to one type of site.[29, 32] In addition, colchicine binding activity in preparations from a water-mould is associated with a component of mol. wt 30,000 daltons which is stable to trichloroacetic acid.[51] Thus, *in vitro* studies have not yet clarified exactly what, or how many, moieties can act as possible target sites for the action of colchicine in plant tissues.

BASIS OF PLANT RESISTANCE TO COLCHICINE

The basis of this seeming resistance of plant cells towards colchicine is as yet undetermined. Reduced cellular permeability towards colchicine is thought to be involved in the relative insensitivity of cells of the Chinese hamster.[41] In this respect, it has been noted that colchicine seems to penetrate coleoptile cells relatively slowly,[52] and that cells deep in the root meristem do not show profound responses to the drug.[9] However, although the analogue colcemid does seem to enter fungal cells more readily than colchicine, differential permeability was thought not to account fully for the greater inhibition of fungal growth by colcemid.[27, 32] A mechanism related to the problem of permeability lies in the possible sequestration of the drug by non-tubulin sites. Binding of the drug by sites possibly located in the glycocalyx has been suggested as a factor in the resistance of cells of the Syrian hamster.[45] If an analogous mechanism is operating in plant tissue, then at least in certain cases, such sites would have to be located internal to the cell wall, since yeast protoplasts also show resistance to colchicine.[40] Non-specific sites with an affinity for colchicine have been demonstrated in membrane fractions from animal cells[65] and in preparations from higher plants.[29]

Alternatively, colchicine may be inactivated, metabolically[58] or otherwise, by some feature of the plant intracellular environment. The physiological effectiveness of colchicine can be lowered by interaction of the drug with Tris buffer.[44] Its antimitotic activity is decreased in the presence of vitamin B or riboflavin,[17] perhaps by a photosensitisation effect. Components in pea tissue also change the potency of the drug *in vivo.*[17] Further investigation of the fate of colchicine in plant tissues may be valuable.

However, while such general mechanisms involving uptake systems and inactivation of colchicine may indeed be operating *in vivo*, they probably do not wholly account for the relative ineffectiveness of the drug in depolymerising plant microtubules. Any proposed mechanism must also account for the resistance of plant microtubules to a broad spectrum of depolymerising agents. This includes a wide range of antimitotic drugs,[17] all required at relatively high concentration to be effective, yet differing in chemical structure and presumably also in their interaction with other cellular components. Furthermore, plant microtubules are also relatively stable towards even physical depolymerising agents such as pressure and low temperatures.[31] It is worth emphasising that while cytoplasmic microtubules of plant cells are generally included in the labile category,[43, 71] observation suggests that not only do they have high stability but also that their stability differs depending upon their location within cells or tissues.[9, 13, 52, 74] In the case of *Ochromonas*, such a situation has been specifically noted, with the two spatially-distinct sets of microtubules within the cell body differing in their sensitivities towards colchicine and pressure.[10]

The basis of the greater stability of plant microtubules towards colchicine is unclear. Differences in the tubulin moiety itself, perhaps reflected in the higher affinity of a soluble component of yeast cells for the colcemid analogue,[27] would have to be of such a nature as to account for the increased stability of the microtubules towards other depolymerising agents. However, the successful copolymerisation of tubulin from such disparate sources as *Chlamydomonas* flagella and brain,[6, 7] suggests that there is no major difference in these tubulin moieties. Similarly, the stable microtubules of flagella, while being themselves unaffected by colchicine, do consist of tubulin with normal colchicine binding activity,[73] again suggesting that the stability does not arise from a change in the tubulin molecule. It is rapidly becoming clear that there are other components chemically associated with isolated and purified microtubules.[7] There are also prominent ultrastructural features, such as side-arms and the so-called 'clear zone', associated with certain microtubules.[34, 48] It remains to be seen whether any of these associated components can be correlated with increased microtubule stability.

CONCLUSIONS

Consideration of the effects of colchicine on plant tissues raises a number of points. Of major importance is the fact that the concentrations of drug required for physiological effectiveness are high enough to cause doubt concerning the basis of many described effects. Experiments designed to delineate plant processes directly dependent upon microtubule function should therefore include a number of control steps: (a) the effects of the high drug concentrations on general ultrastructure and on such areas of metabolism as respiration, protein synthesis and metabolite transport should be ascertained; (b) physiological inhibition and release should be correlated with microtubule disappearance and reappearance respectively; and (c) before conclusions are drawn regarding a possible effect of colchicine specifically on a microtubule-dependent process, equivalent concentrations of lumicolchicine should be shown to be without effect. The use of lumicolchicine is also critical in biochemically distinguishing non-tubulin sites with an affinity for colchicine.

The basis of the apparent stability of many plant microtubules towards colchicine also requires to be established. To this end, quantification of the uptake, transport and possible inactivation of the drug in defined systems would be useful. If, as seems likely, the stability is an inherent property of the microtubule, questions arise relating to the factor responsible and to the possibility that not all the cytoplasmic microtubules of a plant cell are similar in this respect.

We believe such considerations to be important, not only in defining more clearly the roles of microtubules in plant growth and development but also in facilitating isolation of the subunit of plant microtubules for comparison with animal tubulin. The latter aspect, in addition to answering questions concerning the evolutionary origins of this organelle, would also seem to be a prerequisite to determining any common modes of functioning in their various roles.

REFERENCES

1. ALLISON, A.C. The role of microfilaments and microtubules in cell movement, endocytosis and exocytosis. In *Locomotion of Tissue Cells.* Ciba Found. Symp. 14. R. PORTER and R.W. FITZ-SIMMONS (eds.), Elsevier Excerpta Medica, North-Holland (1973) pp. 109–143.

2. ARONSON, J. and INOUÉ, S. Reversal by light of the action of N-methyl, N-desacetyl colchicine on mitosis. *J. Cell Biol.* **45** (1970) 470–477.

3. AUCLAIR, W. and SIEGEL, B. Cilia regeneration in the sea urchin embryo: evidence for a pool of ciliary proteins. *Science* **154** (1966) 913–915.

4. BAMBURG, J.R., SHOOTER, E.M. and WILSON, L. Developmental changes in microtubule proteins of chick brain. *Biochemistry* **12** (1973) 1476–1482.

5. BERLIN, R.D. Temperature dependence of nucleoside membrane transport in rabbit alveolar macrophages and polymorphonuclear leucocytes. *J. biol. Chem.* **248** (1973) 4724–4730.

6. BINDER, L.I., DENTLER, W.L. and ROSENBAUM, J.L. Assembly of chick brain tubulin onto flagellar microtubules from *Chlamydomonas* and sea urchin sperm. *Proc. natn. Acad. Sci. U.S.A.* **72** (1975) 1122–1126.

7. BORISY, G.G., OLMSTED, J.B., MARCUM, J.M. and ALLEN, C. Microtubule assembly *in vitro. Fedn. Proc.* **33** (1974) 167–174.

8. BORISY, G.G. and TAYLOR, E.W. The mechanism of action of colchicine. Colchicine binding to sea urchin eggs and the mitotic apparatus. *J. Cell. Biol.* **34** (1967) 535–548.

9. BRENNAN, J.R. Colchicine induced microtubule degradation: the basis of C-tumour formation in *Phleum pratense. Phytomorphology* **20** (1970) 309–315.

10. BROWN, D.L. and BOUCK, G.B. Microtubule biogenesis and cell shape in *Ochromonas*—II. The role of nucleating sites in shape development. *J. Cell. Biol.* **56** (1973) 360–378.

11. BROWN, R.M. and MONTEZINOS, D. Cellulose microfibrils — visualisation of biosynthetic and orienting complexes in association with plasma membrane. *Proc. natn. Acad. Sci. U.S.A.* **73** (1976) 143–147.

12. BRYAN, J. Definition of three classes of binding sites in isolated microtubule crystals. *Biochemistry* **11** (1972) 2611–2616.

13. BURGESS, J. and NORTHCOTE, D.H. Action of colchicine and heavy water on the polymerisation of microtubules in wheat root meristem. *J. Cell Sci.* **5** (1969) 433–451.

14. BURNS, R.G. ^{3}H-colchicine binding. Failure to detect any binding to soluble proteins from various lower organisms. *Expl. Cell Res.* **81** (1973) 285–292.

15. COYNE, B. and ROSENBAUM, J.L. Flagellar elongation and shortening in *Chlamydomonas*—II. Re-utilisation of flagellar proteins. *J. Cell Biol.* **47** (1970) 777–781.

16. CREASY, W.A., BENSCH, K.G. and MALAWISTA, S.E. Colchicine, vinblastine and griseofulvin. Pharmacological studies with human leucocytes. *Biochem. Pharmac.* **20** (1971) 1579–1588.

17. DEYSSON, G. Antimitotic substances. *Int. Rev. Cytol.* **24** (1968) 99–148.

18. DUSTIN, P. New aspects of the pharmacology of antimitotic agents. *Pharmacol. Rev.* **15** (1963) 449–480.

19. EBNER, H. and STRECKER, H. Über die Wirkung des Colicins *in vivo* auf die alkalische Phosphatase der Rattenleber. *Experientia* **6** (1950) 388–389.

20. EIGSTI, O.J. and DUSTIN, P. *Colchicine in Agriculture, Medicine, Biology and Chemistry.* Iowa State College Press, Ames, Iowa, (1955).

21. FARRELL, K.W. and BURNS, R.G. Inability to detect *Chlamydomonas* microtubule assembly *in vitro*: possible implications to the *in vivo* regulation of microtubule assembly. *J. Cell Sci.* **17** (1975) 669–681.

22. FORBES, E.J. Colchicine and related compounds—XIV. Structure of β and γ lumicolchicine. *J. Chem. Soc.* **iv** (1955) 3864–3870.

23. FOSKET, D.E. Cell division and the differentiation of wound vessel members in cultured stem segments of *Coleus. Proc. natn. Acad. Sci. U.S.A.* **59** (1968) 1089–1096.

24. FREDRICSSON, B. and WIRSEN, C. *In vivo*

effects of colchicine on alkaline phosphatase of rat intestinal epithelium. *Expl. Cell Res.* **10** (1956) 749–751.

25. GABBAY, K.H. and TZE, W.J. Inhibition of glucose induced release of insulin by aldose reductase inhibitors. *Proc. natn. Acad. Sci. U.S.A.* **69** (1972) 1435–1439.

26. GREEN, P.B. Cell morphogenesis. *Annu. Rev. Pl. Physiol.* **20** (1969) 365–394.

27. HABER, J.E., PELOQUIN, J.G. HALVORSON, H.O. and BORISY, G.G. Colcemid inhibition of cell growth and the characterisation of a colcemid binding activity in *Saccharomyces cerevisiae. J. Cell Biol.* **58** (1972) 355–367.

28. HART, J.W. and SABNIS, D.D. Colchicine binding protein from phloem and xylem of a higher plant. *Planta* **109** (1973) 147–152.

29. HART, J.W. and SABNIS, D.D. Colchicine and lumicolchicine binding to components in extracts of higher plants. Manuscript submitted to *J. Cell Biol.*

30. HEATH, I.B. A unified hypothesis for the role of membrane bound enzyme complexes and microtubules in plant cell wall synthesis. *J. Theor. Biol.* **48** (1974) 445–449.

31. HEATH, I.B. The effect of anti-microtubular agents on the growth and ultrastructure of the fungus *Saprolegnia ferax* and their ineffectiveness in disrupting hyphal microtubules. *Protoplasma* **85** (1975) 147–176.

32. HEATH, I.B. Colchicine and colcemid binding components of the fungus *Saprolegnia ferax. Protoplasma* **85** (1975) 177–192.

33. HEPLER, P.K. and FOSKET, D.E. The role of microtubules in vessel member differentiation in *Coleus. Protoplasma* **72** (1971) 213–236.

34. HEPLER, P.K. and PALEVITZ, B.A. Microtubules and microfilaments. *Annu. Rev. Pl. Physiol.* **25** (1974) 309–362.

35. ILAN, J. and QUASTEL, J.H. Effects of colchicine on nucleic acid metabolism during metamorphosis of *Tenebrio moliter* L. and in some mammalian tissues. *Biochem. J.* **100** (1966) 448–457.

36. INOUÉ, S. and SATO, H. Cell motility by labile association of molecules. *J. gen. Physiol.* (Suppl.) **50** (1967) 259–292.

37. JEFFS, R.A. and NORTHCOTE, D.H. The influence of indole-3-acetic acid and sugar on the pattern of induced differentiation in plant tissue culture. *J. Cell Sci.* **2** (1967) 77–78.

38. KUZMICH, M.J. and ZIMMERMAN, A.M. Colcemid action on the division schedule of synchronised *Tetrahymena. Expl. Cell Res.* **72** (1972) 441–452.

39. LAWSON, V.R. and WEINTRAUB, R.L. Interaction of microtubule disorganisers, plant hormones and red light in wheat coleoptile segment growth. *Pl. Physiol.* **55** (1975) 1062–1066.

40. LEDERBERG, S. and STETTEN. G. Colcemid sensitivity of fission yeast and the isolation of colcemid-resistant mutants. *Science* **168** (1970) 485–487.

41. LING, V. and THOMPSON, L.H. Reduced permeability in CHO cells as a mechanism of resistance to colchicine. *J. Cell Physiol.* **83** (1974) 103–116.

42. LINSKENS, H.F. and WULF, N. Über die Trennung und Mitosewirkung der Lumicolchicine. *Naturwissenschaft* **40** (1953) 487–488.

43. MARGULIS, L. Colchicine sensitive microtubules. *Int. Rev. Cytol.* **34** (1973) 333–361.

44. MARGULIS, L., BANNERJEE, S. and WHITE, T. Colchicine-inhibited cilia regeneration: explanation for lack of effect in Tris buffer medium. *Science* **164** (1969) 1177–1178.

45. MINOR, P.D. and ROSCOE, D.H. Colchicine resistance in mammalian cell lines. *J. Cell Sci.* **17** (1975) 381–396.

46. MIZEL, S.B. and WILSON, L. Nucleoside transport in mammalian cells. Inhibition by colchicine. *Biochemistry* **11** (1972) 2573–2578.

47. MOOG, F. Acceleration of the normal and corticoid-induced increase of alkaline phosphatase in the duodenum of the nursling mouse by actinomycin D. puromycin, colchicine and ethionine. *Adv. Enzyme Regul.* **3** (1965) 221–236.

48. NEWCOMB, E.H. Plant microtubules. *Annu. Rev. Pl. Physiol.* **20** (1969) 253–288.

49. O'BRIEN, T.P. Cytology of cell wall formation. *Bot. Rev.* **38** (1972) 87–118.

50. OLMSTED, J.B. and BORISY, G.G. Microtubules. *Annu. Rev. Biochem.* **42** (1973) 507–540.

51. OLSON, L.W. A low molecular weight colchicine-binding protein from the aquatic phycomycete *Allomyces neo-moniliformis. Archs. Microbiol.* **91** (1973) 281–286.

52. PICKETT-HEAPS, J.D. The effects of colchicine on the ultrastructure of dividing wheat cells, xylem wall differentiation and distribution of cytoplasmic microtubules. *Develop. Biol.* **15** (1967) 206–236.

53. PICKETT-HEAPS, J.D. Plant microtubules. In *Dynamic Aspects of Plant Ultrastructure*, A.W. ROBARDS (ed.), McGraw-Hill, London (1974) pp. 219–255.

54. ROBARDS, A.W. and KIDWAI, P. Microtubules and microfibrils in xylem fibres during secondary cell wall formation. *Cytobiologie* **6** (1972) 1–21.

55. ROBERTS, L.W. The initiation of xylem differentiation. *Bot. Rev.* **35** (1969) 201–250.

56. ROBERTS, L.W. and BABA, S. IAA induced xylem differentiation in the presence of colchicine. *Pl. Cell Physiol., Tokyo* **9** (1968) 315–321.

57. SARTORELLI, A.C. and CREASY, W.A. Cancer chemotherapy. *Annu. Rev. Pharmacol.* **9** (1969) 51–72.

58. SCHONHARTING, M., MENDE, G. and SIEBERT, G. Metabolic transformation of colchicine—II. The metabolism of colchicine by mammalian liver microsomes. *Hoppe-Seylers Z. Physiol. Chem.* **355** (1974) 1391–1399.

59. SHARMA, A. and CHATTERJEE, T. *Cytologie* **26** (1961) 12–19. Quoted in reference 64.

60. SHELANSKI, M.L. Methods for the neurochemical study of microtubules. In *Research Methods in Neurochemistry*, N. MARKS and R. RODNIGHT (eds.), Plenum Press, New York (1974) Vol. 2, pp. 281–299.

61. SHERLINE, P., LEUNG, J.T. and KIPNIS, D.M. Binding of colchicine to purified microtubule protein. *J. biol. Chem.* **250** (1975) 5481–5486.

62. SHETERLINE, P., SCHOFIELD, J.G. and MIRA, F. Colchicine binding to bovine anterior pituitary slices and inhibition of growth hormone release. *Biochem. J.* **148** (1975) 435–459.

63. SHIBAOKA, H. Involvement of wall microtubules in gibberellin-promotion and kinetin inhibition of stem elongation. *Pl. Cell Physiol. Tokyo* **15** (1974) 255–263.

64. SIEBERT, G., SCHONHARTING, M., OTT, M. and SURJANA, S. Metabolic transformation of colchicine—III. Inhibition of phosphatases by the metabolite 0^{10}-demethylcolchicine (colchiceine) and their reactivation by divalent cations. *Hoppe-Seylers Z. Physiol. Chem.* **356** (1975) 855–860.

65. STADLER, J. and FRANKE, W.W. Characterisation of the colchicine-binding of membrane fractions from rat and mouse liver. *J. Cell Biol.* **60** (1974) 297–303.

66. TILNEY, L. Origin and continuity of microtubules. In *Origin and Continuity of Cell Organelles*. J. REINERT and H. URSPRUNG (eds.), Springer, New York (1971) pp. 222–260.

67. TURNER, F. The effects of colchicine on spermotogenesis in *Nitella. J. Cell Biol.* **46** (1970) 220–234.

68. WEISENBERG, R.C., BORISY, G.G. and TAYLOR, E.W. The colchicine-binding protein of mammalian brain and its relation to microtubules. *Biochemistry* **7** (1968) 4466–4479.

69. WILSON, L. Properties of colchicine-binding protein from chick embryo brain. Interactions with Vinca alkaloids and podophyllotoxin. *Biochemistry* **9** (1970) 4999–5007.

70. WILSON, L., BAMBURG, J.R., MIZEL, S.B., GRISHAM, L.M. and CRESWELL, K.M. Interaction of drugs with microtubule proteins. *Fedn. Proc.* **33** (1974) 158–166.

71. WILSON, L. and BRYAN, J. Biochemical and pharmacological properties of microtubules. *Adv. Cell Molec. Biol.* **3** (1974) 21–72.

72. WILSON, L. and FRIEDKIN, M. The biochemical events of mitosis. The *in vivo* and *in vitro* binding of colchicine in grasshopper embryos and its possible relation to inhibition of mitosis. *Biochemistry* **6** (1967) 3126–3135.

73. WILSON, L. and MEZA, I. The mechanism of action of colchicine. Colchicine binding properties of sea urchin sperm tail outer doublet tubulin. *J. Cell Biol.* **58** (1973) 709–719.

74. WOODING, F.B.P. P-protein and microtubule systems in *Nicotiana. Planta* **85** (1969) 284–298.

75. WUNDERLICH, F., MULLER, R. and SPETH, V. Direct evidence for a colchicine-induced impairment in the mobility of membrane components. *Science* **182** (1973) 1136–1138.

INSECT RESISTANCE IN CROP PLANTS*

M. D. PATHAK and R. C. SAXENA**

The International Rice Research Institute, Los Baños, Laguna, Philippines

Most species of insect pest have well-developed mechanisms for selection of their host plants and have distinct nutritional requirements. Even the so-called polyphagous insects do not feed on just any plant in their range, while many oligophagous species show marked differences in their ability to feed and survive even on different varieties of one species of the host plant. While such differences in varietal susceptibility have been known for over a century it is only since about the last three decades that general and concerted efforts have been made to utilize them as an effective method for insect control. Probably the main reasons for this lag are a lack of training of most entomologists in plant genetics, the reciprocal lack of training of plant breeders in the entomological sciences and the dearth of research systems which encourage problem-oriented interdisciplinary research.

The increasing complexities of pest control, in recent years, have identified critical need for effective control measures which are relatively free of adverse side effects. The use of insect-resistant varieties is one such promising possibility. Plant varieties that are insect resistant have the unique advantage of providing inherent insect control to the crop; it involves no expense or environmental pollution and is generally compatible with other methods of pest control. Earlier concerns that varietal resistance is frequently associated with poor quality and low yield have generally been proven to be unfounded.

The principles of insect resistance in crop plants and early work in this area have been reviewed by Painter.[101, 102] Various papers have reviewed recent developments.[38, 53, 91, 108] This paper discusses the major aspects of insect resistance in host plant varieties and its role in insect pest control. It uses selected examples as appropriate illustrations but is not intended to be an exhaustive review of the literature on varietal resistance.

RESISTANCE – GENERAL ASPECTS

Resistance or susceptibility of plants to insects involves a cause-and-effect relationship between the insect's response(s) to the plant and, in turn, the plant's reaction(s) to the insect – for example, lack of attraction of insects to a particular plant for oviposition or feeding, and the unsuitability of plants for insects. Painter's[101] definition of varietal resistance to insect pests describes well the actual field conditions: "The relative amount of heritable qualities possessed by the plant which influence the ultimate degree of damage done by the insect. In practical agriculture it represents the ability of a certain variety to produce a larger crop of good quality than do ordinary varieties at the same level of insect populations."

*This Commentary was fully revised and rewritten in November 1979.
**Director-Research and Training Coordinator and Associate Entomologist, respectively.

Painter[101] classified varietal resistance into three categories:

(a) *Non-preference* – when a plant possesses characteristics that make it unattractive to insect pests for oviposition, feeding, or shelter.

(b) *Antibiosis* – when the host plant adversely affects the bionomics of the insects feeding on it.

(c) *Tolerance* – when the damage to the host plant is only slight despite its supporting an insect population of a size sufficient to damage severely susceptible hosts.

Recently, Kogan and Ortman[78] have suggested the term *antixenosis* to replace non-preference. A plant's resistance may be caused by one or more of these factors. Varietal resistance to insect pests is generally a relative quality and under severe infestations even resistant varieties are subject to significant crop losses, which seems a contradiction in terms because resistant varieties are expected to remain at best damage free, and at worst experience only slight insect damage. To preclude this ambiguity, some commercial seed firms have replaced the term "tolerance" for "resistance"; however, tolerance is only one aspect of resistance.

Antibiosis is considered a major aspect of resistance, and, under some circumstances, non-preference can be equally important, particularly where brief infestations cause severe plant damage. One example of the latter is feeding by stem borers which causes white head and insect transmission of virus diseases. In field plantings, non-preferred crop varieties frequently escape infestations and when insects are caged on non-preferred hosts, they lay fewer eggs and smaller populations develop. Thus, both antibiosis and non-preference affect insect populations. Tolerant varieties do not inhibit insect populations and, under certain conditions, may even be responsible for the build up of a population larger in size than one found with susceptible varieties which are killed when a certain level of infestation is reached. However, the ability of tolerant plants to survive insect infestations for longer periods permits a longer time for insects to be exposed to their natural enemies and other physical factors which limit insect populations. Frequently, tolerant varieties are valuable in a programme of integrated insect control.

CAUSES OF RESISTANCE

The non-preference, antibiosis, and tolerance aspects of resistance result from a series of interactions between insects and plants which influence the selection of plants as hosts and the effects of plants on insect survival and multiplication. Establishment of insects on plants involves orientation, feeding, metabolic utilization of ingested food, growth, survival and egg-production, oviposition,[127, 128] and hatching of eggs. The capacity of a plant to cause the interruption of any of these may account for its resistance to insect infestation. Such interruptions may be caused by the biochemical or biophysical characteristics of a plant. Often such responses are due to several factors and few investigations have been able to identify the exact role of these plant characters in the resistance of the host plant to a particular insect. However, the available information on the causes of resistance is broadly summarized as follows:

Physical factors

Many morphological and anatomical characters of plants contribute to the resistance of plants to insect pests, but only in a few cases has resistance been found to be solely due to biophysical characters. Generally, the biophysical characteristics of a plant are complementary to other factors of resistance. For example, tight husks in certain corn varieties help confine the corn earworm, *Heliothis zea*, larvae longer in the husks which increases the probability of cannibalism among the larvae,[101] and forces the larvae to feed on silk channels which, in certain genotypes of corn, contain a lethal biochemical factor.[50] The silk of Zapalote Chico corn contains a feeding deterrent and growth inhibitor to the larvae of corn earworms.[150]

Some of the more important plant biophysical characters that have been recorded associated with resistance to insects are as follows:

Shape, size and colour of plants. The general shape, size, and colour of a plant contribute to its attractiveness to insects. For example, in cotton, red plant colour, in combination with hairiness and an absence of stem glands, was involved in the plant's resistance to the cotton boll weevil, *Anthonomus grandis.*[145] The imported cabbageworm, *Pieris rapae,*[30] also in both field and greenhouse tests for oviposition, preferred other *Brassica* lines to the red cabbage.

Pubescence. Hairiness of plants interferes with oviposition or feeding of certain insects.[161] In cotton, hairiness of foliage increases resistance to leafhoppers, *Empoasca* spp.[34, 72, 101, 136], cotton aphid, *Aphis gossypii,*[72] cotton leafworm, *Spodoptera littoralis*, and spider mites, *Tetranychus* species.[1, 72] Hairiness of cotton flower buds had two features contributing to the plant's resistance to the boll weevil: (1) impeded the weevils feeding on the bracteoles enclosing the flower buds, and (2) exerted antibiotic effects on the larvae.[146] Additionally, pubescence of cotton was found to confuse and impede the movements of pink bollworm larvae, *Pectinophora gossypiella.*[140] Hairiness of wheat leaves provides protection against the cereal leaf beetle, *Oulema melanopus*, by deterring oviposition by the beetle and the feeding of the hatching larvae.[37, 62, 163] Resistance of certain *Phaseolus* bean cultivars to the potato leafhopper, *Empoasca fabae*, has been attributed to specialized hairs or hooked trichomes, which capture or impale both nymphs and adults.[115, 116]

Even with these reports one can only speculate on the major role of plant pubescence in insect resistance. Orientation of hairs rather than their number is considered important for resistance to the potato leafhopper in soybeans,[16] but in many cases the existence of an association between a plant's hairiness and its resistance to insects has not been confirmed. For instance, pubescence of certain accessions of wheat was not correlated to their resistance to the cereal leaf beetle.[130] Rice varieties with hairy leaf surfaces are non-preferred by the striped borer moths, *Chilo suppressalis*, for oviposition, but artificial removal of hairs from the leaves of a resistant variety, TKM 6, did not alter this reaction of the ovipositing moths.[108] On the other hand, hairiness has been reported to increase infestation in cotton by the cotton aphid, spider mites,[117] and the cotton fleahopper, *Pseudatomoscelis seriatus.*[86] Likewise, the leaves of highly susceptible clones of strawberries were found to be more pubescent than those that were less susceptible to the two-spotted spider mite, *Tetranychus urticae.*[75]

Trichomes. Glandular hairs are abundant on the leaves of many solanaceous plants, and their resistance to insects has been attributed to sticky exudates secreted by trichomes which entrap or impede the movements of the pests, e.g. potato leafhopper,[157] potato aphid, *Macrosiphum euphorbiae,* green peach aphid, *Myzus persicae,* and the Colorado beetle, *Leptinotarsa decemlineata*, on *Solanum* species,[45, 46] greenhouse whiteflies, *Trialeurodes vaporarium,*[43] carmine spider mite, *Tetranychus cinnabarinus*, and two-spotted spider mite on *Lycopersicon* and *Solanum* species.[44, 46, 148, 149] In some cases, for example in tobacco, trichome exudates are toxic to insects.[152, 153, 154, 155] Similarly, physiological incompatability rather than physical characters of certain species of *Lycopersicon* which have sparse glandular hairs may contribute to the resistance of these plant species to the cowpea aphid, *Aphis craccivora*, green peach aphid,[70] and potato aphid.[42]

Spines. Sugarcane varieties with high spine density on the lower leaf surface or midrib were found to be less susceptible to attack by the sugarcane top borer, *Scirpophaga nivella*, than varieties with a lower spine density.[158]

Hardness of plant tissue. Mechanical obstruction to feeding or oviposition due to toughness or thickness of epidermal and other tissues of the host plant is important in reducing insect infestations. Toughness of leaf veins[2] and thickness of lamina[156] and palisade cells[5] of cotton affect its resistance to the cotton jassid, *Empoasca devastans*. Thick-hulled varieties of southern peas had fewer ovipositional punctures, eggs, and larvae of the cowpea curculio, *Chalcodermus aeneus*, than thin-hulled varieties.[22] Varietal resistance to the squash vine borer, *Melittia cucurbitae*, in cucurbits,[61] to the shoot and fruit borer, *Leucinodes orbanalis*, in eggplant,[103] and to the stem weevil, *Apion virens*, in red clover,[79] has been attributed to the structure and compactness of vascular bundles in the plants. The solid-stemmed nature of Rescue and other wheat varieties is the major cause of their resistance to the wheat stem sawfly, *Cephus cinctus*,[95, 160] the eggs of which are mechanically damaged and desiccated, and the movement of the hatching larvae is limited.

Resistance to the striped borer was recorded in rice varieties with leaf sheaths wrapped tightly around the stem, closely packed vascular bundles and a thick layer of sclerenchymatous tissue. These characters probably interfere with the boring activity of larvae in the stem.[31, 104, 126] Larvae feeding on rice varieties containing high levels of silica exhibited typical antibiotic effects and had wornout mandibles.[31] Sorghum cultivars resistant to the sorghum shootfly, *Atherigona varia soccata*, had silica bodies in the abaxial epidermis at the base of the first, second and third leaf sheaths[12] of a greater density than those of cultivars with less resistance to the pests.

Hypersensitivity of plant tissues. Larvae of the melon leaf miner, *Liriomyza pictella*, mining in young leaves of snap dragon plants are crushed by the rapidly proliferating wound tissue, and in eggplant, the tissue surrounding the larvae dries up, thus stopping their development.[100]

Chemical factors

Insect resistance in plants is often due to their chemical constituents[7, 63, 89, 102] (Table 1). Differences in these chemicals may be qualitative or quantitative and these chemicals may occur only within certain parts of the plant or during specific stages of plant growth. Determination of these factors is intricate because adequate bioassays are necessary before chemical differences can be identified as the cause of resistance. This is often difficult because of a lack of adequate information on the behavioural and physiological responses of the pests to plant biochemicals. Consequently, the chemicals responsible for resistance have been identified in only a few instances. Beck[7] classified these chemicals as physiological inhibitors and nutritional deficiencies. A review of information on the biochemical bases of plant resistance to insects follows.

Physiological inhibitors and feeding deterrents. Physiological inhibitors responsible for plant resistance have been demonstrated in corn.[6, 8, 9] Klun *et al.*[76] demonstrated that DIMBOA

TABLE 1. Selected examples of the biochemical nature of the resistance of host plants to insect pests

Host plant	Chemical attributed for resistance	Insect pest	Effect on insect	Reference
Alfalfa	trichome exudate	alfalfa weevil	antibiosis	132
	high saponin content	white grub	non-preference, antibiosis	58, 59
	medicagenic acid	pea aphid, potato leafhopper	antibiosis	54, 60
Barley	benzyl alcohol	greenbug	antibiosis	71
Corn	high DIMBOA content	European corn borer	antibiosis	76, 120, 122, 125
	unknown factor in silk	corn earworm	lethal to larvae, antibiosis	50, 150
Cotton	high gossypol content	*ca.* 10 different insect pests	antibiosis, toxic	13, 14, 82, 137
	condensed tannin and heliocides in flower buds	*Heliothis* larvae	antibiosis, toxic	23, 147
Cucumber	absence of cucurbitacin	cucumber beetle	non-preference	29, 133
	presence of cucurbitacin	two-spotted spider mite	non-preference	29
Rice	low asparagine content	brown planthopper	non-preference	141
	benzoic & salicylic acid	striped borer	toxic	96
Strawberry	essential oils	two-spotted spider mite, strawberry	repellence	27, 73
Sweet clover	ammonium nitrate	sweetclover weevil	feeding deterrence	3, 10, 88
Tobacco	Trichome exudate	bird-cherry-oat aphid, green peach aphid, tobacco hornworm	toxic	152, 155
	alkaloids (nicotine, anabasine, nornicotine) in trichome exudate	green peach aphid, tobacco hornworm, tobacco budworm	toxic	47, 153, 154
Tomato	unknown chemical(s)	carmine spider mite, two-spotted spider	repellence	18
Wild potato	alkaloidal glycosides (tomatine, solanine, leptine I) and aglycones (solanidine, demissidine)	potato leafhopper	feeding deterrence, antibiosis	28, 119

(2,4-dihydroxy-7-methoxy-(2H)-1,4-benzoxazine-3(4H)-one) caused the leaves of corn inbreds to be resistant to the first-brood European corn borer, *Ostrinia nubilalis*, larvae and noted a significant correlation between the concentration of DIMBOA in the leaf whorl tissues of several corn inbreds and their degree of resistance to first-brood larvae.[77, 120] Analysis of breeding material for DIMBOA content has even been used as a method of screening for resistance to first-brood corn borer larvae.[125] Recently, DIMBOA was also found to be the primary chemical factor in the non-preference mechanism of corn borer-resistant dent corns.[122] However, many accessions of corn from the tropics have low DIMBOA content but are resistant to the corn borer,[131, 151] and corn inbred B 52 has a low DIMBOA concentration but is resistant to the second brood of the European corn borer.[49] Therefore, analysis for DIMBOA of corn plants may not be an adequate procedure for screening varieties and lines for resistance to corn borer.

DIMBOA was also demonstrated to be a factor in resistance of corn to stalk rot,[11] resistance to wheat stem rust,[32] and resistance of corn to 2-chloro-s-triazene herbicides.[52] Many plant varieties show a resistance to a wide variety of pests and diseases,[107] but the possibility that certain endogenous plant chemicals can impart resistance to several pests and diseases remains to be explored.

Resistance to several insect pests in cotton has been recorded to be related to the content of gossypol, a dimeric sesquiterpenoid (8, 8′-dicarboxaldehyde-1, 1′, 6, 6′,7,7′-hexahydroxy-5, 5′-diisopropyl-3,3′-dimethyl-2, 2′-binaphthalene), which occurs in much higher concentrations in the seed and green parts of glanded than glandless varieties of cotton.[87, 134, 137] Glandless varieties, bred to reduce gossypol concentration in seed to a level non-toxic to non-ruminants, were found to be susceptible to about ten different insect pest species while the glanded or high-gossypol varieties were resistant.[14, 82, 93, 97, 137] Toxic effects of gossypol were demonstrated by spraying it on the cotton aphid, lygus bug, *Lygus hesperus*, salt marsh caterpillar, *Estigmene acrea*, and thurberia weevil, *Anthonomus grandis thurberiae*, and by mixing it in a diet used for rearing the bollworm *Heliothis zea.*[14] A synthetic diet containing 0.1% gossypol significantly reduced the weight of the beet armyworm, *Spodoptera exigua*, cabbage looper, *Trichoplusia ni*, and salt marsh caterpillar larvae.[13] An experimental line of cotton, XG-15, containing 1.7% gossypol in flower buds (2-3 times the amount found in cultivated varieties) inhibited larval growth and suppressed populations of the tobacco budworm, *Heliothis virescens.*[87] However, *Heliothis* resistance in the flower buds of a few wild races of cotton could not be explained by gossypol concentration but was ascribed to certain 'x-factors', among which have been identified the sesquiterpenoids, hemigossypolone and hemigossypolone-7 methyl ether, and the derived terpenoids, heliocides — H_1 to H_4 and B_1 to B_4.[147] Recently, condensed tannin from flower buds of certain cotton lines has been reported to have a strong antibiotic activity against tobacco budworm larvae.[23]

Cucurbitacin is a classical example of the presence or absence of a plant biochemical affecting the resistance of the plant to pests. Cucurbitacins, tetracyclic triterpenoids, are feeding attractants to the spotted cucumber beetle, *Diabrotica undecimpunctata howardi*. Cucurbitacin-free cucumbers are resistant because they are not preferred by the beetle, but they are susceptible to the two-spotted spider mite, while those containing cucurbitacin were mite-resistant.[29] Also, a positive correlation has been shown between the concentration of cucurbitacins and their glycosides in various cucurbit fruits and the number of spotted cucumber beetles attracted to them.[133]

Saponins, another biologically active group of chemicals, occur in many plant species and are toxic to many insects, microorganisms, and even to vertebrates. Saponins of alfalfa, *Medicago sativa*, have been identified as glycosides belonging to the triterpenoid class of compounds.[54, 59, 60, 81] High saponin contents in alfalfa roots, and possibly in red clover, white clover, birdsfoot trefoil, sugar beets, and potatoes, and various herbs and forbs were recorded to have antibiosis effects on the white grub, *Melolontha vulgaris.*[58, 59] One per cent or higher concentrations of saponin in a synthetic diet was found to significantly reduce the survival of the potato leafhopper.[124] Medicagenic acid was the most predominant saponin in the high-saponin DuPuits but was nearly absent in low-saponin Lahontan varieties of alfalfa, and it was found to be toxic to the potato leafhopper and the pea aphid, *Acyrthosiphon pisum*, inhibited growth of the fungus, *Trichoderma viride*, and caused hemolysis of human blood cells.[54, 60] The possible deleterious effects of saponins on vertebrates limits the practicality of developing varieties containing high levels of saponin which could improve the resistance of such plants to pests and diseases.

Resistance in *Nicotiana* species to the green peach aphid, bird-cherry-oat aphid, *Rhopalosiphum padi*, and the pea aphid, was attributed to toxic materials exuded from leaf hairs;

the toxic effects resembled nicotine poisoning.[155] Analysis of these exudates showed the presence of alkaloids, such as nicotine, anabasine, and nornicotine;[153] these were contact toxic to adults and nymphs of the green peach aphid and larvae of the tobacco hornworm, *Manduca sexta.*[154] The toxicity of tobacco plants to tobacco hornworm larvae was reduced when the plants were washed to remove the exudates.[152] In flue-cured varieties of tobacco, differences in the degree of leaf damage caused by the tobacco budworm were attributed to differences in their nicotine contents,[47] and the alkaloids present in the trichome secretions of several *Nicotiana* species caused these species to be resistant to the two-spotted spider mite.[112]

Resistance to the potato leafhopper in certain wild *Solanum* species was attributed to the feeding-deterrent and survival-reducing effects of the solanaceous alkaloidal glycosides tomatine, solanine and leptine I, and the aglycones solanidine and demissidine.[28] Recently, further evidence of a causal role for potato glycoalkaloids in leafhopper resistance has been provided.[119]

Ethanol extracts of tomato plants served as repellents to the two-spotted spider mite and the carmine spider mite.[18] Resistance to the two-spotted spider mite was attributed to both repellency and antibiosis caused by essential oils present in much higher concentrations in the resistant tomato line PI 251303 than in the susceptible line PI 143524. Two fractions of these oils were toxic to mites and were found to be sesquiterpenoids.[111] Also, differences were found in the composition of essential oils in the leaves of clones of strawberries resistant and susceptible to the two-spotted spider mite and the strawberry spider mite, *Tetranychus turkestani.*[73] Resistance of the strawberry plants to these mites appeared to be caused by a repellence factor in the leaves.[27, 123]

Biochemicals that are responsible for plants' resistance to insects are not always complex molecules like DIMBOA, gossypol and others, but could be simple compounds. For example – benzoic acid and salicylic acid found in rice plants were reported to be toxic to striped borer larvae.[96] Benzyl alcohol, another simple compound, has been reported to be responsible for the resistance of Omugi barley to the greenbug, *Schizaphis graminum*. It is readily absorbed by barley and sorghum plants rendering susceptible plants resistant to the greenbug,[71] and thereby, serves as an excellent example of the potential application of the knowledge of biochemical basis of plant resistance to an insect pest.

Ammonium nitrate was isolated by Akeson *et al.*[3] from a resistant sweetclover species, *Melilotus infesta*, and was identified as a feeding deterrent for the sweetclover weevil, *Sitona cylindricollis.* Apparently, the "nitrate" ion was responsible for the deterrent activity *in vivo.*[3, 10, 88]

Nutritional deficiencies. The importance of deficiencies in host plants of nutritional factors indispensible to insect pests is not fully understood, although such factors have been associated with the resistance of some plants to a few insects. Varieties of pea that were resistant to the pea aphid contained deficient levels of amino acids and thus were less nutritious for the insect than the susceptible varieties.[4]

Insects appear to respond positively to various nutritional plant compounds including sugars, amino acids, polypeptides, and vitamins.[7, 74] These compounds may serve as feeding stimulants or as cofactors and synergists to more specific stimulants. The concentration of a water-soluble feeding stimulant, which was isolated from southern peas,[17] had a direct and positive effect on the amount of feeding by the cowpea curculio.[26] Additionally, positive correlations were found between the level of total carbohydrates and the nitrogen concentration in the hull of different varieties of cowpea and the number of feeding punctures made in their hull by the cowpea curculio.[21]

Resistance of the rice variety Mudgo to the brown planthopper, *Nilaparvata lugens*, was suggested to be due to its lower asparagine content as compared to that of susceptible varieties. Mudgo plants either lacked a feeding stimulant or possessed a feeding deterrent as insects caged on Mudgo died because of a lack of sustained feeding.[141]

GENETICS OF RESISTANCE

Knowledge of plant genetics relative to insect resistance can contribute to the development of plant varieties with a broad base of insect resistance which may reduce the development of insect biotypes adapted to resistant varieties. Also, plant genetics can be applied to the development of isogenic lines which are used to identify the mechanism of resistance, to multi-line plant populations, and others.

Varietal resistance to many of the insect pests of plants is of polygenic nature which generally imparts a moderate level of resistance, but little has been studied on this type of inheritance in plants. Most of the known high levels of resistance are monogenic. The completeness of the information on the number of genes involved in insect resistance depends on the thoroughness with which the germ plasm has been screened.

Pathak[108] summarized the current information on the genetics of plant resistance to insects. Since then two additional genes affecting rice plant resistance to the brown planthopper[80] and four additional genes affecting the resistance of rice to the green leafhopper, *Nephotettix virescens*,[121, 138] have been identified. Also, two duplicate factor genes have been associated with the resistance of the Seneca variety of wheat to Race E of the Hessian fly, *Mayetiola destructor.*[113] The resistance of wheat varieties to the wheat stem sawfly is caused by the solidness of the stem which is genetically controlled by one or more dominant, recessive, or complementary genes, depending on the parents used, the crosses and the ploidy involved.[159] The resistance of some wheat varieties to the greenbug was once attributed to a single pair of recessive genes,[25] but subsequent findings indicated a mode of inheritance more complex than monogenic resistance.[118] The resistance of barley to the greenbug is attributed to a single dominant gene.[40, 41, 139] The resistance of oat varieties to the greenbug was caused by a single pair of genes,[39] while a single incompletely dominant gene,[162] or dominant genes at more than one locus,[51] account for sorghum resistance to this plant pest.

Inheritance of resistance in corn to the corn leaf aphid, *Rhopalosiphum maidis*, is determined by many genes with varying degrees of dominance and additivity.[35]

Diallele systems of crossing can be used in case of polygenic resistance to increase further the existing levels of insect resistance. For example, in rice, a significant increase in resistance to striped borer and yellow borer, *Tryporyza incertulas*, was achieved by intercrossing moderately resistant rice varieties.[66, 67, 68]

INSECT BIOTYPES

The development of insect biotypes capable of surviving on resistant plant limits the use of resistant varieties. Wild populations of many insect species are believed to consist of individuals genetically diverse in their ability to survive on different plant varieties, but their number is limited because of some inherent handicap. However, continuous planting of resistant plant varieties screens out the latter, thereby promoting the development of a new biotype of plant pest. Also, the possibility of the development of new pest biotypes through mutation should not be excluded.

The development of biotypes or biological races of pathogenic fungi is frequent, and in some cases, such as wheat rust, *Puccinia graminis*, several hundred races of the fungus have been identified. The southern corn leaf blight epidemic of 1970 in the United States[98] exemplifies the magnitude of the problem which may result from large areas being planted to a single plant genotype, such as the Texas cytoplasm which was carried by the majority of corn hybrids in America at that time, and the development of a virulent race of a fungal pathogen, such as *Helminthosporium maydis*. However, biotypes are generally less common among insect pests, primarily because of the insect's own complex physiology and because the resistance of plants to pests is often governed by several independent factors. Apparently, insect biotypes are more likely to develop on plants which have monogenic resistance than on plants with polygenic resistance. Thus, the genetic base of at least major crop plants should be broadened to include pest-resistant characteristics to equip the crop with inherent protection from polyphagous insects.[55, 99, 108]

In a few cases biotypes capable of surviving on resistant plants may be larger and more vigorous insects, such as the pea aphid, but generally biotypes are adapted to the effects of specific genes for resistance. A knowledge of insect biotypes can help plant breeders to identify diverse sources of resistance and to breed for multigenic resistance.

Differential host varieties are commonly used to detect biotypes. A list of some of the commonly known biotypes is given in Table 2. Six of these eight insects with known biotypes are aphids. Since most aphids commonly reproduce parthenogenetically and have a relatively short life cycle, an aphid that can survive on a resistant plant can build up a new biotype within one or two crop seasons. A destructive new Biotype C of the greenbug threatened serious losses to the U.S. sorghum crop in 1968,[56] but sources of resistance to this Biotype were identified within a rela-

Table 2. Insect biotypes involved in plant resistance (Pathak[106] *)

Insect	Common name	Crops	Biotypes (number)
Acyrthosiphon pisum	pea aphid	peas, alfalfa	9
Amphorophora rubi	raspberry aphid	raspberry	3
Aphis fabae	bean aphid	beans	2
Mayetiola destructor	Hessian fly	wheat	9
Nilaparvata lugens	brown planthopper	rice	3
Rhopalosiphum maidis	corn leaf aphid	corn, sorghum	5
Schizaphis graminum	greenbug	wheat, barley, sorghum	3
Therioaphis trifolii (= *maculata*)	spotted alfalfa aphid	alfalfa	8

*Revised

tively short time in sorghum and barley.[166] Also, intergeneric crossing to transfer greenbug resistance from Argentine rye via an octoploid hybrid *Triticale* 'Gaucho' to winter wheat susceptible to greenbug populations appears to be feasible because only one of the pairs of the rye chromosomes bears the genes for resistance.[167]

Four biotypes of the corn leaf aphid differing in their ability to infest different hosts, geographical distribution, and toleranceto environmental conditions were recognized.[19, 110] In 1972 a new biotype was identified on the basis of its ability to attack *Triticum timopheevi*, a cereal species previously considered resistant, and on its ability to reproduce well at higher temperatures.[164]

The stability of the resistance of a crop variety is known to vary among areas. Wheat varieties grown in Kansas and California in the U.S.A. have been resistant to the Hessian fly since 1946.[102] However, in Indiana, intensive cultivation of wheat varieties containing the H_3 gene for resistance caused a shift from the predominance of the fly Biotype A to Biotype B.[57] Additional biotypes were isolated in the greenhouse and field populations and now there are nine biotypes (GP, A, B, C, D, E, F, G, and L) of the Hessian fly.[142] While most wheat varieties now being grown in Indiana carry resistance to the common biotypes of the Hessian fly, there is a continued need to monitor the occurrence of virulent biotype populations and corresponding programmes for development of effective resistant wheat cultivars.

As mentioned previously in this paper, rice varieties resistant to the brown planthopper are being grown over large areas of Asia but development of biotypes capable of surviving on resistant plants appears to be a major threat to the success of these varieties. Two additional biotypes were identified, each adapted to a specific gene for brown planthopper resistance, when field-collected insects were reared in the greenhouse on rices having different resistance to this pest. The original Biotype plus these two more recently identified biotypes were named Biotypes 1, 2, and 3. High-yielding improved rice varieties resistant to each of these three biotypes have been developed. In 1975 Biotype 2 was reported to be present at a few locations in the Philippines. In certain parts of India, Sri Lanka, and the Solomon Islands the brown planthopper appears to be different from any of the three biotypes identified at the International Rice Research Institute (IRRI).

ROLE OF RESISTANT VARIETIES IN INSECT CONTROL

Resistant varieties of plants offer an inexpensive preventive measure, generally compatible with other methods of pest control. Depending upon the level of resistance, they can either be used as the principal method of control or can be combined with other methods in developing appropriate pest management systems. Whenever possible, the use of resistant varieties appears to be the most practical method of pest control in that: it requires less technological inputs, is less costly, and causes no environmental pollution problems.

Resistant varieties effect cumulative reductions in pest populations. Insects infesting varieties possessing an antibiosis type of resistance exhibit a high mortality, lay fewer eggs, and have slower rates of growth and, in general, smaller body size. The same is true for insects feeding on non-preferred varieties of plants but to a lesser degree. Even normal adults from susceptible hosts when caged on non-preferred plants frequently lay a smaller number of eggs than they do on preferred hosts; for example, an ovipositional arrestant for boll weevil is present in certain cotton varieties.[92]

In his review of twenty-one different insect species, Pathak[106] noted that, in many of these cases, within one generation of the insect, the population on susceptible plants became 5-10 times larger than on resistant plants.

In one of the studies of the actual development of populations of several rice insect pests on resistant and susceptible varieties at IRRI, the population of the striped borer within 120 days or approximately four generations became 30-fold larger on the susceptible rice variety Sapan Kwai than on the resistant rice Chianan 2.[109] Differences in the rate of population build-up on resistant and susceptible varieties are of even greater magnitude for leafhopper and planthopper pests of rice. Because varietal resistance to these insects is high, insects caged on resistant plants frequently die within the first generation, whereas they multiply rapidly on susceptible varieties.[108]

Because of the microenvironment in fields planted with modern heavy-tillering rice varieties and treated with higher rates of fertilizers, this is apparently well suited to the brown planthopper which inhabits the basal parts of the rice plant; the population of this pest increased markedly in several areas in the Philippines, for example around IRRI (Fig. 1). The first brown planthopper-resistant rice variety, IR26, was released for cultivation in early 1973, and at present, resistant varieties are being planted in over 90% of this area of the Philippines. This may have

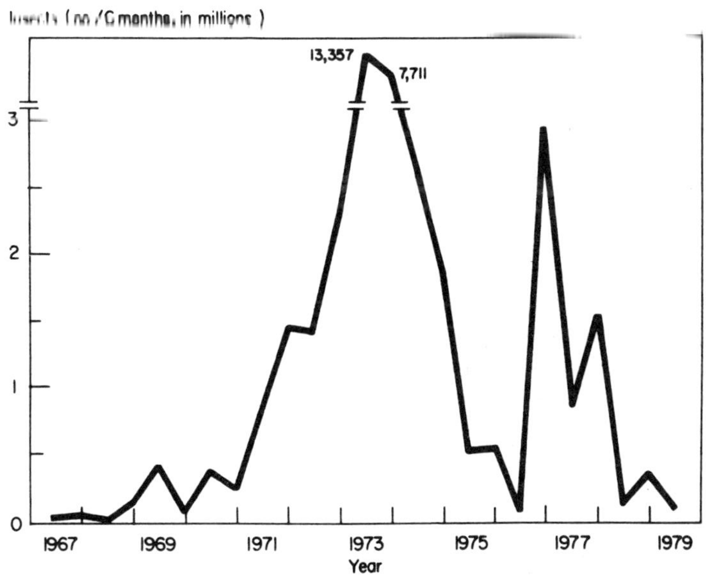

FIG. 1. Abundance of brown planthopper based on catches in three light traps at IRRI (V.A. Dyck, unpublished data). Intensive cultivation of resistant varieties is considered a major factor affecting the decline of brown planthopper numbers since mid-1973.

contributed, in large part, to a subsequent decline in the insect's abundance. A slight increase in the brown planthopper population in 1976, as evidenced by somewhat high catches in light traps at IRRI, may be because of the development of Biotype 2, which is capable of surviving on IR26 and sibling varieties. Since then, introduction and large-scale cultivation of new rice varieties such as IR32, IR36, etc. (carrying a different source of resistance), has effectively curbed the pest incidence in the Philippines.

The level of resistance of the rice varieties resistant to the brown planthopper, presently grown over large areas in Asia, is very high (Fig. 2). In certain areas, however, where brown planthoppers appear to be of different biotypes, rice varieties developed as resistant at IRRI are susceptible.

FIG. 2. Both replications of a resistant line showed no visible damage while two other varieties (susceptible) were killed by the brown planthopper, IRRI, 1972.

The cultivation of resistant wheat varieties has reduced the field populations of the Hessian fly and the wheat stem sawfly, from a very serious problem (amounting to annual crop losses of 10 million dollars in the U.S. alone)[38] to a situation of that of minor pests.[84, 85, 102] By 1974 nearly 16 million acres in the U.S.A. were planted to varieties resistant to Hessian fly, and by 1977 more than twenty-eight resistant varieties had been released.[36] Similarly, alfalfa varieties resistant to the spotted alfalfa aphid and corn varieties resistant to the European corn borer help, to a major extent, control these two economically important plant pests.[24, 48, 84, 114]

Rice variety IR8 is resistant to the rice delphacid, *Sogatodes orizicola*, in Colombia but is susceptible to the *hoja blanca* virus transmitted by this insect. Fields of IR8 rice remained virtually virus free, apparently because of its resistance to the vector, while other rice varieties which are susceptible to this pest and growing in adjacent fields were heavily infected.[69] Similarly, many rice varieties and breeding lines resistant to leafhoppers and planthoppers have exhibited field resistance to virus diseases because of their resistance to the vectors, even though these rices are susceptible to the virus.

Where a number of insect pest species infest a crop, varietal resistance can be integrated with alternate control methods to attain a higher degree of control efficiency than can be attained by each procedure alone. This approach has been found to be highly practical for rice grown in tropical and subtropical Asia where infestations of a crop by about a dozen insect pest species is general. However, presently rices carry varietal resistance only to stem borers, gall midge, leafhopper and planthopper pests.

Host plant resistance can be an integral part of the total pest management approach. Insects feeding on resistant plants are often less vigorous, smaller, and less likely to be able to survive environmental inclemencies, such as high or cold temperature, parasites, and predators. A predator may devour more small-sized insects from resistant plants than larger insects from susceptible hosts. Also significant attrition in terms of exhaustion and disorientation can be expected from the restless behaviour of a pest on a resistant plant.

Cotton bollworm larvae hatching from eggs placed on plant apices remained longer because of the hairiness of the cotton plant, and were thus more exposed to predators on the terminal growth of Frego-type resistant cotton than those on susceptible DPL-smooth leaf variety.[83] Consequently, predators killed twice as many larvae on the resistant varieties as on susceptible varieties. Early maturing smooth cotton genotypes, which are resistant to cotton fleahopper, bollworm, tobacco budworm, and other pests, supported as many or more parasites and predators at any one time as the susceptible hirsute genotype, indicating that resistance by a plant to a pest need not affect the abundance or activity of a predator of the pest.[135] Similarly, greenbug parasites *Aphelinus asychis* and *Lysiphlebus testaceipes* and varietal resistance of barley and sorghum to the greenbug were complementary and resulted in less infestation by this plant pest.[33, 144]

Varietal resistance in rice and insecticidal treatments were found at IRRI to be complementary, providing a more complete pest-control program than the use of either method alone. The level of borer infestation of varieties with moderate resistance to rice stem borers receiving only one or two insecticidal treatments was similar or lower than that observed with the susceptible varieties receiving two-fold more insecticidal treatments.[64, 65, 105] Other such examples include sweet potato against the sweet potato flea beetle, *Chaetocnema confinis*;[15] cabbage against the cabbage looper and the imported cabbageworm;[20] and sweet corn hybrids resistant to the corn earworm.[94, 165]

Varietal resistance and cultural practices can be used synergistically in pest management programs. Planting of a resistant variety around a susceptible one may prevent a migrating pest from reaching the susceptible hosts. Conversely, a susceptible variety planted around a non-preferred variety may serve as a trap crop for the pest. Trials in Louisiana, U.S.A., showed that when 10 to 20% of a field was planted with green cotton around a red cotton, which is non-preferred by boll weevils, the insects concentrated in the green rows.[90] Insecticidal sprays were needed only in the green portion of the field, thus protecting the pest's natural enemies in the larger red portion of the field. Commercial lines of red cotton are now available for cultivation.

CONCLUSION

The use of insect-resistant host plant varieties has a number of advantages: because the resistance is inherent in the plants no extra costs are involved, it does not impair the quality of the environment, and it is generally compatible with other methods of pest control. Genetic resistance of crop plants to insect pests is generally combined with other desirable plant characteristics, such as high yields and good quality. This method of insect control has been shown to be practical and is recently receiving greater attention by scientists throughout the world.

While the exact mechanism of resistance is not always known, the stability of resistance against the development of new biotypes deserves top priority. Identification of sources of resistant germ plasm and breeding for resistance constitute the major thrusts and lead to the use of resistant and susceptible varieties in basic studies on insect host–plant relationships to gain a better understanding of the causes of resistance. Information on the stability of resistance in economically important crops is urgently needed.

REFERENCES

1. ABUL-NASR, S. The susceptibility of different varieties of cotton plants to infestation with insect and mite pests. *Bull. Soc. ent. Egypt* **44** (1960) 143–156.

2. AFZAL, M. and GHANI, M.A. Studies on the cotton jassid (*Empoasca devastans* Distant) in the Western Punjab — XIII. Method of cotton breeding for jassid resistance. *J.N.Y. ent. Soc.* **56** (1948) 209–218.

3. AKESON, W.R., HASKINS, F.A. and GORZ, H.J. Sweetclover weevil feeding deterrent B: isolation and identification. *Science* **163** (1969) 293–294.

4. AUCLAIR, J.L. Aphid feeding and nutrition. *Annu. Rev. Entomol.* **8** (1963) 439–490.

5. BATRA, G.R. and GUPTA, D.S. Screening of varieties of cotton for resistance to the jassid. *Cotton Grow. Rev.* **47** (1970) 285–291.

6. BECK, S.D. The European corn borer, *Pyrausta nubilalis* (Hübner), and its principal host plant — VI. Host plant resistance to larval establishment. *J. Insect Physiol.* **1** (1957) 158–177.

7. BECK, S.D. Resistance of plants to insects. *Annu. Rev. Entomol.* **10** (1965) 207–232.

8. BECK, S.D. and SMISSMAN, E.E. The European corn borer, *Pyrausta nubilalis*, and its principal host plant — IX. Biological activity of chemical analogs of corn resistance factor A (6-methoxybenzoxazolinone). *Ann. ent. Soc. Am.* **54** (1961) 53–61.

9. BECK, S.D. and STAUFFER, J.F. The European corn borer, *Pyrausta nubilalis* (Hübner), and its principal host plants — III. Toxic factor influencing larval establishment. *Ann. ent. Soc. Am.* **50** (1957) 166–170.

10. BELAND, G.L., AKESON, W.R. and MANGLITZ, G.R. Influence of plant maturity and plant part on nitrate content of the sweetclover weevil-resistant species *Melilotus infesta*. *J. Econ. Entomol.* **63** (1970) 1037-1039.

11. BEMILLER, J.N. and PAPPELIS, A.J. 2,4-Dihydroxy-7-methoxy-1,4-benzoxazin-3-one-glucoside in corn — I. Relation of water-soluble 1-butanol-soluble glucoside fraction content of pith cores and stalk rot resistance. *Phytopathology* **55** (1965) 1237–1240.

12. BLUM, A. Anatomical phenomena in seedlings of sorghum varieties resistant to the sorghum shootfly (*Atherigona varia soccata*). *Crop Sci.* **8** (1968) 388–391.

13. BOTTGER, G.T. and PATANA, R. Growth, development, and survival of certain Lepidoptera fed gossypol in the diet. *J. Econ. Entomol.* **59** (1966) 1166–1168.

14. BOTTGER, G.R., SHEEHAN, E.T. and LUKEFAHR, M.J. Relation of gossypol content of cotton plants to insect resistance. *J. Econ. Entomol.* **57** (1964) 283–285.

15. BRETT, C.H. and SULLIVAN, M.J. Sweet-potato flea beetle control. Vegetable insects Annual Report, N.C. State University, Raleigh, 1969.

16. BROERSMA, D.B., BERNARD, R.L. and LUCKMANN, W.H. Some effects of soybean pubescence on populations of the potato leaf-hopper. *J. Econ. Entomol.* **65** (1972) 78–82.

17. CANERDAY, T.D. and CHALFANT, R.B. An arrestant and feeding stimulant for the cowpea curculio, *Chalcodermus aeneus* (Coleoptera: Curculionidae). *J. Ga. ent. Soc.* **4** (1969) 59–64.

18. CANTELO, W.W., BOSWELL, A.L. and ARGAUER, R.J. *Tetranychus* mite repellant in tomato. *Environ. Entomol.* **3** (1974) 128–130.

19. CARTIER, J.J. and PAINTER, R.H. Differential reactions of the two biotypes of the corn leaf aphid to resistant and susceptible varieties, hybrids and selections of sorghums. *J. Econ. Entomol.* **49** (1956) 498–508.

20. CHALFANT, R.B. and BRETT, C.H. Inter-relationship of cabbage variety, season, and insecticide on control of the cabbage looper and the imported cabbageworm. *J. Econ. Entomol.* **60** (1967) 687–690.

21. CHALFANT, R.B. and GAINES, T.P. Cowpea curculio correlation between chemical composition of the southern pea and varietal resistance. *J. Econ. Entomol.* **66** (1973) 1011–1013.

22. CHALFANT, R.B., SUBER, E.G. and CANERDAY, T.D. Resistance of southern peas to the cowpea curculio in the field. *J. Econ. Entomol.* **65** (1972) 1679–1682.

23. CHAN, B.G., WAISS, A.C., Jr. and LUKEFAHR, M. Condensed tannin, an antibiotic chemical from *Gossypium hirsutum*. *J. Insect Physiol.* **24** (1978) 113–118.

24. CHIANG, H.C. Host variety as an ecological environmental factor in the population dynamics of the European corn borer, *Ostrinia nubilalis*. *Ann. ent. Soc. Am.* **61** (1968) 1521–1523.

25. CURTIS, B.C., SCHLEHUBER, A.M. and WOOD, E.A. Genetics of greenbug resistance in two strains of common wheat. *Agron. J.* (1960) 599–602.

26. CUTHBERT, F.P. and DAVIS, B.W. Factors contributing to cowpea curculio resistance in southern peas. *J. Econ. Entomol.* **65** (1972) 778–781.

27. DABROWSKI, Z.T., RORIGUEZ, J.G. and CHAPLIN, C.E. Studies in the resistance of strawberries to mites – IV. Effect of season on preference or non-preference of strawberries to *Tetranychus urticae*. *J. Econ. Entomol.* **64** (1971) 806–809.

28. DAHLMAN, D.L. and HIBBS, E.T. Responses of *Empoasca fabae* (Cicadellidae: Homoptera) to tomatine, solanine, leptine I, tomatidine, solanidine, and demissidine. *Ann. ent. Soc. Am.* **60** (1967) 732–740.

29. DECOSTA, C.P. and JONES, C.M. Cucumber beetle resistance and mite susceptibility controlled by the bitter gene in *Cucumis sativus* L. *Science* **172** (1971) 1145–1146.

30. DICKSON, M.H. and ECKENRODE, C.J. Variation in *Brassica oleracea* resistance to cabbage looper and imported cabbageworm in the greenhouse and field. *J. Econ. Entomol.* **68** (1975) 757–760.

31. DJAMIN, A. and PATHAK, M.D. Role of silica in resistance to Asiatic borer, *Chilo suppressalis* (Walker), in rice varieties. *J. Econ. Entomol.* **60** (1967) 347–351.

32. ELNAGHY, M.A. and LINKO, P. The role of 4-0-glucosyl-2,4-dihydroxy-7-methoxy-1,4-benzoxazin-3-one in resistance of wheat to stem rust. *Physiologia Pl.* **15** (1962) 764–771.

33. ESMAILI, M. and WILDE, G. Behavior of the parasite *Aphelinus asychis* in relation to the greenbug and certain hosts. *Environ. Entomol.* **1** (1972) 266–268.

34. EVANS, D.E. Jassid populations on three hairy varieties of cotton. *Emp. Cotton Grow. Rev.* **42** (1965) 211–217.

35. EVERLY, R.T. Establishment and development of corn leaf aphid populations on inbred and single cross dent corn. *Proc. N. Cent. Br. ent. Soc. Am.* **22** (1967) 80–84.

36. GALLUN, R.L. Genetic basis of Hessian fly epidemics. *Ann. N.Y. Acad. Sci.* **287** (1977) 223–229.

37. GALLUN, R.L., ROBERTS, J.J., FINNEY, R.E. and PATTERSON, F.L. Leaf pubescence of field grown wheat: a deterrent to oviposition by the cereal leaf beetle. *J. Environ. Qual.* **2** (1973) 333–334.

38. GALLUN, R.L., STARKS, K.J. and GUTHRIE, W.D. Plant resistance to insects attacking cereals. *Annu. Rev. Entomol.* **20** (1975) 337–357.

39. GARDENHIRE, J.H. Inheritance of resistance in oats. *Crop. Sci.* **4** (1964) 443.

40. GARDENHIRE, J.H. Inheritance and linkage studies on greenbug resistance in barley (*Hordeum vulgare* L.) *Crop Sci.* **5** (1965) 28–29.

41. GARDENHIRE, J.H. and CHADA, H.L. Inheritance of greenbug resistance in barley. *Crop Sci.* **1** (1961) 349–352.

42. GENTILE, A.G. and STONER, A.K. Resistance in *Lycopersicon* and *Solanum* species to the potato aphid. *J. Econ. Entomol.* **61** (1968) 1152–1154.

43. GENTILE, A.G., WEBB, R.E. and STONER, A.K. Resistance in *Lycopersicon* and *Solanum* to greenhouse whiteflies. *J. Econ. Entomol.* **61** (1968) 1355–1357.

44. GENTILE, A.G., WEBB, R.E. and STONER, A.K. *Lycopersicon* and *Solanum* spp. resistant to the carmine and the two-spotted spider mite. *J. Econ. Entomol.* **62** (1969) 834–836.

45. GIBSON, R.W. Glandular hairs providing resistance to aphids in certain wild potato species. *Ann. appl. Biol.* **68** (1971) 113–119.

46. GIBSON, R.W. and TURNER, R.H. Insect-trapping hairs on potato plants. *PANS* **22** (1977) 272–277.

47. GIRARDEAU, J.H., GAINES, T.P. and GOLDEN, B.A. Possible causes of differences in tobacco budworm damage to flue-cured tobacco varieties. *J. Econ. Entomol.* **66** (1973) 470–472.

48. GUTHRIE, W.D. European corn borer investigations. European Corn Borer Laboratory, Ankeny, Iowa (1969).

49. GUTHRIE, W.D., RUSSELL, W.A. and JENNINGS, C.W. Resistance of maize to second-brood European corn borers. *Proc. 26th Corn–Sorghum Res. Conf. Am. Seed Trade Assoc.* (1971) 165–179.

50. GUTHRIE, W.D. and WALTER, E.V. Corn earworm and European corn borer resistance in sweet corn inbred lines. *J. Econ. Entomol.* **54** (1961) 1248–1250.

51. HACKEROTT, H.L., HARVEY, T.L. and ROSS, W.M. Greenbug resistance in sorghums. *Crop Sci.* **9** (1969) 656–658.

52. HAMILTON, R.H. A corn mutant deficient in 2,4-dihydroxy-7-methoxy-1,4-benzoxazin-3-one with an altered tolerance of atrazine. *Weeds* **12** (1964) 27–30.

53. HANOVER, J.W. Physiology of tree resistance to insects. *Annu. Rev. Entomol.* **20** (1975) 75–95.

54. HANSON, C.H., PEDERSON, M.W., BERRANG, B., WALL, M.E. and DAVIS, Jr., K.H. The saponins in alfalfa cultivars. In MATCHES, A.G. (Ed.) *Anti-quality Components of Forages.* Coop. Sci. Soc. America, Madison, Wisc. Special Pub. No. 4 (1973) 33–52.

55. HARRIS, M.K. Allopatric resistance: Searching for sources of insect resistance for use in agriculture. *Environ. Entomol.* **4** (1975) 661–669.

56. HARVEY, T.L. and HACKEROTT, H.L. Recognition of a greenbug biotype injurious to sorghum. *J. Econ. Entomol.* **62** (1969) 776–779.

57. HATCHETT, H.J. and GALLUN, R.L. Frequency of Hessian fly, *Mayetiola destructor*, races in field populations. *Ann. ent. Soc. Am.* **61** (1968) 1446–1449.

58. HORBER, E. Isolation of components from the roots of alfalfa (*Medicago sativa* L.) toxic to the white grubs (*Melolontha vulgaris* F.). In *Proc.*

12th Int. Congr. Entomol. London (1964) pp. 540–541.

59. HORBER, E. Alfalfa saponins significant in resistance to insects. In RODRIGUEZ, J.G. (Ed.) *Insect* and *Mite Nutrition.* North-Holland, Amsterdam (1972), pp. 611–628.

60. HORBER, E., LEATH, K.T., BERRANG, B., MARCARIAN, V. and HANSON, C.H. Biological activities of saponin components from Dupuits and Lahontan alfalfa. *Entomologia exp. app.* **17** (1974) 410–424.

61. HOWE, W.L. Factors affecting the resistance of certain cucurbits to the squash borer. *J. Econ. Entomol.* **42** (1949) 321–326.

62. HOXIE, R.P., WELLSO, S.G. and WEBSTER, J.A. Cereal leaf beetle response to wheat trichome length density. *Environ. Entomol.* **4** (1975) 365–370.

63. HSIAO, T.H. Chemical basis of host selection and plant resistance in oligophagous insects. *Entomologia exp. app.* **12** (1969) 777–788.

64. INTERNATIONAL RICE RESEARCH INSTITUTE. Annual Report for 1967. Los Baños, Philippines (1968) 213–218.

65. INTERNATIONAL RICE RESEARCH INSTITUTE. Annual Report for 1968. Los Baños, Philippines (1969) 221–246.

66. INTERNATIONAL RICE RESEARCH INSTITUTE. Annual Report for 1975. Los Baños, Philippines (1976) 101–110.

67. INTERNATIONAL RICE RESEARCH INSTITUTE. Annual Report for 1976. Los Baños, Philippines (1977) 53–61.

68. INTERNATIONAL RICE RESEARCH INSTITUTE. Annual Report for 1977. Los Baños, Philippines (1978) 61–72.

69. JENNINGS, P.R. and PINEDA, A.T. Screening rice for resistance to the planthopper, *Sogatodes orizicola* (Muir). *Crop Sci.* **10** (1970) 687–689.

70. JOHNSON, B. The influence on aphids of the glandular hairs on tomato plants. *Pl. Path.* **5** (1956) 130–132.

71. JUNEJA, P.S., GHOLSON, R.K., BURTON, R.L. and STARKS, K.J. The chemical basis for greenbug resistance in small grains — I. Benzyl alcohol as a possible resistance factor. *Ann. ent. Soc. Am.* **65** (1972) 961–964.

72. KAMEL, S.A. and ELKASSABY, F.Y. Relative resistance of cotton varieties in Egypt to spider mites, leafhoppers, and aphids. *J. Econ. Entomol.* **58** (1965) 209–212.

73. KEMP, T.R., STOLTZ, L.P., SMITH, Jr., W.T., and CHAPLIN, E.C. The composition of the essential oils of leaves of strawberry cultivar "Citation". *Proc. Am. Soc. hort. Sci.* **93** (1968) 334–339.

74. KENNEDY, J.S. Mechanisms of host plant selection. *Ann. appl. Biol.* **56** (1965) 317–322.

75. KISHABA, A.N., VOTH, V., HOWLAND, A.F., BRINGHURST, R.S. and TOBA, H.H. Two-spotted mite resistance in California strawberries. *J. Econ. Entomol.* **65** (1972) 117–119.

76. KLUN, J.A., TIPTON, C.L. and BRINDLEY, T.A. 2,4-dihydroxy-7-methoxy-1, 4-benzoxazin-3-one (DIMBOA), an active agent in the resistance of maize to the European corn borer. *J. Econ. Entomol.* **60** (1967) 1529–1533.

77. KLUN, J.A., GUTHRIE, W.D., HALLAUER, A.R. and RUSSELL, W.A. Genetic nature of the concentration of 2,4-dihydroxy-7-methoxy 2H-1,4-benzoxazin-3 (4H)-one and resistance to the European corn borer in a diallele set of eleven maize inbreds. *Crop Sci.* **10** (1970) 87–90.

78. KOGAN, M. and ORTMAN, E.F. Antixenosis — a new term to define Painter's "nonpreference" modality of resistance. *Bull. ent. Soc. Am.* **24** (1978) 87–90.

79. KOKORIN, A.N. (The anatomical and morphological structure of stems of clover and the resistance of the plant to the stem weevil, *Apion virens* Hbst.) *Trudy Vsesoyuznogo Naucho-issledovatel'skogo Instituta Zaschity Rastenii* **37** (1973) 144–153.

80. LAKSHMINARYANA, A. and KHUSH, G.S. New genes for resistance to the brown planthopper in rice. *Crop Sci.* **17** (1977) 96–100.

81. LEATH, K.T., DAVIS Jr. K.H., WALL, M.E. and HANSON, C.H. Vegetative growth responses of alfalfa pathogens to saponin and other extracts from alfalfa (*Medicago sativa* L.) *Crop Sci.* **12** (1972) 851–856.

82. LEIGH, T.F. and HYER, A.H. Susceptibility of glandless cotton to attack by *Diabrotica* beetle. *J. Econ. Entomol.* **64** (1971) 319.

83. LINCOLN, C., DEAN, G., WADDLE, B.A., YEARIAN, W.C., PHILIPS, J.R. and ROBERTS, L. Resistance of Frego-type cotton to boll weevil and bollworm. *J. Econ. Entomol.* **64** (1971) 1326–1327.

84. LUGINBILL, P. Developing resistant plants — the ideal method of controlling insects. USDA, ARS, Prod. Res. Rep. 111 (1969) 14pp.

85. LUGINBILL, P. and KNIPLING, E.F. Suppression of wheat stem sawfly with resistant wheat. USDA, ARS, Prod. Res. Rep. 107 (1969) 9pp.

86. LUKEFAHR, M.J., COWAN, C.B. and HOUGHTALING, J.E. Field evaluations of improved cotton strains resistant to the cotton fleahopper. *J. Econ. Entomol.* **63** (1970) 1101–1103.

87. LUKEFAHR, M.J. and HOUGHTALING, J.E. Resistance of cotton strains with high gossypol content to *Heliothis* spp. *J. Econ. Entomol.* **62** (1969) 588–591.

88. MANGLITZ, G.R., GORZ, H.J., HASKINS, F.A., AKESON, W.R. and BELAND, G.L. Interaction between insects and chemical components of sweetclover. *J. Environ. Qual.* **5** (1976) 347–352.

89. MAXWELL, F.G. Host plant resistance to insects nutritional and pest management relationships. RODRIGUEZ, J.G. (Ed.) *Insect* and *Mite Nutrition*. North Holland, Amsterdam (1972) 702pp.

90. MAXWELL, F.G. Plant resistance to insects. *Bull. ent. Soc. Am.* **23** (1977) 199–203.

91. MAXWELL, F.G., JENKINS, J.N. and PARROTT, W.L. Resistance of plants to insects. *Adv. Agron.* **24** (1972) 187–265.

92. MAXWELL, F.G., JENKINS, J.N., PARROTT, W.L. and BUFORD, W.T. Factors contributing to resistance and susceptibility of cotton and other hosts to the boll weevil, *Anthonomus grandis. Entomologia exp. app.* **12** (1969) 801–810.

93. MAXWELL, F.G., LAFEVER, H.N. and JENKINS, J.N. Bister beetles on glandless cotton. *J. Econ. Entomol.* **58** (1965) 792–793.

94. McMILLIAN, W.W., WISEMAN, B.R., WIDSTROM, N.W. and HARRELL, E.A. Resistant sweet corn hybrid plus insecticide reduce losses from corn earworms. *J. Econ. Entomol.* **65** (1972) 229–231.

95. McNEAL, F.H., WALLACE, L.E., BERG, M.A. and McGUIRE, C.F. Wheat stem sawfly resistance with agronomic and quality information in hybrid wheats. *J. Econ. Entomol.* **64** (1971) 939–941.

96. MUNAKATA, K. and OKAMOTO, D. Varietal resistance to rice stem borers in Japan. In *The Major Insect Pests of Rice Plant. Proc. Symp. Int. Rice Res. Inst., Los Baños.* Johns Hopkins, Baltimore (1964) pp. 419–430.

97. MURRAY, J.C., VERHALEN, L.M. and BRYAN, D.E. Observation on the feeding preference of the striped blister beetle, *Epicauta vittata* (Fabricius), to glanded and glandless cotton. *Crop Sci.* **5** (1965) 189.

98. NATIONAL ACADEMY OF SCIENCES. National Research Council Committee on genetic vulnerability of major crops, Agric. Board, Div. Biol. & Agric., *Natl. Res. Council* (1972) 307pp.

99. NIELSON, M.W. and DON, H. Interaction between biotypes of the spotted alfalfa aphid and resistance in alfalfa. *J. Econ. Entomol.* **67** (1974) 368–370.

100. OATMAN, E.R. Host range studies of the melon leaf miner *Liriomyza pictella* (Thompson). *Ann. ent. Soc. Am.* **52** (1959) 739–741.

101. PAINTER, R.H. *Insect Resistance in Crop Plants.* MacMillan, New York (1951) p. 520.

102. PAINTER, R.H. Resistance of plants to insects. *Annu. Rev. Entomol.* **3** (1958) 267–290.

103. PANDA, N., MAHAPATRA, A. and SAHOO, M. Field evaluation of some brinjal varieties for resistance to shoot and fruit-borer (*Leucinodes orbanalis* Guen.). *Indian J. agric. Sci.* **41** (1971) 597–601.

104. PATANAKAMJORN, S. and PATHAK, M.D. Varietal resistance of rice to the Asiatic rice borer, *Chilo suppressalis* (Lepidoptera: Crambidae), and its association with various plant characters. *Ann. ent. Soc. Am.* **60** (1967) 287–292.

105. PATHAK, M.D. Varietal resistance to rice stem borers at IRRI. In *The Major Insect Pests of the Rice Plant. Proc. Symp. Int. Rice Res. Inst., Los Baños*, 1964. Johns Hopkins, Baltimore (1967) pp. 405–418.

106. PATHAK, M.D. Genetics of plants in pest management. In RABB, R.L. and GUTHRIE, F.E. (Eds.) *Concepts of Pest Management*, N.C. State Univ. Raleigh (1970) pp. 138–157.

107. PATHAK, M.D. Resistance to insect pests in rice varieties. In *Int. Rice Res. Inst., Rice Breeding Symposium. Los Baños,* Philippines (1972) pp. 325–341.

108. PATHAK, M.D. Utilization of insect–plant interrelationships in pest control. In PIMENTAL, D. (Ed.) *Insects, Science and Society.* Academic Press, London and New York (1975) pp. 121–148.

109. PATHAK, M.D., ANDRES, F. GALACGAC, N. and RAROS, R. Resistance of rice varieties to striped rice borers. *Int. Rice Res. Inst. Tech. Bull.* **11** (1971) 69pp.

110. PATHAK, M.D. and PAINTER, R.H. Differential amounts of materials taken up by from biotypes of corn leaf aphids from resistant and susceptible sorghums. *Ann. ent. Soc. Am.* **51** (1958) 250–254.

111. PATTERSON, C.G., KNAVEL, D.E., KEMP, T.R. and RODRIGUEZ, J.G. Chemical basis for resistance to *Tetranychus urticae* Koch in tomatoes. *Environ. Entomol.* **4** (1975) 670–674.

112. PATTERSON, C.G., THURSTON, R. and RODRIGUEZ, J.G. Two-spotted spider mite resistance in *Nicotiana* species. *J. Econ. Entomol.* **67** (1974) 341–343.

113. PATTERSON, F.L. and GALLUN, R.L. Inheritance of resistance of Seneca wheat to Race E of Hessian fly. *Proc. Int. Wheat Genet. Symp. 4th Colombia, MO.* (1973) pp. 445–449.

114. PESHO, G.R., DICKE, F.F. and RUSSELL, W.A. Resistance of inbred lines of corn (*Zea mays*) to the second brood of the European corn borer (*Ostrinia nubilalis* (Hübner)). *Iowa State J. Sci.* **40** (1965) 85–98.

115. PILLEMER, E.A. and TINGEY, W.M. Hooked trichomes: a physical plant barrier to a major agricultural pest. *Science* **193** (1976) 482–484.

116. PILLEMER, E.A. and TINGEY, W.M. Hooked trichomes and resistance of *Phaseolus vulgaris* to *Empoasca fabae* (Harris). *Entomologia exp. app.* **24** (1978) 83–94.

117. POLLARD, D.G. and SAUNDERS, J.H. Relations of some cotton pests to jassid resistant sakel. *Emp. Cotton Grow. Rev.* **33** (1956) 197–202.

118. PORTER, K.P. and DANIELS, N.E. Inheritance and heritability of greenbug resistance in a common wheat cross. *Crop Sci.* **3** (1963) 116–118.

119. RAMAN, K.V., TINGEY, W.M. and GREGORY, P. Potato glycoalkaloids: effect on survival and feeding behavior of the potato leafhopper. *J. Econ. Entomol.* **72** (1979) 337–341.

120. REED, G.L., BRINDLEY, T.A. and SHOWERS, W.B. Influence of resistant corn leaf tissue on the biology of the European corn borer. *Ann. ent. Soc. Am.* **65** (1972) 658–662.

121. REZAUL KARIM, A.N.M. Varietal resistance of rice to green leafhopper, *Nephotettix virescens* (Distant): sources, mechanism, and genetics of resistance. Ph.D. thesis, University of the Philippines at Los Baños (1978) 163pp.

122. ROBINSON, J.F. KLUN, J.A. and BRINDLEY, T.A. European corn borer: a nonpreference mechanism of leaf feeding resistance and its relationship to 1,4-benzoxazin-3-one concentration in dent corn tissue. *J. Econ. Entomol.* **71** (1978) 461–465

123. RODRIGUEZ, J.G., DABROWSKI, Z.T., STOLTZ, L.P., CHAPLIN, C.E. and SMITH, Jr. W.O. Studies on resistance of strawberries to mites — II. Preference and nonpreference responses of *Tetranychus urticae* and *T. turkestani* to water-soluble extracts of foliage. *J. Econ. Entomol.* **64** (1971) 383–387.

124. ROOF, M.E., HORBER, E. and SORENSEN, E.L. Bioassay technique for the potato leafhopper, *Empoasca fabae* (Harris). *Proc. North Central Branch Meeting Ent. Soc. Amer.* **27** (1972) 140–143.

125. RUSSELL, W.A., GUTHRIE, W.D. KLUN, J.A. and GRINDELAND, R. Selection for resistance in maize to first-brood European corn borer by using leaf feeding damage of the insect and chemical analysis for DIMBOA in the plant. *J. Econ. Entomol.* **68** (1975) 31–34.

126. SASAMOTO, K. Resistance of the rice plant supplied with silicate and nitrogenous fertilizers to the rice stem borer, *Chilo suppressalis* Walker. *Proc. Fac. Liberal Arts and Educ. Yamanashi Univ., Japan* **3** (1961) 1–73.

127. SAXENA, K.N. Patterns of insect-plant relationships determining susceptibility or resistance of different plants to an insect. *Entomologia exp. app.* **12** (1969) 751–766.

128. SAXENA, K.N., GANDHI, J.R. and SAXENA, R.C. Patterns of relationship between certain leafhoppers and plants — I. Responses to plants. *Entomologia exp. app.* **17** (1974) 303–318.

129. SCHALK, J.M. and RATCLIFFE, R.H. Evaluation of the United States Department of Agriculture program on alternative methods of insect control: Host plant resistance to insects. *FAO Plant Prot. Bull.* **25** (1977) 9–14.

130. SCHILLINGER, J.A. Larval growth as a method of screening *Triticum* sp. for resistance to the cereal leaf beetle. *J. Econ. Entomol.* **59** (1966) 1163–1166.

131. SCRIBER, J.M. TINGEY, W.M., GRACEN, V.E. and SULLIVAN, S.L. Leaf-feeding resistance to the European corn borer in genotypes of tropical (low-DIMBOA) and U.S. inbred (high-DIMBOA) maize. *J. Econ. Entomol.* **68** (1975) 823–826.

132. SHADE, R.E., THOMPSON, T.E. and CAMPBELL, W.R. An alfalfa weevil larval resistance mechanism detected in Medicago. *J. Econ. Entomol.* **68** (1975) 399–404.

133. SHARMA, G.C. and HALL, C.V. Relative attractance of spotted cucumber beetle to fruits of fifteen species of Cucurbitaceae. *Environ. Entomol.* **2** (1973) 154–156.

134. SHAVER, T.N. and GARCIA, J.A. Gossypol content of cotton flower buds. *J. Econ. Entomol.* **66** (1973) 327–329.

135. SHEPARD, M., STERLING, W. and WALKER, Jr. J.K. Abundance of beneficial arthropods on cotton genotypes. *Environ. Entomol.* **1** (1972) 117–121.

136. SIKKA, S.M., SAHNI, V.M. and BUTANI, D.K. Studies on jassid resistance in relation to hairiness of cotton leaves. *Euphytica* **15** (1966) 383–388.

137. SINGH, I.D. and WEAVER, Jr. J.B. Growth and infestation of boll weevils on normal-glanded, glandless, and high-gossypol strains of cotton. *J. Econ. Entomol.* **65** (1972) 821–824.

138. SIWI, B.H. and KHUSH, G.S. New genes for resistance to the green leafhopper in rice. *Crop Sci.* **17** (1977) 17–20.

139. SMITH, O.D., SCHLEHUBER, A.M. and CURTIS, B.C. Inheritance studies of greenbug (*Toxoptera graminum* Rond.) resistance in four varieties of winter barley. *Crop Sci.* **2** (1962) 489–491.

140. SMITH, R.L., WILSON, R.L. and WILSON, F.D. Resistance of cotton plant hairs to mobility of first instars of the pink bollworm. *J. Econ. Entomol.* **68** (1975) 679–683.

141. SOGAWA, K. and PATHAK, M.D. Mechanisms of brown planthopper resistance in Mudgo variety of rice (Hemiptera: Delphacidae). *Appl. Entomol. Zool.* **5** (1970) 145–158.

142. SOSA, O. Jr. Biotype L, ninth biotype of the Hessian fly. *J. Econ. Entomol.* **71** (1978) 458–460.

143. SOTO, P.E. Ovipositional preference and antibiosis in relation to resistance to a sorghum shootfly. *J. Econ. Entomol.* **67** (1974) 265–267.

144. STARKS, K.J., MUNIAPPAN, R. and EIKENBARY, R.D. Interaction between plant resistance and parasitism against the greenbug on barley and sorghum. *Ann. ent. Soc. Am.* **65** (1972) 650–655.

145. STEPHENS, S.G. Sources of resistance of cotton strains to the boll weevil and their possible utilization. *J. Econ. Entomol.* **50** (1957) 415–418.

146. STEPHENS, S.G. and LEE, H.S. Further studies on the feeding and oviposition preferences of the boll weevil (*Anthonomus grandis*). *J. Econ. Entomol.* **54** (1961) 1085–1090.

147. STIPANOVIC, R.D., BELL, A.A. and LUKEFAHR, M.J. Natural insecticides from cotton (*Gossypium*). In HEDIN, P.A. (Ed.) *Host Plant Resistance to Pests.* ACS Symposium Series **62** (1977) 197–214.

148. STONER, A.K. Selecting tomatoes resistant to spider mites. *J. Am. Soc. hort. Sci.* **95** (1970) 78–80.

149. STONER, A.K., FRANK, J.A. and GENTILE, A.G. The relationship of glandular hairs on tomatoes to spider mite resistance. *Proc. Am. Soc. hort. Sci.* **93** (1968) 532–538.

150. STRAUB, R.W. and FAIRCHILD, M.L. Laboratory studies of resistance in corn to the corn earworm. *J. Econ. Entomol.* **63** (1970) 1901–1903.

151. SULLIVAN, S.L., GRACEN, V.E. and ORTEGA, A. Resistance of exotic maize varieties to the European corn borer *Ostrinia nubilalis* (Hübner). *Environ. Entomol.* **3** (1974) 718–720.

152. THURSTON, R. Toxicity of trichome exudates of *Nicotiana* amd *Petunia* species to tobacco hornworm larvae. *J. Econ. Entomol.* **63** (1970) 272–274.

153. THURSTON, R., SMITH, W.T. and COOPER, B.P. Alkaloid secretion by trichomes of *Nicotiana* species and resistance to aphids. *Entomologia exp. app.* **9** (1966) 428–432.

154. THURSTON, R., PARR, J.C. and SMITH, W.T. The phylogeny of *Nicotiana* and resistance to insects. *4th Int. Tobacco Sci. Congr. Proc., The Natl. Tobacco Board Greece, Athens* (1966) 424–430.

155. THURSTON, R. and WEBSTER, J.A. Toxicity of *Nicotiana gossei* Domin to *Myzus persicae* (Sulzer). *Entomologia exp. app.* **5** (1962) 233–238.

156. TIDKE, P.M. and SANE, P.V. Jassid resistance and morphology of cotton leaf. *Indian Cotton Grow. Rev.* **16** (1962) 324–327.

157. TINGEY, W.M. and GIBSON, R.W. Feeding and mobility of the potato leafhopper impaired by glandular trichomes of *Solanum berthaultii* and *S. polyadenium*. *J. Econ. Entomol.* **71** (1978) 856–858.

158. VERMA, S.C. and MATHUR, P.S. The epidermal characters of sugarcane leaf in relation to insect pests. *Indian J. agric. Sci.* **20** (1950) 387–389.

159. WALLACE, L.E. and McNEAL, F.H. Stem sawflies of economic importance in grain crops in the United States. USDA. *Tech. Bull.* 1350 (1966) 50pp.

160. WALLACE, L.E., McNEAL, F.H. and BERG, M.A. Minimum stem solidness required in wheat for resistance to the wheat stem sawfly. *J. Econ. Entomol.* **66** (1973) 1121–1123.

161. WEBSTER, J.A. Association of plant hairs and insect resistance, an annotated bibliography USDA., *Misc. Publ.* 1297 (1975) 18pp.

162. WEIBEL, D.E., STARKS, K.J., WOOD, Jr. E.A. and MORRISON, R.D. Sorghum cultivars and progenies rated for resistance to greenbugs. *Crop Sci.* **12** (1972) 334–336.

163. WELLSO, S.G. Cereal leaf beetle: larval feeding, orientation, development, and survival on four small grain cultivars in the laboratory. *Ann. ent. Soc. Am.* **66** (1973) 1201–1208.

164. WILDE, G. and FEESE, H. A new corn leaf aphid biotype and its effect on some cereal and small grains. *J. Econ. Entomol.* **66** (1973) 570–571.

165. WISEMAN, B.R., HARRELL, E.A. and McMILLIAN, W.W. Continuation of tests of resistant sweet corn hybrid plus insecticides to reduce losses from corn earworm. *Environ. Entomol.* **2** (1973) 919–920.

166. WOOD, Jr. E.A. Non-preference, fecundity, and longevity of three greenbug biotypes cultured on greenbug tolerant sorghum species. USDA., *Spec. Rept. W.* 303 (1969) 7pp.

167. WOOD, Jr. E.A., SEBESTA, E.E. and STARKS, K.J. Resistance of "Gaucho" *Triticale* to *Schizaphis graminum*. *Environ. Entomol.* **3** (1974) 720–721.

SI UNITS IN PUBLICATIONS IN PLANT SCIENCE

L.D. INCOLL, S.P. LONG and M.R. ASHMORE

Department of Plant Sciences, University of Leeds,
Leeds LS2 9JT; Department of Biology, University of Essex,
Colchester CO4 3SQ; Botany Department, Imperial College Field Station,
Silwood Park, Ascot, SL5 7PY, England

INTRODUCTION

What do a mval, a μE and a tr/mn mean to you? Does a quintal/ha mean the same on both sides of the Atlantic? Do you describe the concentration of solutes in solvents by meq/l, by ppm, by %, by normality, by molarity or by all of these? Do you think it really matters what units you use in your publications?

These units are amongst the hundreds which we have found in 1976 issues of journals publishing papers in plant science. Few of us could boast of an understanding of all of these units and indeed the context in which many were used revealed that the authors too often did not wholly understand the units that they themselves had used. Until recently, plant scientists have felt it necessary to communicate the results of their research only with scientists in their own discipline and so have tended to develop their own conventions for units. In recent years there has been an increase in joint studies involving disciplines both within plant science, for example physiology and pathology, and outside plant science, for example plant physiology and micrometeorology. Such interdisciplinary studies are hindered by the lack of a common vocabulary.

Ideally we would all wish to reduce this diversity and improve communication. It was with these objectives in mind that the Système International d'Unités (SI) was proposed in its present form in 1958 and adopted in 1960 by all signatories of the Metre Convention, which by 1972 included all the countries which publish scientific journals. Whilst it is not our intention to describe the system in detail since there are already many publications about it (see Bibliography), it has the following features:

1. It is easily understood;
2. It is metric;
3. It has an economy of units, for example only one unit, the joule, is used for all the different forms of the physical quantity, energy;
4. Six of its seven base units, from which the other units may be derived, have internationally accepted standards of high precision (the exception is the candela; however, this unit and its derived units, the lumen and lux, have no relevance to plant science);

5. It has a set of prefixes which avoid the use of exponents and allow the expression of numbers in sizes between 1 and 1000 which are readily understood and amenable to verbal communication;
6. It is coherent, i.e. a physical quantity of unit size can be derived from other units each of unit size.

Although other systems of units have some of these features which to us seem essential for both international and interdisciplinary communication of results of modern quantitative research in plant science, not one possesses them all.[14] However, SI has its faults, which have been discussed elsewhere[12] and in plant science, as in other disciplines, there are cases where SI appears unsuitable. For example, what can replace the milliequivalent in the expression of cation exchange capacity of soils? We therefore decided to assess the usage of SI today, sixteen years after its official recognition, in order to identify those areas of plant science in which its adoption has caused particular problems, and to suggest how these might best be resolved.

POLICY OF JOURNALS

We identified the particular cases where problems have arisen by surveying the first 1976 number of 81 journals, published in 13 different countries and covering all aspects of plant science. While carrying out this survey, it became clear to us that individual journals vary greatly in the editorial policy, if any, that they have adopted toward units.

Thus of the 81 journals surveyed, only two journals (*Vegetatio* and *Aust. J. Exp. Agric. Animal Husb.*) specifically stated in their "Instructions to Authors" that units *must* follow SI. Twenty-nine journals "encouraged", "preferred", "recommended" or "requested" the use of SI. Fifteen recommended the use of the metric system and 35 had no stated editorial policy. The recommendations of several journals, e.g. *Plant Physiology* and *Plant & Cell Physiology*, would prevent authors from using SI units and abbreviations for certain quantities. It is apparent from our correspondence with editors that the great majority of them are reluctant to actively impose a particular policy, preferring changes to occur gradually through changes in the practice of individual authors. Our experience leads us to believe that many authors given this option will allow the short-term difficulties to outweigh the long-term advantages of SI.

PROBLEMS

Presentation of units

Strict adherence to the conventions of SI is essential to the success of the system. We have found many errors in presentation which at first sight may appear trivial, but to scientists in other disciplines or of other nationalities may produce ambiguities.

The symbols for units and prefixes are frequently wrong, for example sec for second, v for volt, K for kilo, ° or C for °C, day for day, hr for hour and min. for minute. It may seem pedantic to insist on the correct abbreviation, but each symbol has only one meaning in SI[36] thus preventing any ambiguities which may be unknown to the author. For example, while it might seem logical to use sec as an abbreviation for the English second, it is not logical in many other languages, such as

German – Sekunde, and Spanish – Segundo. The trigonometric function secant is also abbreviated to sec.

Ambiguity also arises from the use of more than one solidus in units. For example, net photosynthetic flux of CO_2 per unit area of leaf is often erroneously expressed as mg/dm^2/h which mathematically could mean $\frac{mg}{dm^2 \cdot h}$ or $\frac{mg.h}{dm^2}$. Such ambiguities could be avoided by always using negative indices instead of solidi. Thus in our example, the correct expression mg dm^{-2} h^{-1} has only one interpretation.

The separation of digits with a comma is also confusing. To most English-speaking scientists "The distance from A to B is 2,500 m" means that A is 2½ km from B; yet most European scientists, accustomed to using the comma as a decimal indicator, would conclude that A is only 2½ m from B. When using SI, multiples of 10^3 should be separated only by spaces, i.e. 2 500. This is only a problem when numbers are being used without units or are physical constants. For physical quantities of magnitude greater than 10^3 or less than 10^{-3} the SI set of prefixes saves the writing of numbers in full.

We commonly encountered prefixed units such as the cm in the denominators of physical quantities. This practice can produce a number of valid units for the same physical quantity. Thus mW cm^{-2}, W m^{-2}, μW mm^{-2}, mW m^{-2}, W cm^{-2} and even GW pm^{-2} are valid units for irradiance. While the relation between W m^{-2} and mW m^{-2} is immediately obvious, that between W m^{-2} and mW cm^{-2} is not. Throughout this Commentary we have adhered to the recommendation of the International Organisation for Standardization (ISO)[8, 20] that a prefix should only be attached to a unit in the numerator.

Recommendations: that the SI conventions on symbols and prefixes be adhered to; that negative powers be always used in place of solidi; that spaces be used instead of commas to separate groups of three digits; that prefixed units be confined to numerators.

Concentrations of substances in solutions and mixtures

Most published expressions of concentration are ambiguous and often unintelligible. What, for example, is a "6% glutaraldehyde solution"? Is the solvent water? Is it a solution in which glutaraldehyde constitutes 6% of the total weight or 6% of the total volume? Could it be a solution which contains 6 g of glutaraldehyde to every 94 cm^3 of solution, or perhaps one which contains 6 cm^3 of glutaraldehyde in every 100 g of solution? When twelve research biologists of different disciplines were asked this question, they produced eight different answers. Plant scientists express mass concentration not only by the % but also by other direct ratios, viz. ppm, vpm, pphm, ppb and ppt. Frequently two or three of these methods can be found in the same sentence, thus we can have air containing "21% oxygen, 4 ppm CO_2 and 4000 ppt CO" or a list of solutions "10% HCl, 4:4:7 butanol–acetic–water, and 50 ppm IAA". Although qualifying a ratio by (v/v), (w/w), (w/v) or (v/w) decreases ambiguity, v and w are not internationally understood abbreviations, e.g. the French equivalent of (w/v) is (p/v). There is a similar profusion of ways of expressing amount-of-substance concentration. For example, the same concentration of KCl in water was described by one author as 10 μmol ml^{-1} and by another in the same journal as 10 mM.

The simple answer to this apparent complexity is to express all concentrations as either mass per unit volume or amount of substance per unit volume in SI units only, as others have already suggested[31].

The numerator of expressions of concentration is either mass (kg) or amount of substance (mol). The choice between these two appeared to be purely arbitrary in the majority of articles covered in our survey. However, as the physiological activity of a substance is normally proportional to the concentration of molecules or ions present and not to the mass concentration, the mole would seem far more relevant for the majority of biological research. We suggest that plant scientists should follow the recommendation previously given to medical research workers[4, 45] that amount-of-substance concentrations should be given rather than mass concentrations, at least for aqueous solutions. Mass concentration should be used only where the exact molecular weight of a substance is unknown, where the purity of a substance is low and unknown, or where the solution has no physiological significance and thus the use of amount of substance instead of mass would be an unnecessary complication.

How should the denominator, volume, be expressed? The SI unit for volume is m^3 and not the litre which is only temporarily accepted as a special name for the submultiple dm^3. As only base units should be used in the denominator[8, 20], the denominator should always be m^3. The wisdom of this recommendation will be apparent in calculations of quantities derived from concentration such as osmotic pressure and diffusive conductance (see later). The use of m^3 may seem impractical in some cases. However, a system of units based on the m^3 is no different from one based on the litre. Reconsidering our example, the 6% glutaraldehyde solution becomes 60 kg m^{-3}, glutaraldehyde in water. But 60 kg $m^{-3} \equiv$ 60 g dm^{-3} (60 g l^{-1}) $\equiv$ 60 mg cm^{-3} (60 mg ml^{-1}) $\equiv$ 60 μg mm^3 (60 μg μl^{-1}) i.e. from a practical point of view, whatever the actual volume to be prepared for an experiment the number has stayed the same. By the same arguments, 4 ppm of CO_2 in air becomes 8 mg m^{-3}, CO_2 in air and 10 mM KCl becomes 10 mol m^{-3}, KCl in water.

Recommendations: that the concentrations of substances in solutions or mixtures be confined to either mass concentration (kg m^{-3}) or amount-of-substance concentration (mol m^{-3}); that the substance and solvent or diluent be always named; that dm^{-3} be permitted for a transitional period if followed by a unit using m^{-3} in parenthesis to facilitate the adoption of the latter.

The equivalent

The equivalent (e, eq, equiv. or Eq) and normality (N) have provided yet another means of describing the solute contents of solutions. However, these terms are also abandoned under SI. Some may feel that the absence of the equivalent, which could be considered as a unit derived from the mole, is an inadequacy in SI. The main argument for the use of the equivalent has been that the number of equivalents of B required to react completely with a given amount of A would be the same as the number of equivalents of C, D, E or any other compound which reacts with A. The number of moles would not necessarily be constant but would vary according to the valency of the reactant and thus could only be determined from an inspection of the equation for the reaction concerned. However, many substances have more than one valency and thus the stoichiometry of many reactions must still be examined before the correct equivalent weight can be used. The use of molecular weights provides a consistency in the treatment of chemical reactions which the use of equivalent weights cannot provide.

It is interesting to note that in both pure chemistry and medicine, disciplines in which the equivalent was used as much as, if not more than, in plant science, the abandonment of the equivalent has been recommended for both teaching and research. Our enquiries suggest that equivalents have not been used in the teaching of pure chemistry in both secondary and tertiary education for the last few years. Indeed a review of current research publications suggests that plant science is the last stronghold of the equivalent.

One usage of the equivalent which has not been covered in the preceding discussion is its use in describing the ion exchange capacity of soils and ion exchange resins. At present the cation exchange capacity of a soil is described most commonly by meq per 100 g of oven-dry soil. Although this expression is not accepted in SI, the system does provide a method of expressing cation exchange capacity. Cation exchange capacity is defined as the number of exchange sites. The mole in SI is used to describe a defined number of elementary entities which include ions and electrons. Each cation exchange site represents a negative charge in the absence of a counterion, and therefore cation exchange capacity could validly be described by moles of negative charge, or of electrons, per unit mass of soil. Thus, a cation exchange capacity of 10 meq/100 g would be described as 100 mmol kg^{-1}. The same method has been adopted for describing the ion exchange capacity of polyvalent macromolecules[34].

In practice, the cation exchange capacity is determined by the exchange of one species of cation. However, experimental and theoretical evidence[16, 18] shows that cation exchange in soils does not always obey the law of equivalents. It would, therefore, seem both more acceptable and more useful to express cation exchange capacity in moles of cation. For example, if the species of test cation was ammonium "a cation exchange capacity of 10 meq/100 g" would be described as an exchange capacity for ammonium ions of 100 mmol kg^{-1}.

Recommendation: that the equivalent should be replaced in all its usages by the mole.

Light: its terminology and units

A newcomer to the literature of plant science would be forced to conclude that the subject of light was not considered important by plant scientists, such is the apparent lack of understanding of concepts and profusion of terms and units. For example, in our survey, a unit was rarely paired with its correct physical quantity. Thus, of eight authors who measured light in $\mu E\ m^{-2}\ s^{-1}$, five called the measured quantity light intensity, one said it was irradiance, one said it was photon flux and only one correctly said it was a photon flux density. We are by no means the first to draw attention to this confusion[27, 32, 41, 42, 43].

All scientists working with green plants are expected to state in their methods the conditions under which they grew their plants. So what units and terms for light should they, as well as those for whom light is an experimental variable, use?

The most commonly used term is light intensity which is deeply entrenched in scientific jargon. Light, or luminous, intensity is the luminous flux emitted by a point source into unit solid angle. Its unit is the candela (cd) a base unit of SI. This physical quantity is quite different from the flux density which is the flux of energy or matter through unit area. Plant scientists are more often concerned with interception or absorption by areas of leaf than with emission from point sources. Yet intensity is commonly and inappropriately used with the photometric units for luminous flux density and illuminance – lumen m^{-2} or lux (cd sr m^{-2}) and lumen ft^{-2} or ft candle (cd sr ft^{-2}), or with the radiometric units for radiant flux density and irradiance — W m^{-2}.

The metric units of photometry and radiometry are part of SI. Unfortunately, the inclusion of the candela in SI has given it and its derived photometric units an undeserved aura of respectability, particularly in plant science. These photometric units constitute a psychophysical system based on the spectral responses of 52 pairs of American eyes measured in 1923[12] and thus have no relevance to plants.

Thus, if we are concerned with the energy balance of plants, the physical quantities of irradiance or radiant flux density should be used, with units of W m^{-2} and with the measured range of wavelengths. Irradiance is the radiant power incident on a surface divided by the area of that surface whereas radiant flux density (W m^{-2}) is the radiant power incident on a small sphere divided by the cross-sectional area of that sphere[21]. In practice, it is almost always irradiance rather than radiant flux density, as defined above, which is measured.[25] Thus the term irradiance is preferred.

We must emphasise that irradiance is concerned with energy, and people should ask themselves whether it is energy which they want to express. We feel that more often it is the photon content of the radiation which is important. The primary photochemical act in photosynthesis depends on the number of photons trapped and not on the energy of those photons. It therefore seems logical to express incident or absorbed photosynthetically active light in relation to its photochemical action[27], i.e. as the number of photons m^{-2} s^{-1}, the photon flux density in the appropriate waveband. The minor imperfections of measuring photon flux density have been succinctly stated by McCree[27] in a Commentary which we would recommend to all plant scientists. Sceptics might ask the question, paraphrasing McCree[26], "Are we ready to abandon the lux?". To us this is like refusing to abandon a square wheel because the round one is not a perfect circle.

Recommendation: that the term light intensity and all the units and terminology of the photometric system be banned from plant science journals and replaced by irradiance (W m^{-2}) or photon flux density, as appropriate.

The einstein

There is now available a commercial version of a sensor which is calibrated to measure photon flux density[5]. As is common with new measuring devices, the unit used by the designers and manufacturers, the microeinstein per metre^2 per second, has set the standard. However, there are two definitions of the einstein (symbol — E[10]) in the literature. The first definition, which appears in dictionaries of units[10, 14, 24] and in books on photochemistry, photobiology and radiation physics[33, 39, 40, 43] can be restated as the quantity of radiant energy in Avogadro's number of quanta, i.e. Nhv joules/mole. The second defines the einstein as Avogadro's number of quanta or photons, i.e. as a dimensionless constant[35, 38, 39, 44]. It is used by the designers and manufacturers of one Quantum Sensor in their calibration procedure[5]. We take the first definition to be correct on historical precedent and in the absence of any evidence of redefinition. In our view the einstein has been used inappropriately in the unit for photon flux density. Furthermore, the einstein is not an SI or metric unit and it cannot be derived coherently from base units.

Do we need a new unit for Avogadro's number of photons? Two courses are open, the first to use the absolute number of photons m^{-2} s^{-1} as particle physicists would, the second to use the mol m^{-2} s^{-1}. The perils of the first are well demonstrated in *Aust. J. Pl. Physiol.* **3**(1) 1976 where sunlight of from 66 to 230 X 10^{18} quanta cm^{-2} s^{-1} was reported. Such values seem impossible to us, even at the Great Barrier Reef where they were recorded, as they are around 1000 X full sunlight. The einstein has been used to cope with the large number of photons involved. For example, full sunlight has more than 10^{21} photons m^{-2} s^{-1} in the waveband 400–700 nm. The current definition of the SI base unit the mole is "the amount of substance of a system which contains as many elementary entities as there are atoms in 0.012 kilogram of carbon 12 (Note: when the mole is used, elementary entities must be specified and may be atoms, molecules, ions, electrons, other particles or specified groups of particles)",[36] i.e. Avogadro's number of particles. The photon is an elementary particle so that the einstein could be simply replaced by the mole and the unit for photon flux density would become mol m^{-2} s^{-1}.

Neither of these options cause any difficulties in measurement because numerically 6.02 X 10^{23} photons m^{-2} s^{-1} $\equiv$ 1 mol m^{-2} s^{-1} ($\equiv$ 1 E m^{-2} s^{-1}) and so no change in calibration of already existing sensors will be required. We prefer the second because in practice it gives numbers, in the range 0–2000 μmol m^{-2} s^{-1}, which are easy to remember and to talk about.

Another advantage of our proposed unit comes in deriving quantum yield in terms which photobiologists and photochemists as well as plant physiologists and ecologists would understand, i.e. mole photons/mole reactant – a dimensionless ratio.

Recommendations: that the einstein should not be used in the unit for photon flux density and that E m^{-2} s^{-1} be replaced by mol m^{-2} s^{-1}.

Photosynthetic rate and other mass flux densities

Flux densities (flux/area) can take many forms and are governed by many different physical laws. We will consider one flux density, photosynthetic rate, because we found that its unit, and those of its associated measures, varied most. The principles we elucidate should, however, be readily applicable to other cases.

In our survey, we found that twelve different combinations of units were used to describe photosynthetic rate as a CO_2 flux density. These ranged from "μl/cm^2/min" and "μmol/cm^2.min" to "kg m^{-2} s^{-1}" and "mg dm^{-2} h^{-1}, with over 60% of authors using the latter. Why mg dm^{-2} h^{-1}? Many authors, perhaps, use these units simply because everyone else does, or because these are the units they are familiar with. Another justification for the use of mg dm^{-1} h^{-1} is that measurements of CO_2 flux density are often made over periods of an hour or more and not over periods of seconds. However, these same authors often measure irradiance over exactly the same periods but they do not report these as J dm^{-2} h^{-1}, but as W m^{-2} (J m^{-2} s^{-1}).

Which units should be used for CO_2 flux density? Since only base units may be used in denominator[20], m^2 should be used for area and s for time. Additionally, while h and min are accepted in SI as multiples of s they are not decimal multiples and so they make conversions between different units of flux density very difficult. Thus, whilst measurements reported at a conference in μg m^{-2} s^{-1} will be almost immediately meaningful to someone who is used to working in mg m^{-2} s^{-1} and vice versa, a pocket calculator would be needed to convert measurements which were reported in mg dm^{-2} h^{-1} or μg cm^{-2} min^{-1}.

There are many who feel that flux densities should be expressed on the basis of the time scale and area on which they were actually measured. For example, it may seem odd to express the net assimilation rate (NAR) of a crop in units of mg m^{-2} s^{-1}. We would propose two ways in which these conflicting arguments can be reconciled. First, when units based on the ha, wk, etc. are used, they should be accompanied by the equivalent value with SI base units in the denominator, i.e. the NAR of the crop was 10 μg m^{-2} s^{-1} (605 kg ha^{-1} wk^{-1}). Alternatively, the period of averaging should be specified, i.e. the NAR of the crop was 10 μg m^{-2} s^{-1}averaged over a week.

As the quantity of CO_2 forms the numerator in the unit of CO_2 flux density, sub-multiples of the SI base units are permitted. The main problem in the presentation of the quantity of CO_2 lies in the choice of appropriate physical quantity. Volume, mass and amount of substance are all currently used. Of these, volume seems the least satisfactory because a rate reported for one air temperature is not directly comparable with a rate at another air temperature. The choice between amount of substance or mass should depend on the purposes of the study. If it is a study which relates photosynthetic rate to the biochemistry or cell physiology of the plant then amount-of-substance flux density of CO_2 would be the more appropriate. If the study is one which relates photosynthetic rate to dry matter yield of the plant, crop or community, then mass flux density of CO_2 would be more appropriate.

The units of measurement of photosynthetic rate should be coherent with those of related parameters; most commonly these are irradiance, photon flux density, temperature, CO_2 concentration in air, and CO_2 diffusive resistances and conductances. The pitfalls and unsuitability of many of the units in current use are well illustrated by the authors who created a term, which they called "Carboxylation Efficiency" (CE), for comparing genotypes of one species[3]. "Carboxylation Efficiency" was defined as the slope, at the CO_2 compensation point, of photosynthetic rate plotted against the concentration of CO_2 in air, the units of "Carboxylation Efficiency" being mg CO_2 dm^{-2} hr^{-1}/ppm! Had these authors used SI units such as mg m^{-2} s^{-1} for CO_2 flux density and mg m^{-3} for the concentration of CO_2 in air they would have seen that their "Carboxylation Efficiency" had the units mg m^{-2} s^{-1}/mg m^{-3}, i.e. it was simply the total conductance of the leaf to CO_2 (m s^{-1}) at the CO_2 compensation point.

Table 1a compares the units of photosynthetic rate and associated measures most commonly used at present with SI units for the expression of these same terms based on the preceding arguments. It includes CO_2 concentration in air as a mass concentration (mg m^{-3}) or as an amount-of-substance concentration (μmol m^{-3}), photon flux density (μmol m^{-2} s^{-1}) and irradiance (W m^{-2}), which have all been discussed elsewhere in this commentary. The units most frequently used at present (column 1), include three different units of time, three different units of length, two of mass, two of amount of substance, and one ratio. The recommended units (column 2) on the other hand use only one unit for each of time, length, mass and amount of substance. The sub-multiples chosen for the numerator of each expression give numbers of sensible size.

One advantage of the recommended set of units is the ease with which quantities derived from measured quantities can be calculated. For example, the calculation of diffusive conductance (k), from values of CO_2 flux density (F) and CO_2 concentration difference (ΔC) using Fick's Law of Diffusion, F = k.ΔC, can easily be done using the units in column 2. However, when the units of column 1 are used in this calculation, a slide rule or pocket calculator and a set of tables presenting the density of gaseous CO_2 at different temperatures would be required.

The CO_2 diffusive conductance is exceptional in Table 1 in that the figures are outside the range 0.1–1000. With a minimal loss to simplicity, CO_2 diffusive conductance could be expressed in mm s^{-1} in place of m s^{-1}, giving the range of typical conductances of 0.1–20.0. The CO_2 diffusive resistance is absent from Table 1a not because of an oversight, but because we believe that its reciprocal, CO_2 diffusive conductance should be preferred.

Recommendations: that the units for photosynthetic rate and associated measures, in Table 1a, column 2, be used exclusively.

Table 1. Recommended units for photosynthetic rate, transpiration rate and associated measurements. Column 1: the units most commonly used at present; column 2: a set of units based on SI. The figures with each set of units indicate typical ranges for these measurements

	Measurement	1	2
(a)	Photosynthetic rate (CO_2 mass flux density)	0–100 mg dm^{-2} h^{-1}	0–2.80 mg m^{-2} s^{-1}
	(CO_2 amount-of-substance flux density)	0–0.38 μmol cm^{-2} min^{-1}	0–63.6 μmol m^{-2} s^{-1}
	CO_2 concentration in air (mass concentration)	0–2000 ppm	0–3640 mg m^{-3}
	(amount-of-substance concentration)		0–826 μmol m^{-3}
	CO_2 diffusive conductance	0.01–2.00 cm s^{-1}	0.0001–0.02 m s^{-1} (0.1–20 mm s^{-1})
	Irradiance	0–500 W m^{-2}	0–500 W m^{-2}
	Photon flux density	0–2500 μE m^{-2} s^{-1}	0–2500 μmol m^{-2} s^{-1}
	Quantum yield	0.01–0.12 mol E^{-1}	0.01–0.12 (dimensionless)
	Quantum requirement	8–100 E mol^{-1}	8–100 (dimensionless)
(b)	Transpiration rate (H_2O mass flux density)	0–3.6 g dm^{-2} h^{-1}	0–100 mg m^{-2} s^{-1}
	H_2O concentration in air	30–100 % r.h.	0–30 g m^{-3}
	H_2O vapour pressure	1–100 mbar	0.1–10 kPa

Water relations

Soil scientists currently describe the state of water in the soil in units ranging from % oven dry wt. and pF to mm and bar, while meteorologists and plant scientists describe the state of water in the atmosphere in units ranging from mbar and mm Hg to % r.h. and g m^{-3}. Thus an obstacle both to communication between disciplines and to the study of water relations exists.

The state of water in plant tissues and in soils can be described by its water potential, that is the free energy content per unit volume relative to that of pure, free water. This potential has the units of J m^{-3}, the dimensions of which are the same as those of pressure. Thus the state of water in all of its sites, from the soil through the plant and into the atmosphere may be described by multiples of the SI unit for pressure, which is the pascal (Pa), the name for N m^{-2}. Despite the adoption of the pascal in 1971 its use today remains rare. Since the pascal is part of SI, it is based on a single internationally recognised set of standards and it is coherent with all other units of SI. Therefore the conversions needed to reach, for example, the equivalent free energy units are simple (1 Pa = 1 J m^{-3}, 1 MPa = 1 MJ m^{-3}). On the other hand, the bar, which is almost always used, is part of no coherent system of units; it is not, as is sometimes stated[30] the cgs unit of pressure. Thus, no conversions can be made from units based on the bar without either locating or knowing the required conversion factor. For these reasons, the pascal should replace the bar. In practice this would mean the use of the MPa (= 10 bar) for the expression of water potentials and the kPa (= 10 mbar) for the expression of atmospheric water vapour pressures[30]. The kPa should in our opinion also be used for atmospheric pressures in meteorology in place of the beloved mbar. The numbers which result are not very different and no less convenient, e.g. 0–5.0 MPa (0–50 bar) for water potentials; 0.1–10 kPa (1–100 mbar) for water vapour pressures and vapour pressure deficits; and 88–108 kPa (880–1080 mbar[28]) for atmospheric pressures.

Although the water vapour content of air is commonly described by its partial pressure a number of authors continue to use relative humidity. The relative humidity by itself conveys no useful information and is irrelevant, even when it is accompanied by air temperature; reference to tables must be made before a term meaningful in plant physiology, such as water vapour pressure or absolute humidity can be calculated.

Another error commonly found in reports of studies in plant water relations is in the graphical presentation of water potentials. Since water potential is defined relative to that of pure free water, plant water potentials will most commonly be negative. This can lead to some confusion, especially in the minds of those unfamiliar with the terminology of water relations, since we all tend to look at the size of the numbers involved and ignore the negative sign. Any such confusion is undoubtedly enhanced by the all too frequent practice of plotting water potentials on graphs as if the values were positive. Thus the lowest water potentials are found at the topmost or rightmost ends of the axes, where the largest values would be expected. This practice is in opposition to all graphical conventions, and we feel that it should no longer be tolerated in scientific publications.

The foregoing discussion has shown that the vast range of units currently used to describe the state of water in all sites of the hydrological cycle may be replaced by multiples of a single SI unit. An equally large range of units was found in current use for the description of fluxes of water between these sites. Thus whilst one author studying the hydrological cycle in a forest may express evapotranspiration in kg ha^{-1} wk^{-1}, another studying short-term changes in leaf transpiration may use units of mg cm^{-2} s^{-1}. Evaporation and transpiration rates, like photosynthetic rate, are mass flux densities and the same arguments can be applied in deriving the appropriate units which are submultiples of kg m^{-2} s^{-1} (Table 1b).

The mm h^{-1}, a unit favoured by some meteorologists and hydrologists, is not a mass flux density. This unit is not coherent with, and not readily converted to, others used in water relations research. We think that evapotranspiration can be expressed as a mass flux density, rather than as depth of water per unit time, without any loss of information.

Recommendations: that the state of water in soils, plants and the atmosphere be described only in SI units, the most useful units being the MPa for soil and plant water potentials and the kPa for water vapour pressures and atmospheric pressures; that use of the bar and its submultiples be banned in future Plant Science publications; that the units for transpiration rate and associated measures in Table 1b, column 2, be used exclusively.

CONCLUDING REMARKS

Plant science journals in the range covered by *Current Advances in Plant Science* are generally distinguished by the vagueness and brevity of their instructions to contributors. Instructions as thorough, up-to-date and comprehensive as those of the *Biochem. J.* are rare. There is nothing to match the IUPAP[23] and IUPAC[22] manuals. Attempts to standardise the use of quantities and units are almost negated by their publication in expensive books[44] or in unpublicised pamphlets[19]. Very often, symbols for physical quantities, units for physical quantities and other abbreviations are presented in mixed lists which are also often full of mistakes and out of date. Despite the adoption in mid-1975 of new symbols for radioactivity, Bq (the becquerel) and for absorbed dose, Gy (the gray) and of new prefixes peta-(P) and exa-(E) for 10^{15} and 10^{18} , these have only been detailed twice to date, by the Royal Society[37] and by Australian journals published by C.S.I.R.O. We recommend that an up-to-date summary of SI be placed in the first issue of each journal every year as in *Aust. J. Pl. Physiol.*, that instructions to contributors be made much more comprehensive as in *Biochem. J.* and that "do-nots" as well as "dos" be given, as in *Photosynthetica.*

There has been little evidence of willingness of plant science journals to adapt or to take concerted action as the medical journals have[15]. There are many instances of editorial boards and authors of books acting either in complete ignorance of or defiance of such bodies as the ISO and Conférence Générale des Poids et Mesures (CGPM), organisations of professional scientists whose business is metrology and whose aims are to simplify, codify and unify. It may be that editors find a concern for units to be the "last straw" in their already busy lives and they may find that they lack the expertise to arbitrate on units. If so then a sub-editor may be needed whose sole task would be to scan papers on acceptance for correct use of quantities and units. The work of such a sub-editor and his authors would be considerably simplified if a cheap manual with contents like the *Council of Biology Editors Style Manual*[11] but valid internationally, updated annually and with the same status and intellectual backing as the publications of ISO, CGPM, IUPAP and IUPAC, was readily available.

The advantages of SI outweigh its disadvantages. In advocating its universal adoption by editors, reviewers and authors we are sure that more effort must be made to explain the logic behind it.

BIBLIOGRAPHY

This bibliography includes publications relevant to the use of SI in various parts of the world. Those publications which we have found to be especially useful are marked with an asterisk.

1. ADAMS, H.F.R. *SI Metric Units: An Introduction.* McGraw-Hill, Ryerson, Toronto (1974).

*2. ANDERTON, P. and BIGG, P.H. *Changing to the Metric System. Conversion Factors, Symbols and Definitions.* HMSO, London (1972).

3. AUGUSTINE, J.J., STEVENS, M.A., BREIDENBACH, R.W. and PAIGE, D.F. Genotypic variation in carboxylation of tomatoes. *Pl. Physiol.* **57** (1976) 325–333.

4. BARON, D.N. SI units. *Brit. med. J.* **1** (1974) 509–512.

5. BIGGS, W.W., EDISON, A.R., EASTIN, J.D., BROWN, K.W., MARANVILLE, J.W. and CLEGG, M.D. Photosynthesis light sensor and meter. *Ecology* **52** (1971) 125–131.

6. BLACKMAN, D.R. *S.I. Units in Engineering.* Macmillan, Melbourne (1969).

7. British Standards Institution, London. BS3763. *The International System (SI) Units* (1970).

8. British Standards Institution, London. PD5686. *The Use of SI Units* (1972).

9. CHISWELL, B. and GRIGG, E.C.M. *SI Units.* Wiley, Sydney (1971).

10. CLASON, W.E. *Elsevier's Lexicon of International and National Units.* Elsevier, Amsterdam (1964).

11. Council of Biology Editors, Committee on Form and Style. *CBE Style Manual.* 3rd Edition. AIBS, Washington DC (1972).

*12. DANLOUX-DUMESNILS, M. *The Metric System. A Critical Study of Its Principles and Practice* (Translation by GARRETT, A. and ROWLINSON, J.S.). The Athlone Press, London (1969).

13. DRAZIL, J.V. *Dictionary of Quantities and Units.* Leonard Hill, London (1971).

*14. DRESNER, S. *Units of Measurement. An Encyclopedic Dictionary of Units both Scientific and Popular and the Quantities they Measure.* Harvey, Miller & Metcalf, Aylesbury, England (1971).

15. ELLIS, G. *Units, Symbols and Abbreviations. A Guide for Biological and Medical Editors and Authors.* The Royal Society of Medicine, London (1971).

16. HELMY, A.K. Cation-exchange stoichiometry. *Soil Sci.* **95** (1963) 204–205.

17. HESLOP, R.B. and WILD, G.M. *SI Units in Chemistry.* Applied Science Publishers, London (1971).

18. HESSE, P.R. *A Textbook of Soil Chemical Analysis.* Murray, London (1971) 91–92.

19. International Association for Plant Physiology. *IAPP Tentative Recommendations of Terminology, Symbols and Units in Plant Physiology.* Bloms, Lund (1972).

20. International Organisation for Standardization. *International Standard ISO* 1000 (1973) 2.

21. International Organisation for Standardization. *International Standard ISO* 31/VI. Quantities and units of light and related electromagnetic radiations (1973).

*22. International Union of Pure and Applied Chemistry (IUPAC). *Manual of Symbols and Terminology for Physicochemical Quantities and Units.* 1973 revised edn. Butterworths, London (1975).

23. International Union of Pure and Applied Physics (IUPAP). *Symbols, Units and Nomenclature in Physics.* Document U.I.P. 11 (S.U.N. 65-3) (1965).

24. JERRARD, H.G. and McNEILL, D.B. *A Dictionary of Scientific Units.* 3rd edn. Chapman & Hall, London (1972).

25. KUBIN, S. Measurement of radiant energy. In *Plant Photosynthetic Production. Manual of Methods,* SESTAK, Z., CATSKY, J. and JARVIS, P.G. (Eds.) Junk, The Hague (1971) 702–763.

26. McCREE, K.J. Are we ready to abandon the footcandle? *Pl. Physiol.* **51** (1973a) Suppl. 20.

*27. McCREE, K.J. A rational approach to light measurements in plant ecology. Commentary in *Plant Sciences* No. 5. *Current Adv. Pl. Sci.* (1973b) 39–43.

28. McWHIRTER, R. and McWHIRTER, N. *The Guiness Book of Records.* Guiness Superlatives, London (1975) 67.

29. MARGERISON, D. *SI Units: An Explanation of the International System of Units.* Koch-Light Laboratories, Colnbrook (1973).

30. MONTEITH, J.L. and WEATHERLY, P.E. Units and symbols. In *A discussion on water relations of plants. Phil. Trans. R. Soc. Lond. B.* 273 (1976) 434.

31. MORRIS, J.G. *A Biologist's Physical Chemistry,* 2nd Edn. Edward Arnold, London (1974) 14–25.

32. NORRIS, K.H. Evaluation of visible radiation for plant growth. *Annu. Rev. Pl. Physiol.* **19** (1968) 490–499.

33. NOYES, W.A. and LEIGHTON, P.A. *The Photochemistry of Gases.* Reinhold, New York (1941) 13–14.

34. OOSAWA, F. *Polyelectrolytes.* Dekker, New York (1971) 12.

35. OWEN, J.A., EDWARDS, R.G. and COLLER, B.A.W. *The Mole in Medicine and Biology.* Livingstone, Edinburgh (1970) 6.

*36. PAGE, C.H. and VIGOUREUX, P. (Eds.) *SI. The International System of Units.* Translation of *Le Systeme International d'Unites.* HMSO, London; U.S. Govt. Printing Office, Washington (1973).

*37. Royal Society, London. *Quantities, Units and Symbols.* A report of the Symbols Committee. 2nd Edn. (1975).

38. SESTAK, Z., CATSKY, J. and JARVIS, P.G. (Eds.) *Plant Photosynthetic Production. Manual of Methods.* Junk, The Hague (1971) 30, 416.

39. SMITH, H. *Phytochrome and Photomorphogenesis.* McGraw-Hill, London (1975) 8.

40. TURRO, N.J. *Molecular Photochemistry.* Benjamin, New York (1965) 4–5.

41. TYLER, J.E. Lux vs quanta. *Limnol. Oceanogr.* **18** (1973) 810.

42. WITHROW, R.B. Radiant energy nomenclature. *Pl. Physiol.* **18** (1943) 476–487.

43. WITHROW, R.B. and WITHROW, A.P. Generation, control and measurement of visible and near visible radiant energy. In A. HOLLAENDER (Ed.), *Radiation Biology* Vol. 3, McGraw-Hill, New York (1956) 126–138.

*44. WORTHINGTON, E.B. (Ed.) *International Biological Programme 1. The Evolution of IBP.* Cambridge University Press, Cambridge (1975) 203–255.

45. YOUNG, D.S. Standardized reporting of laboratory data. *New Engl. J. Med.* 290 (1974) 368–373.

NOTE: Reference 27, McCree (1973b), has also been published in SMITH, H. (Ed.) *Commentaries in Plant Science.* Pergamon, Oxford (1976) 45–50.

ADDENDUM

Water relations

In our recommendations for water relations, we may have implied that the state of water should be expressed in terms of pressure rather than free energy. This was not our intention. Our concern was that if pressure-based units are to be used, the correct SI unit is the pascal.

Additional bibliography

A further seven publications have come to our attention, two of which warrant special mention. In one,[52] recent international progress towards developing a uniform terminology for describing all radiations is summarized and proposals are put forward for criticism. The other[48] provides a means of representing units and prefixes uniformly when sets of characters lack letters such as Ω and μ or are limited to one case, for example on the output from terminals to computers. These representations of SI are not intended to replace the international symbols or to be printed in publications.

ADDITIONAL BIBLIOGRAPHY

46. GREEN, M.H. *Metric Conversion Handbook.* Chemical Publishing Co., New York (1978).

47. HESSE, P.R. SI units and nomenclature in soil science. *Food and Agriculture Organisation Soils Bulletin*, no.28. F.A.O., Rome (1975).

48. International Organisation for Standardization. *International Standard ISO 2955.* Information processing — Representations of SI and other units for use in systems with limited character sets (1974) *Note.* Also published in reference 49.

49. International Organisation for Standardization. *ISO Standards Handbook 2. Units of Measurement* (Handbook of International Standards for Units of Measurement) ISO, Genève (1979).

50. LOWE, D.A. *Progress in Standardization: 2* A guide to international recommendations on names and symbols for quantities and on units of measurement. (Supplement to Vol.52 of the *Bulletin of the World Health Organisation*) W.H.O., Geneva (1975).

51. McGLASHAN, M.L. Physico-chemical Units and Quantities: The Grammar and Spelling of Physical Chemistry (Royal Institute of Chemistry Monographs for Teachers No.15) Royal Institute of Chemistry, London (1971).

52. RUPERT, C.S. and LATARJET, R. Toward a nomenclature and dosimetric scheme applicable to all radiations. *Photochem. Photobiol.* **28** (1978) 3–5.

OXYGEN – A PHYSIOLOGICAL ELECTRON ACCEPTOR IN PHOTOSYNTHESIS?

J.F. ALLEN

Botany School, South Parks Road, Oxford OX1 3RA, England

It has been known for some time that isolated, illuminated chloroplasts are able to use oxygen as a Hill oxidant, that is, as an electron acceptor in the 'light' reactions of photosynthesis. This photosynthetic reduction of oxygen results in a net consumption of oxygen by the chloroplast suspension, since hydrogen peroxide rather than water is usually the product.[20, 34, 35] Oxygen uptake by illuminated chloroplasts is termed the 'Mehler reaction'. The role of the superoxide anion, O_2^-, in this process is reviewed in detail elsewhere[3], but since the participation of superoxide is of some relevance to a consideration of possible roles of oxygen reduction in photosynthesis, a short and schematic account of the steps involved will now be given.

The first part of the Mehler reaction (and of the Hill reaction) is transfer of reducing equivalents ($2e^-$) along the photosynthetic chain, making them available for reduction of a suitable mediator:

$$H_2O \xrightarrow{\text{PHOTOSYSTEMS}} \tfrac{1}{2}O_2 + 2H^+ + 2e^-$$

In the second part (unique to the Mehler reaction) the mediator (which is 'autoxidisable') passes on reducing equivalents to oxygen:

$$2e^- + O_2 + 2H^+ \xrightarrow{\text{MEDIATOR}} H_2O_2$$

This second reaction is now known to proceed by any one of the three routes depicted in Fig. 1. The identity of the mediator that is involved determines which mechanism of the three applies, though in all cases the overall process may be represented by the sum of the above equations, which is:

$$H_2O + \tfrac{1}{2}O_2 \xrightarrow{2e^-} H_2O_2$$

In its simplest form, photosynthetic oxygen reduction is mediated by a compound ('R' in

Fig. 1a) which transfers electrons singly from the chloroplast to oxygen. Examples of such a compound are the viologen dyes (e.g. methyl viologen), flavin mononucleotide, adrenochrome, anthraquinone and phenazine methosulphate. Superoxide (univalently reduced oxygen) may subsequently dismutate either spontaneously or catalytically:

$$2O_2^{-} + 2H^{+} \longrightarrow O_2 + H_2O_2$$

though it may instead be reduced, e.g. by ascorbate[4, 17] or manganous ions[32], in which case net oxygen uptake is increased. Such effects of ascorbate or Mn^{2+} may be reversed by addition of the enzyme superoxide dismutase, which restores the dismutation reaction and with it the original stoichiometry. This 'single-step' oxygen reduction is important chiefly for electron transport studies *in vitro*[5], while the cytotoxicity of superoxide may be the basis of the herbicidal effects of paraquat.[18]

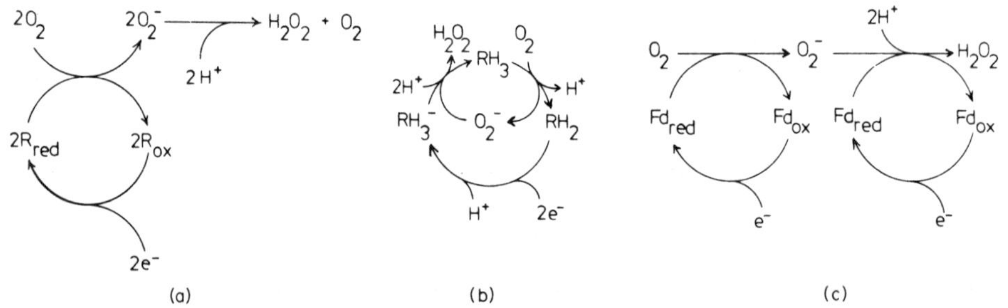

Fig. 1. Three ways of reducing oxygen in photosynthesis. (a) Single-step reduction followed by superoxide dismutation. (b) Two-step reduction by a mediator which has three valence states. (c) Two-step reduction by ferredoxin.

A second mechanism of chloroplast oxygen reduction, first described by Elstner and Heupel[15], is shown in Fig. 1b. Here a mediator such as adrenalin (RH_3^{-} in Fig. 1b) or dopamine accepts two electrons per molecule from the photosynthetic chain, subsequently donating one electron to oxygen and the other to superoxide. Three valence states of the mediator participate, and the overall reaction is achieved without dismutation of superoxide. Indeed, addition of superoxide dismutase now interrupts the catalytic cycle and hence causes inhibition of both oxygen uptake and photosynthetic electron transport.

The third route of oxygen reduction is the one most likely to be of importance *in vivo*, since it is mediated by an electron carrier, ferredoxin, which is native to photosynthetic systems. As shown in Fig. 1c, this mechanism also involves reduction of both oxygen and superoxide[1], though it differs from that of Fig. 1b in that with ferredoxin only two valence states of the mediator are involved. Addition of superoxide dismutase does not completely inhibit the process, since its presence competitively replaces only one of the ferredoxin-oxidising reactions (superoxide reduction) while the other (reduction of oxygen to superoxide) continues. The relatively high affinity of reduced ferredoxin for superoxide results in a much lower steady-state superoxide concentration than would prevail in the methyl viologen-mediated Mehler reaction, and to this extent the mechanism of reduced ferredoxin's reduction of oxygen to hydrogen peroxide may be regarded as one which which is physiologically 'safe'.

PSEUDOCYCLIC PHOTOPHOSPHORYLATION

'Pseudocyclic' photophosphorylation accompanies non-cyclic electron transport where oxygen is the terminal electron acceptor, and in the presence of an appropriate mediator it is easily observed *in vitro*.[7] It resembles cyclic phosphorylation in that it will occur even where no electron acceptor has been added to an aerobic suspension of chloroplasts, and in the presence of catalase net oxygen exchange does not take place.

It has generally been assumed that ferredoxin will reduce oxygen in chloroplasts only when NADP is absent.[8] This assumption is questioned by the observation[2] that in broken, washed chloroplasts, addition of catalase stimulates net oxygen evolution, an effect which can be understood to result from production of hydrogen peroxide by ferredoxin autoxidation which is occurring even where NADP is the major terminal electron acceptor. The proportion of total electron flux which accounts for reduction of oxygen to hydrogen peroxide in this system can be calculated from the observed rates of net oxygen evolution in the presence and absence of catalase, and at ferredoxin concentrations saturating for electron transport as such, not less than 15% of total electron flow is to oxygen rather than to NADP.[2]

Furthermore, circumstances may arise *in vivo* in which the rate of purely non-cyclic phosphorylation cannot keep pace with demands for ATP made by both CO_2-fixation and biosynthetic processes. Under these circumstances reoxidation of NADPH in the Calvin cycle would be the factor limiting to electron transport, and electrons could consequently be diverted from NADP to oxygen. Even without considering additional biosynthesis, the accepted pathway of CO_2-fixation (in C_3 plants at least) requires ATP and NADPH in the ratio 3ATP:2NADPH, and so an absolute stoichiometry of non-cyclic phosphorylation ($P/2e^-$) of less than 1.5 would create a need for an ancillary reaction producing ATP but not NADPH. Though the $P/2e^-$ ratio of non-cyclic phosphorylation is in dispute[6, 22, 25, 44], even the highest estimates ($P/2e^- = 2$) would not allow non-cyclic phosphorylation by itself to account for the ATP-requirement of CO_2-fixation in C_4 plants (ATP:NADPH = 5:2).[9]

The two chief candidates for the possible ancillary reaction are pseudocyclic and cyclic phosphorylation. Examples of cyclic phosphorylation which appear to depend on the presence of oxygen will be the concern of the next section. Evidence for the occurrence *in vivo* of each of the three main types of photophosphorylation has been discussed by Simonis and Urbach.[42]

Among the various pieces of evidence that pseudocyclic phosphorylation occurs in relatively intact photosynthetic systems, Egneus *et al.*[13] have demonstrated an enhancement by catalase addition of net oxygen evolution in intact, CO_2-fixing spinach chloroplasts. Such chloroplasts were also shown by these investigators to exhibit a light-dependent uptake of ^{18}O-oxygen. Both effects were absent from intact, oxygen-evolving chloroplasts when phosphoglycerate replaced CO_2 as the terminal oxidant; for phosphoglycerate reduction ATP and NADPH are required in equimolar amounts, and so no ancillary ATP synthesis would be required.

Kaiser[30] has also obtained evidence that intact, CO_2-fixing chloroplasts produce hydrogen peroxide, and that added hydrogen peroxide inhibits CO_2-fixation. Kaiser concludes that the hydrogen peroxide which is produced by intact chloroplasts causes inhibition of CO_2-fixation by bringing about depletion of Calvin cycle intermediates, since either catalase or added intermediates (ribose-5-phosphate, fructose-6-phosphate or dihydroxyacetone phosphate) will increase the rate of CO_2-fixation in a reconstituted chloroplast system, while addition of the same intermediates has little effect when the catalase activity of the chloroplast preparation is high.

Huber and Edwards[26] describe a net oxygen evolution by chloroplasts extracted from mesophyll protoplasts of the C_4 plant *Digitaria sanguinalis.* Oxygen evolution accompanied CO_2 fixation in the presence of pyruvate, and, on addition of cyanide, was replaced by oxygen uptake. The authors regard this effect of cyanide as resulting from an inhibition of catalase with a consequent inhibition of release of oxygen from hydrogen peroxide. Cyanide in their experimental system apparently had no effect on CO_2-fixation itself. They suggest that in C_4 photosynthesis ATP is supplied by both non-cyclic and pseudocyclic phosphorylation.

Raven and Glidewell[38] have shown that light-driven active phosphate transport in whole cells of the green alga *Hydrodictyon africanum* occurs even in the presence of the metabolic inhibitors cyanide and antimycin A, and in the absence of CO_2. In this situation pseudocyclic phosphorylation is likely to be the only remaining source of ATP; oxygen is required for phosphate uptake to take place, and the process is inhibited by DCMU. Glidewell and Raven[19] have also shown that ^{18}O-oxygen uptake accompanies phosphate transport in *Hydrodictyon*.

Using a technique based on fluorescence of scopoletin, Patterson and Myers[36] have demonstrated a light-dependent and DCMU-sensitive production and 'excretion' of hydrogen peroxide by whole cells of a strain of *Anacystis nidulans*. The hydrogen peroxide production was faster under conditions where CO_2-fixation itself is suppressed, though it occurred to a significant extent even when CO_2-fixation was allowed to proceed. They concluded that 'a low potential reductant is produced more rapidly than it can be used in the normal pathway to CO_2 reduction, and, instead, reacts with oxygen'.

More recently, Radmer and Kok[37] have used mass spectrometry to monitor both uptake and evolution of oxygen together with associated CO_2 uptake in suspensions of *Scenedesmus obliquous, Chlorella vulgaris* and *Anacystis nidulans.* They found that during a period of 0.5 min immediately following the onset of illumination (corresponding to the lag-phase of CO_2-assimilation), evolution and uptake of oxygen proceeded simultaneously at a rate only 20–30% lower than the steady-state rate of photosynthesis which was ultimately reached (180 μmole/mg chl. per hr). All three exchange reactions were sensitive to the electron transport inhibitor DCMU. Even after the maximal rate of CO_2 uptake had been reached, oxygen uptake accounted for a small proportion of total oxygen exchange, and for a much higher proportion (up to 50%, when net oxygen evolution ceases to be observed) in the presence of inhibitors of CO_2-assimilation such as iodoacetamide, KCN and FCCP. The persistence of oxygen uptake in the presence of uncoupler or inhibitors of the Calvin cycle seems to suggest that a fairly direct photoreduction of oxygen, such as that which occurs via ferredoxin, is taking place in this system. Radmer and Kok[37] conclude that rapid photoreduction of oxygen may be required in these algae to bring about the onset of CO_2-fixation, and that oxygen competes with CO_2 for electrons from the photosynthetic chain in such a way that high rates of electron transport occur under circumstances where CO_2-fixation cannot proceed.

Any photosynthetic reduction of oxygen *in vivo* clearly cannot occur in the same manner as the reaction *in vitro* that is catalysed by the herbicide methyl viologen; here toxicity results from continuous photosynthetic production of superoxide with consequent peroxidation of lipids and damage to membrane integrity. The two-step oxygen reduction which is catalysed by ferredoxin is, however, quite acceptable as the basis of a truly physiological pseudocyclic photophosphorylation.

OXYGEN AND CYCLIC PHOTOPHOSPHORYLATION

A number of artificial cofactors are known to support cyclic phosphorylation in isolated chloroplasts under anaerobic conditions. This is also true of ferredoxin, the only known cyclic cofactor which is also a natural chloroplast component. Anaerobic conditions are unlikely, however, to prevail *in vivo*, and so the recent description[6] of an aerobic cyclic phosphorylation which requires ferredoxin at much lower concentrations than does anaerobic cyclic phosphorylation considerably strengthens the claim that cyclic phosphorylation can occur *in situ* in a leaf. According to Arnon and Chain[6] aerobic cyclic phosphorylation requires only the same concentration of ferredoxin (10 μM) as does non-cyclic phosphorylation.

One explanation of this requirement for oxygen of cyclic phosphorylation at low ferredoxin concentrations is that oxygen may act as an oxidant and so regulate ('poise') the system in such a way that an optimum net oxidation–reduction state of the components is achieved. This interpretation is favoured by Arnon and Chain[6], who showed that DCMU inhibits aerobic cyclic phosphorylation in 'photosystem I' light. Such inhibition is likely to result from 'over-oxidation' of the cyclic chain's components, and in the absence of DCMU a slow input of electrons from photosystem II could occur, counteracting the tendency towards over-oxidation which would be caused by ferredoxin's reaction with oxygen.

Inhibition of cyclic electron flow *in vivo* could occur for a similar reason. With a continuous flow of electrons into photosystem I from photosystem II, and in situations where ATP is required and hence pyridine nucleotide is largely reduced, over-reduction would be expected to prevent optimum cyclic electron flow. Oxidation of ferredoxin by oxygen could then provide an 'escape' for electrons and a release of this inhibition. 'Poising' in chloroplasts has been discussed in detail by Grant and Whatley.[21] A PMS-catalysed cyclic phosphorylation which is subject to inhibition by over-reduction under anaerobic conditions has also been described.[45] Poising and hence resumption of phosphorylation could be achieved in this case by addition of oxygen or ferricyanide.

Huber and Edwards[27] have obtained evidence that oxygen, because of its reduction by ferredoxin, causes over-oxidation of the cyclic pathway's components in isolated, intact mesophyll chloroplasts of *Digitaria sanguinalis* under far-red light. Using pyruvate-dependent CO_2-fixation (resulting from the β-carboxylation of C_4 photosynthesis) as a measure of cyclic phosphorylation in this system, they observed an inhibition of CO_2-fixation by oxygen. DCMU also had an inhibitory effect under these conditions, presumably by preventing a supply of electrons from photosystem II, as in the analogous experiment (but with broken spinach chloroplasts incorporating ^{32}P-phosphate) of Arnon and Chain.[6] In white light, however, DCMU stimulated pyruvate-dependent CO_2-fixation in the experiments of Huber and Edwards, an effect which they attributed to a relief by DCMU of over-reduction of the cyclic pathway. If this explanation is correct, and if oxygen does indeed provide an escape route for electrons and hence an alternative antidote to over-reduction, an increase in oxygen concentration should itself stimulate cyclic phosphorylation and associated CO_2-fixation in white light, when both photosystems are active. This oxygen-dependent poising could clearly be of importance in the regulation of any cyclic phosphorylation which occurs *in vivo* in the mesophyll chloroplasts of C_4 plants.

Since cyclic electron transport can result in synthesis of ATP but not of NADPH, cyclic phosphorylation, like pseudocyclic, can provide a solution to the problem, described in the previous section, concerning the stoichiometry of production of ATP and NADPH. In the context of 'poising' it becomes particularly apparent that the operation of neither mode (cyclic or pseudocyclic) of phosphorylation precludes the simultaneous operation of the other.

An alternative explanation for oxygen-enhancement of ferredoxin-mediated cyclic phosphorylation is possible, though as yet it commands no direct experimental support. The basis of this is the number of examples that are known of what can be presumed to be cyclic electron transport in which the O_2/O_2^- couple functions as an electron-carrying component.

In a study of the kinetics of oxidation of plastocyanin by photosystem I, Wood and Bendall[47] found it necessary to add superoxide dismutase to detergent-treated chloroplasts in the presence of methyl viologen in order to suppress reduction of plastocyanin by superoxide. In the absence of superoxide dismutase, a superoxide-mediated cyclic electron transport must presumably have occurred. An analogous reaction has been proposed for *Rhodospirillum rubrum* reaction centres by Boucher and Gingras.[10] Here added horse heart cytochrome c functioned both as an electron donor to the bacterial photosystem and, by virtue of its reduction by superoxide, as an electron acceptor. Takahama and Nishimura[43] taking malondialdehyde formation as a measure of membrane lipid peroxidation in spinach chloroplast fragments, concluded that singlet oxygen (the peroxidising agent) is produced as a result of oxidation of superoxide by components of the inter-photosystem electron transport chain. If this conclusion is correct, and if superoxide is produced when oxygen is reduced by photosystem I, a cyclic system with O_2/O_2^- as a component must have been set up.

Since plastocyanin is now thought to be located near the inner surface of the thylakoid membrane (see Trebst[44]), superoxide, produced at the outer surface of the same membrane (or even in the medium), would not be expected to function as a link for this kind of cyclic electron transport except in detergent-treated or otherwise damaged chloroplasts or chloroplast fragments. This may help to explain an inhibition by ferredoxin of net oxygen uptake in *Euglena* chloroplasts, and in digitonin-treated spinach chloroplast fragments, but not in untreated spinach chloroplasts, as described by Elstner *et al.*[16] Elstner *et al.* attributed the inhibition to a ferredoxin-mediated cyclic electron transfer via cytochrome 552 (in *Euglena* an analogue of plastocyanin).

In vivo, however, the permeability barrier of the thylakoid membrane would prevent a reaction of superoxide with plastocyanin, and, even if it were to occur, such a cyclic pathway would be a wasteful one in the sense that it would not contribute to the transmembrane proton concentration gradient which is held to be the link between electron transport and synthesis of ATP. A contribution to photophosphorylation by a superoxide-mediated cyclic electron transport would occur only if the proton-translocating step involving plastoquinone were to form part of the cycle. The thylakoid would not represent a permeability barrier to such a process, since the site of plastoquinone reduction (and of proton binding) is located, like the site of oxygen reduction by photosystem I, on the outer surface of the membrane.[44, 29] Cytochrome b_{563} is thought to mediate electron transfer from ferredoxin in the cyclic pathway.[41] The possibility that O_2/O_2^- may also represent a link in this chain cannot yet be ruled out on either thermodynamic or kinetic grounds, though experimental support for this idea has yet to emerge.

SUPEROXIDE AND NET TRANSLOCATION OF PROTONS

Following the experiments of Junge and Auslander[29] the net proton-translocating capacity of illuminated thylakoids can be seen to result from the existence of two sites of proton binding at the outer surface of the thylakoid and two sites of proton production at the inner surface. In these experiments absorbance changes of the dye cresol red were used to measure pH changes at the outer surface of the thylakoid. The autoxidisable electron acceptor benzyl viologen was used, and proton-binding at photosystem I would be expected to result from the formation of undissociated hydrogen peroxide which follows dismutation of the superoxide anion.

Schmid[40] has reported that this proton-binding step seems not to occur in chloroplasts which which have been EDTA-treated in order to remove the chloroplast coupling factor. In such chloroplasts Schmid found only a net acidification following a single-turnover flash, whereas in chloroplasts not treated with EDTA an initial alkalinization at the outer surface of the thylakoid is followed by a relatively slow return to the initial pH. Schmid[40] proposed that EDTA-treatment removes an intrinsic chloroplast superoxide dismutase, and thereby 'makes O_2^- a rather stable end product of the electron transport'. The proton uptake associated with the dismutation is thus avoided, and net acidification results. In support of this explanation, addition of superoxide dismutase to EDTA-treated chloroplasts was shown to restore the kinetics of proton uptake which were found in chloroplasts which had not been treated with EDTA.

On a millisecond time-scale it may be not unreasonable to regard superoxide as 'rather stable', though Schmid's results show that the effect of acidification persists for several seconds after illumination. It would be of interest to know if a similar effect could be observed with cofactors which autoxidise by a two-step mechanism.

PHOTORESPIRATION

The reductive pentose phosphate pathway of photosynthetic CO_2-fixation[11] is a process which does not itself require oxygen, even if the ATP which it consumes may in certain circumstances have been manufactured in an oxygen-requiring photophosphorylation. The same is true of the additional steps of CO_2-fixation found in C_4 plants.[24] However, in photorespiration, a phenomenon associated particularly with C_3 plants, oxygen is consumed while CO_2 is released from compounds derived directly from intermediates of the CO_2-fixation pathway itself. This subject has been reviewed by Jackson and Volk[28], Black[9] and Zelitch[49].

An oxygen-consuming reaction which occurs readily at least *in vitro*, and which recently has become popular as an explanation of photorespiration, is the ribulose diphosphate oxygenase reaction[33], the products of which are phosphoglycerate and phosphoglycollate. The reaction is catalysed by the same protein (ribulose diphosphate carboxylase) which brings about the initial step of CO_2-fixation, and the product of this carboxylation reaction is phosphoglycerate alone. The oxygenase reaction's mechanism is generally thought to involve oxygen itself[33], though it has been suggested[46] that superoxide is required for oxidation of a sulphydryl group at the active site of the enzyme. No inhibition of the oxygenase reaction by superoxide dismutase has yet been reported, and an 'apparent' superoxide dismutase (which is heat-stable) may even be associated with the active purified enzyme.[39] It should be mentioned that there is by no means general acceptance that ribulose diphosphate oxygenase contributes appreciably to photorespiration. Zelitch[50], for example, attaches importance to a disparity between the pH optima of the carboxylase and oxygenase reactions, and to the lack of evidence that phosphoglycollate is produced by ^{14}C labelling of intact, photorespiring tissue.

Another process involving oxygen consumption and production of glycollate was proposed by Coombs and Whittingham.[12] Here hydrogen peroxide is generated in a Mehler reaction and leads to production of glycollate from a transketolase intermediate. Kaiser's demonstration[30] that hydrogen peroxide is inhibitory to net CO_2-fixation by intact chloroplasts has already been mentioned. If the Coombs–Whittingham explanation of photorespiration and the 'Warburg effect' (inhibition of photosynthesis by increasing oxygen concentration) is correct, then an increase by hydrogen peroxide of chloroplast glycollate production should occur.

Kirk and Heber[31] have recently published data showing that hydrogen peroxide is inhibitory to chloroplast glycollate production, while addition of catalase brings about a stimulation of glycollate production even without previous addition of hydrogen peroxide to the chloroplast suspension. A number of the observations of Kirk and Heber[31] suggest that glycollate is produced in intact chloroplasts directly from ribulose diphosphate, with the proportion of carbon fixed in glycollate increasing both with oxygen concentration and with pH in the range pH 7.2–8.2. One piece of evidence which in particular seems to rule out a transketolase intermediate as a source of glycollate in intact chloroplasts is the inhibition by the uncoupler FCCP of formation of glycollate from dihydroxyacetone phosphate.[31] This suggests that production of glycollate does not occur at any point between dihydroxyacetone phosphate and the phosphoribulokinase step of the Calvin cycle, but instead occurs after the phosphoribulokinase reaction (the first ATP-requiring step encountered by carbon flowing from dihydroxyacetone phosphate).

The fate of glycollate and the source of photorespiratory CO_2 is also a controversial topic. A contribution of photosynthetic oxygen reduction here may take the form of a requirement for hydrogen peroxide for oxidative decarboxylation of glyoxylate to formate[14, 48], though even if this reaction does participate in photorespiration it is possible that it does so in leaf peroxisomes rather than in chloroplasts themselves.[23] Elstner and Heupel[14] have demonstrated decarboxylation of a number of α-keto acids by chloroplasts in the presence of an autoxidisable electron acceptor, and with NADP (in the presence of ferredoxin) as electron acceptor the decarboxylation commenced as soon as all the NADP present had been reduced. Photorespiration by such a mechanism could be linked to any pseudocyclic phosphorylation which may occur *in vivo*, and so hydrogen peroxide generated in a Mehler reaction may yet be shown to play an important role in photorespiration, even if, as suggested by Kirk and Heber[31] it plays no part in synthesis of glycollate itself.

Acknowledgement

The author is indebted to Professor F.R. Whatley, F.R.S., for advice and helpful criticism of the manuscript, and to the U.K. Science Research Council for a Postdoctoral Research Fellowship.

REFERENCES

1. ALLEN, J.F. A two-step mechanism for the photosynthetic reduction of oxygen by ferredoxin. *Biochem. biophys. Res. Commun.* **66** (1975) 36–43.

2. ALLEN, J.F. Oxygen reduction and optimum production of ATP in photosynthesis. *Nature* **256** (1975) 599–600.

3. ALLEN, J.F. Superoxide and photosynthetic reduction of oxygen. In A.M. MICHELSON, J.M. McCORD, and I. FRIDOVICH (Eds). Superoxide

and superoxide dismutases. Academic Press, New York, in press.

4. ALLEN, J.F. and HALL, D.O. Superoxide reduction as a mechanism of ascorbate-stimulated oxygen uptake by isolated chloroplasts. *Biochem. biophys. Res. Commun.* **52** (1973) 856–862.

5. ALLEN, J.F. and HALL, D.O. The relationship of oxygen uptake to electron transport in photosystem I of isolated chloroplasts. *Biochem. biophys. Res. Commun.* **58** (1974) 579–585.

6. ARNON, D.I. and CHAIN, R.K. Regulation of ferredoxin-catalysed photosynthetic phosphorylation. *Proc. natn. Acad. Sci. USA* **72** (1975) 4961–4965.

7. ARNON, D.I., LOSADA, M., WHATLEY, F.R., TSUJIMOTO, H.Y., HALL, D.O. and HORTON, A.A. Photosynthetic phosphorylation and molecular oxygen. *Proc. natn. Acad. Sci. USA* **47** (1961) 1314–1334.

8. ARNON, D.I., TSUJIMOTO, H.Y. and McSWAIN, B.D. Ferredoxin and photosynthetic phosphorylation *Nature* **214** (1967) 562–566.

9. BLACK, C.C. Photosynthetic carbon fixation in relation to net CO_2 uptake. *Annu. Rev. Plant Physiol.* **24** (1973) 253–286.

10. BOUCHER, R. and GINGRAS, G. The photogeneration of superoxide by isolated photoreaction center from *Rhodospirillum rubrum. Biochem. biophys. Res. Commun.* **67** (1975) 421–426.

11. CALVIN, M. and BASSHAM, J.A. *The photosynthesis of Carbon Compounds*. Benjamin, New York (1962).

12. COOMBS, J. and WHITTINGHAM, C.P. The mechanism of inhibition of photosynthesis by high partial pressures of oxygen in *Chlorella. Proc. Roy. Soc. Lond. Ser. B.* **164** (1966) 511–520.

13. EGNEUS, H., HEBER, U., MATTHIESEN, U. and KIRK, M. Reduction of oxygen by the electron transport chain of chloroplasts during assimilation of carbon dioxide. *Biochim. biophys. Acta* **408** (1975) 252–268.

14. ELSTNER, E.F. and HEUPEL, A. On the decarboxylation of α-keto acids by isolated chloroplasts. *Biochim. biophys. Acta* **325** (1973) 182–188.

15. ELSTNER, E.F. and HEUPEL, A. Involvement of the superoxide free radical ion in photosynthetic oxygen reduction. *Z. Naturforsch Sect. C.* **29** (1974) 559–563.

16. ELSTNER, E.F., WILDNER, G.F. and HEUPEL, A. Oxygen activation by isolated chloroplasts from *Euglena gracilis*. Ferredoxin-dependent function of a fluorescent compound and photosynthetic electron transport close to photosystem I. *Archs. Biochem. Biophys.* **173** (1976) 623–630.

17. EPEL, B.L. and NEUMANN, J. The mechanism of the oxidation of ascorbate and Mn^{2+} by chloroplasts. The role of the radical superoxide. *Biochim. biophys. Acta* **325** (1973) 520–529.

18. FARRINGTON, J.A., EBERT, M. and FLETCHER, K. Bipyridylium quaternary salts and related compounds — V. Pulse radiolysis studies of the reaction of paraquat radical with oxygen. Implications for the mode of action of bipyridyl herbicides. *Biochim. biophys. Acta* **314** (1973) 372–381.

19. GLIDEWELL, S.M. and RAVEN, J.A. Measurement of simultaneous oxygen evolution and uptake in *Hydrodictyon africanum. J. exp. Bot* **26** (1975) 479–488.

20. GOOD, N. and HILL, R. Photochemical reduction of oxygen in chloroplast preparations — II. Mechanisms of the reaction with oxygen. *Archs. Biochem. Biophys.* **57** (1955) 355–366.

21. GRANT, B.R. and WHATLEY, F.R. Some factors affecting the onset of cyclic photophosphorylation. In GOODWIN, T.W. (Ed.) *Biochemistry of Chloroplasts* Vol. 2. Academic Press, New York. (1967) pp.505–521.

22. HALL, D.O. The coupling of photophosphorylation to electron transport in isolated chloroplasts. In BARBER, J. (Ed.) *The Intact Chloroplast: Structure and Function*. ASP Biological and Medical Press, Amsterdam (1976).

23. HALLIWELL, B. and BUTT, V.S. Oxidative decarboxylation of glycollate and glyoxylate by leaf peroxisomes. *Biochem. J.* **138** (1974) 217–224.

24. HATCH, M.D. and SLACK, C.R. Photosynthetic CO_2-fixation pathways. *Annu. Rev. Plant Physiol.* **21** (1970) 141–162.

25. HEBER, U. Stoichiometry of reduction and phosphorylation during illumination of intact chloroplasts. *Biochim. biophys. Acta* **305** (1973) 140–152.

26. HUBER, S. and EDWARDS, G. The effect of oxygen on CO_2-fixation by mesophyll protoplast extracts of C_3 and C_4 plants. *Biochem. biophys. Res. Commun.* **67** (1975) 28–34.

27. HUBER, S.C. and EDWARDS, G.E. Studies on the pathway of cyclic electron flow in mesophyll chloroplasts of a C_4 plant. *Biochim. biophys. Acta* **449** (1976) 420–433.

28. JACKSON, W.A. and VOLK, R.J. Photorespiration. *Annu. Rev. Plant Physiol.* **21** (1970) 385–432.

29. JUNGE, W. and AUSLANDER, W. The electric generator in photosynthesis of green plants.— I. Vectorial and protolytic properties of the electron transport chain. *Biochim. biophys. Acta* **333** (1973) 59–70.

30. KAISER, W. The effect of hydrogen peroxide on CO_2-fixation of isolated intact chloroplasts. *Biochim. biophys. Acta* **440** (1976) 476–482.

31. KIRK, M.R. and HEBER, U. Rates of synthesis and source of glycolate in intact chloroplasts. *Planta* **132** (1976) 131–141.

32. KONO, Y., TAKAHASHI, M. -A. and ASADA, K. Oxidation of manganous pyrophosphate by superoxide radicals and illuminated spinach chloroplasts. *Archs. Biochem. Biophys.* **174** (1976) 454–462.

33. LORIMER, G.H. and ANDREWS, T.J. Plant photorespiration — an inevitable consequence of the existence of atmospheric oxygen. *Nature* **243** (1973) 359–360.

34. MEHLER, A.H. Studies on reactions of illuminated chloroplasts — I. Mechanism of the reduction of oxygen and other Hill reagents. *Archs. Biochim. Biophys.* **33** (1951) 65–77.

35. MEHLER, A.H. Studies on reactions of illuminated chloroplasts — II. Stimulation and inhibition of the reaction with molecular oxygen. *Archs. Biochem. Biophys.* **33** (1951) 339–351.

36. PATTERSON, C.O.P. and MYERS, J. Photosynthetic production of hydrogen peroxide by *Anacystis nidulans. Pl. Physiol.* **51** (1973) 104–109.

37. RADMER, R.J. and KOK, B. Photoreduction of O_2 primes and replaces CO_2-assimilation. *Pl. Physiol.* **58** (1976) 336–340.

38. RAVEN, J.A. and GLIDEWELL, S.M. Sources of ATP for active phosphate transport in *Hydrodictyon africanum*; evidence for pseudocyclic photophosphorylation *in vivo. New Phytol.* **75** (1975) 197–204.

39. RYAN, F.J. and TOLBERT, N.E. Ribulose diphosphate carboxylase-oxygenase — III. Isolation and properties. *J. biol. Chem.* **250** (1975) 4229–4233.

40. SCHMID, R. Deactivation of superoxide dismutase on EDTA-treated chloroplasts. *FEBS Lett.* **60** (1975) 98–102.

41. SCHURMANN, P., BUCHANAN, B.B. and ARNON, D.I. Role of cyclic photophosphorylation in photosynthetic carbon dioxide assimilation by isolated chloroplasts. *Biochim. biophys. Acta* **267** (1972) 111–124.

42. SIMONIS, W. and URBACH, W. Photophosphorylation *in vivo. Annu. Rev. Plant Physiol.* **24** (1973) 89–114.

43. TAKAHAMA, U. and NISHIMURA, M. Effects of electron donor and acceptors, electron transfer mediators, and superoxide dismutase on lipid peroxidation in illuminated chloroplast fragments. *Pl. Cell Physiol. Tokyo* **17** (1976) 111–118.

44. TREBST, A. Energy conservation in photosynthetic electron transport of chloroplasts. *Annu. Rev. Plant Physiol.* **25** (1974) 423–458.

45. WHATLEY, F.R. Some effects of oxygen in photosynthesis by chloroplast preparations. In *Photosynthetic Mechanisms of Green Plants.* Natl. Acad. Sci. — Natl. Res. Council, Washington D.C. (1963) pp.243–251.

46. WILDNER, G.F. and HENKEL, J. Specific inhibition of the oxygenase activity of ribulose-1,5-bisphosphate carboxylase. *Biochem. biophys. Res. Commun.* **69** (1976) 268–275.

47. WOOD, P.M. and BENDALL, D.S. The kinetics and specifity of electron transfer from cytochromes and copper proteins to P700. *Biochim. biophys. Acta* **387** (1975) 115–128.

48. ZELITCH, I. The photoxidation of glyoxylate by envelope-free spinach chloroplasts and its relation to photorespiration. *Archs. Biochem. Biophys.* **150** (1972) 698–704.

49. ZELITCH, I. Pathways of carbon fixation in green plants. *Annu. Rev. Biochem.* **44** (1975) 123–145.

50. ZELITCH, I. The biochemistry of photorespiration. In SMITH, H. (Ed.) *Commentaries in Plant Science.* Pergamon Press, Oxford (1976) pp. 51–61.

RIBULOSE BISPHOSPHATE CARBOXYLASE ACTIVITY IN TERRESTRIAL PLANTS: SIGNIFICANCE OF O_2 AND CO_2 DIFFUSION

J.A. RAVEN

Department of Biological Sciences, University of Dundee, Dundee DD1 4HN, Scotland

INTRODUCTION

Ribulose bisphosphate carboxylase–oxygenase [RuBPc-o; 3-phospho-D-glycerate carboxylase (dimerising), E.C. 4.1.1.39] is the only carboxylase in O_2-evolving plants which participates in a reaction sequence leading to a net conversion of CO_2 into reduced organic compounds. The reaction it catalyses (RuBPc activity) is:

$$\text{Ribulose bisphosphate} + CO_2 + H_2O \rightarrow 2.3\text{-phosphoglycerate.}$$

RuBPc-o thus has a key role in photosynthesis and primary productivity. Two of its properties have, however, been construed as not well adapted to the present composition of the atmosphere. One relates to the $K_{1/2}$ for CO_2, which is rather larger than the 10 μM CO_2 found in water in equilibrium with air.[43, 72, 78, 79] This means that, even in a cell with the same CO_2 concentration as the surrounding air, the full potential of this enzyme cannot be used. The other 'non-adaptive' characteristic is the ability of the enzyme to catalyse not only the RuBPc reaction but also an oxygenase reaction (RuBPo activity):

$$\text{Ribulose bisphosphate} + O_2 \rightarrow 3\text{-phosphoglycerate} + 2\text{-phosphoglycolate.}$$

The RuBPc and RuBPo activities are competitive, so that the rate of CO_2 fixation by the enzyme from a 10 μM CO_2 solution is lower in the presence of O_2 than in its absence.[1, 42] This reduces the rate of CO_2 fixation in gross photosynthesis in air compared with a low O_2, but otherwise similar, atmosphere. The quantity of reduced C available for cell growth is further limited by the intractable nature of the glycolate which results from the action of a specific phosphatase on the phosphoglycolate produced by the oxygenase. This is either excreted, as in some algae; or scavenged back to compounds which can be used for cell growth, at the expense of C loss (as CO_2) and further energy input.[19, 20] The ratio of oxygenase to carboxylase activities of the enzyme[1, 42] is 0.3–0.5 in a solution in equilibrium with air, i.e. containing 10 μM CO_2. The adverse effect of O_2 on net photosynthesis is exacerbated by the existence of reactions other than RuBPo which convert sugar phosphate into glycolate.[19, 20]

These apparent deficiencies in the key carboxylase of photosynthesis are even more marked if the properties of the enzyme are considered in relation to the CO_2 and O_2 concentrations in the chloroplast of a cell photosynthesising in air.[60, 68, 69, 78, 79] The chloroplast is separated from the bulk air by gas- and liquid-phase diffusion pathways; the net flux of CO_2 into the cell, and of O_2 out of the cell, along these pathways which impose a resistance to gas movement, is driven by concentration differences for the two gases. This means that the CO_2 concentration inside a chloroplast obtaining its CO_2 by diffusion from the air is smaller than the CO_2 concentration in water in equilibrium with air (10 μM). Conversely, the O_2 concentration in the photosynthesising chloroplast is greater than the 250 μM in water in equilibrium with air.

This paper attempts to quantify diffusion effects on the *in vivo* reactions of RuBP with CO_2 and O_2, and other O_2 effects on photosynthesis in C3 and C4 terrestial angiosperms, using an analysis of CO_2 and O_2 diffusion[60–62, 68, 69] arrived at independently by Samish[68, 69] and by Raven (unpublished analysis).

GAS DIFFUSION IN PHOTOSYNTHETIC TISSUES

(1) *Samish's analysis*

The basic equation for CO_2 diffusion in photosynthesis is:

$$M_{CO_2} = \frac{[CO_2]_o - [CO_2]_c}{R_{CO_2}}, \tag{1}$$

where M_{CO_2} = net CO_2 influx in photosynthesis [nmole (cm^2 cell surface)$^{-1}$s^{-1}];

$[CO_2]_o$ = CO_2 concentration in bulk air (nmole cm^{-3});
$[CO_2]_c$ = CO_2 concentration in the chloroplast (nmole cm^{-3});
R_{CO_2} = resistance to CO_2 diffusion, (s cm^{-1}).

The maximum value of R_{CO_2} can be computed for any set value of M_{CO_2} by taking $[CO_2]_c$ as being equal to the CO_2 concentration at the CO_2 compensation point, i.e. assuming that the diffusion resistance R_{CO_2} is much higher than the chemical resistance to CO_2 fixation.[54, 60, 69]

Samish[68, 69] has used an analogous equation for O_2:

$$M_{O_2} = \frac{[O_2]_c - [O_2]_o}{R_{O_2}} \tag{2}$$

where M_{O_2} = net O_2 efflux in photosynthesis [nmole (cm^2 cell surface)$^{-1}$s^{-1}];

$[O_2]_c$ = O_2 concentration in the chloroplast (nmole cm^{-3});
$[O_2]_o$ = O_2 concentration in the bulk air (nmole cm^{-3});
R_{O_2} = resistance to O_2 diffusion, (s cm^{-1}).

Equations (1) and (2) may be used to compute a value for $[O_2]_o$. M_{O_2} can be measured, and $[O_2]_o$ is known. Samish[68, 69] has derived a value for R_{O_2} by assuming that the relation between the diffusion coefficient D, and the diffusion resistance, R, is the same for both gases. Thus $R_{CO_2} = k/D_{CO_2}$, and $R_{O_2} = k/D_{O_2}$ (where k is a constant), whence

$$R_{O_2} = \frac{R_{CO_2} \cdot D_{CO_2}}{D_{O_2}} \tag{3}$$

Since gas transport in terrestial plants involves both the gas and the aqueous phases, the analysis must take account of the different gas concentrations in the gas and liquid phases, and the different ratio of D_{CO_2}/D_{O_2} in the gas and aqueous phases. A *maximal* value for R_{O_2} is obtained by using the gas-phase ratio (0.93) rather than the liquid-phase value of 0.78.

Since R_{O_2} is up to 0.93 of R_{CO_2}, then, from equations (1) and (2), $([O_2]_c - [O_2]_o) = 0.93([CO_2]_o - [CO_2]_c)$ when $M_{CO_2} = M_{O_2}$. Since $[CO_2]$ is 10 μM, $([CO_2]_o - [CO_2]_c)$ cannot exceed 10 μM: neither can $([O_2]_c - [O_2]_o)$. Thus the air-equilibrium value of 250 μM cannot be increased above 260 μM in the chloroplast: a negligible increase in the O_2 concentration in terms of the kinetics of RuBPc-o. Thus, although the absolute concentration differences for the two gases are the same, the much greater background level of O_2 means that the diffusion resistance only influences RuDPc-o behaviour in terrestial C3 plants in relation to the CO_2 concentration and its influence of CO_2/O_2. Before considering the influence of changes in this ratio due to diffusion resistances on RuDPc-o in more detail, some attention will be given to the assumptions underlying the derivation of equation (3).

(2) *Lipid bilayers, solutions and cells*

Lipid bilayers. The combination of high lipid solubility[11] of O_2 and CO_2 and their high diffusion coefficient in lipids[17] suggest that the permeability of lipid bilayers to these gases is high. Direct measurements are impossible due to unstirred layer effects; values as high as 10 cm s^{-1} are likely (Refs. 23, 32, 54, 60, 80; cf. Refs. 14, 18, 71); low temperatures phase-change of membrane lipids may reduce permeability 5-fold.[8, 17, 40] In the case of inorganic carbon the trans-membrane flux can be increased by facilitated or active transport of HCO_3^-, provided carbonic anhydrase is present on the side of the membrane to which CO_2 is supplied or from which CO_2 is removed.[15, 16, 48, 49, 58, 60] Such a mechanism is probably not important in the plasmalemma or the chloroplast envelope of terrestial plants.[60, 67, 80]

Aqueous phase of cells. The apparent diffusion constant for CO_2 and O_2 in cell walls, cytoplasm and chloroplast stroma is likely to be reduced up to 50% compared with its value in free solution due to the presence of macromolecules acting as a barrier.[60] However, for both CO_2 and O_2 there is the possibility of facilitated diffusion. For CO_2, the contribution of the hydrated species H_2CO_3 and HCO_3^- to diffusion requires the presence of the enzyme carbonic anhydrase, since the uncatalysed reaction is too slow to contribute to CO_2 supply during photosynthesis; this enzyme is present in many photosynthetic tissues, and can double the apparent diffusion coefficient for CO_2 at cytoplasmic pH values of 7–8.[22, 23, 32, 34, 35, 48, 60, 70, 77] For O_2, facilitation of diffusion can occur *via* soluble molecules which can bind O_2 and diffuse, or membrane associated O_2-binding agents which have sufficient mobility to transfer O_2 in the plane of the membrane by transfer from carrier to carrier. Carriers of the first type include haemoglobin and myoglobin,[23, 35, 37, 81] which are absent from photosynthetic tissues; the second type is exemplified by cytochrome P_{450} on endoplasmic reticulum, which is present in many plants.[7, 10, 21, 44–47, 76]

Cells and tissues. Direct measurements of CO_2 and O_2 diffusion in plant cells and tissues are uncommon; the available data suggest values similar to those of the two gases in water.[5, 6, 31, 75] This accords with predicted influence of the barriers to, and facilitators of, diffusion in plant tissues discussed above, and suggests that the assumptions of Samish[68, 69] as to the relative diffusion coefficients of CO_2 and O_2 are substantially correct.

CO_2 AND O_2 DIFFUSION IN C3 PHOTOSYNTHESIS

In terrestial plants using the C3 pathway, when stomata are wide open and the boundary layer resistance is low, the resistance to CO_2 fixation which can be attributed to the gas- and liquid-phase diffusion path is less than half of the total (diffusive plus chemical) resistance.[32, 54, 59, 60] This conclusion is based on measured rates of net photosynthesis, and estimates of the gas-phase diffusion resistance (from rates of transpiration) and the liquid-phase diffusion resistance (from the length of the diffusion path measured in electron micrographs, and estimates of D_{CO_2} of the kind described above).

From equation (1), this conclusion about the diffusive resistance to CO_2 fixation means that the value of $[CO_2]_c$ is at least 5 μM during steady-state photosynthesis, while the value of $[O_2]_c$ is not more than 255 μM [equation (1)]. As was mentioned earlier, the fact that $[CO_2]_c$ is less than 10 μM, while $[O_2]_c$ is above 250 μM, means that analyses of the rates of RuBPc and RuBPo operation *in vivo* based on measurements of the rates *in vitro* in 10 μM CO_2 and 250 μM O_2 (e.g. Ref. 42) require adjustment to take the *in vivo* values of $[O_2]_c$ into account. Such an adjustment of the valuable analysis of Laing, Ogren and Hagemann[42] makes the agreement of *in vitro* and *in vivo* measurements less exact, in that the *in vitro* properties of the enzyme predict a larger O_2 inhibition of photosynthesis, and a larger synthesis of glycolate, than is observed *in vivo*. The discrepancy is even greater if the possibility of glycolate synthesis not involving RuBPo is admitted (see Introduction). The equation used by Laing, Ogren and Hagemann[42] is:

$$\frac{v_o}{v_c} = \frac{V_o \cdot K_c \cdot [O_2]_e}{V_c \cdot K_o \cdot [CO_2]_e} \tag{4}$$

where v_o = observed rate of RuBPo activity [nmole O_2 (mg enzyme)$^{-1}$s^{-1}];

v_c = observed rate of RuBPc activity [nmole CO_2 (mg enzyme)$^{-1}$s^{-1}];
V_o = RuBPo activity at optimal O_2 [nmole O_2 (mg enzyme)$^{-1}$s^{-1}];
V_c = RuBPc activity at optimal CO_2 [nmole CO_2 (mg enzyme)$^{-1}$s^{-1}];
K_o = half-saturation constant of RuBPo (nmole O_2 cm^{-3});
K_c = half-saturation constant of RuBPc (nmole CO_2 cm^{-3});
$[O_2]_e$ = concentration of O_2 at active site of RuBPo (nmole cm^{-3});
$[CO_2]_e$ = concentration of CO_2 at active site of RuBPc (nmole cm^{-3});

In an air-equilibrated solution, the ratio v_o/v_c for RuBPc-o *in vitro* is 0.3–0.5.[1, 42] This corresponds to 60–100% of the reduced C produced from the photosynthetic carbon reduction cycle being glycolate. Most estimates of the *in vivo* fraction of glycolate production are less than 50% of net C fixation.[40, 50, 51, 62] This corresponds to v_o/v_c of 0.25 or less under the *in vivo* conditions of about 7 μM CO_2 in the chloroplast and 253 μM O_2, i.e. diffusion contributing about 1/3 to the total resistance to CO_2 fixation [equations (1) and (2)].

The 'inadequacies' of RuBPc mentioned in the Introduction relate to the low rate of RuBPc, and the high ratio of RuBPo/RuBPc, at natural levels of O_2 and CO_2. Granted that the kinetic properties of the enzyme (V_o, K_o, V_c and K_c) are unchanged, the maximisation of net photosynthesis in the units used in equations (1) and (2), i.e. nmoles (cm^2 cell surface area)$^{-1}$ s^{-1} has two main requirements. One is that the quantity of enzyme per unit of cell surface should be high; a high value of (mg enzyme) (cm^2 cell surface)$^{-1}$ means a high capacity for CO_2 fixation (nmoles cm^{-2} s^{-1}). The other requirement relates to the realisation of this capacity by keeping $[CO_2]_e$ [equation (4); $[CO_2]_c$ of equation (1)] as high as possible, with $[O_2]_e$ [equation (4); $[O_2]_c$ of equation (2)] as low as possible, bearing in mind the concentrations of these gases in air. Unfortunately, these two requirements are mutually incompatible.

The low turnover number of RuBPc, together with its large molecular weight, means that an increased quantity of enzyme per unit area of cell surface involves an increase in the thickness of the layer of stroma (in which the enzyme is the major soluble protein) in the periphery of the leaf cell.[60, 64, 65] The extra enzyme cannot be accommodated by increasing the concentration of the enzyme in a constant thickness of stroma since this would soon lead to precipitation of the enzyme, with considerable loss of activity.[12, 25, 33] The extra thickness of stroma means that the additional molecules are further from the source of CO_2 in the intercellular spaces. This additional length of aqueous diffusion path leads to a proportionate increase in the values of R_{CO_2} [equation (1)], and R_{O_2}, [equation (2)]. Thus $[CO_2]_e$ [equation (4)] is decreased, while $[O_2]_e$ [equation (4)] is increased by a similar amount. This means that the additional RuBPc molecules are working at a low $[CO_2]_e$ and with a high ratio $[O_2]_e/[CO_2]_e$, i.e. with a low rate of carboxylation and a high rate of oxygenation. In view of the large energy and material cost of the synthesis of RuBPc[64–66] there must be strong selective pressure against synthesis of additional RuBPc molecules which can only make such a small contribution to net photosynthesis.[46, 65]

This analysis shows that the properties of RuBPc/o are such that the photosynthetic capacity on a cell surface basis is not fully expressed during photosynthesis in air with diffusive entry of CO_2 and exit of O_2. One way in which more of the RuBPc capacity can be used is by increasing $[CO_2]_e$ [equation (4)]; this may be achieved in certain algae by active HCO_3^- influx across the chloroplast envelope,[67] and in some angiosperms by the use of the carboxylic acid 'CO_2 pump'. This latter mechanism is discussed in more detail below.

CO_2 AND O_2 DIFFUSION IN C4 PHOTOSYNTHESIS

A commonly accepted view of C4 photosynthesis[28] is shown in Fig. 1, with the C4 acids acting as a 'CO_2 pump' which increases the CO_2 concentration in the bundle sheath chloroplasts containing the RuBPc. This lowers the ratio $[O_2]_e/[CO_2]_e$ [equation (4)], and hence decreases v_o/v_c, thus accounting for many features of C4 photosynthesis, such as a higher rate of photosynthesis in air on a unit RuBPc activity basis, the much reduced rate of glycolate synthesis and the much lower O_2 inhibition of net photosynthesis. The initial carboxylation in this scheme is cata-

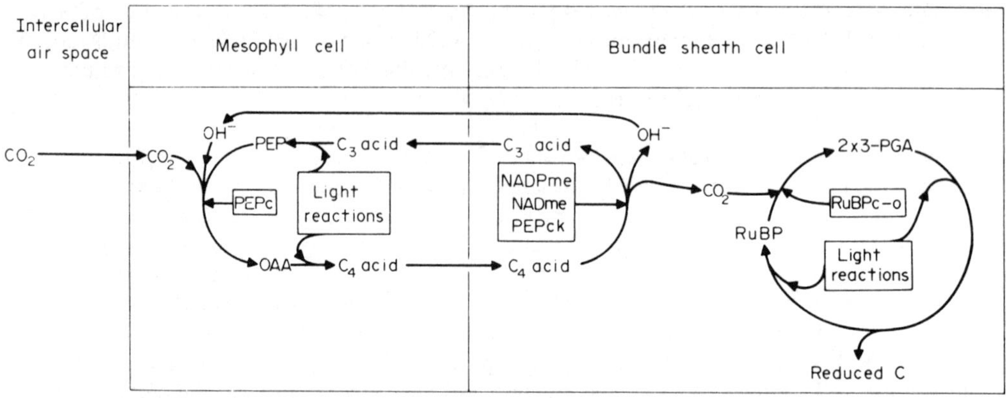

Fig. 1. C_4 Photosynthesis. The spatial distribution of reactions shown here is supported by most[26, 28, 30] but not all[13, 36] workers in the field. The need for OH^- transfer between bundle sheath and mesophyll was pointed out by Dr. F.A. Smith (personal communication). The shuttling of reductant and high-energy phosphate between mesophyll and bundle sheath[38, 39] is not indicated in this scheme.

Abbreviations (other than those used elsewhere in the text): OAA = oxaloacetate; PEP = phosphoenolpyruvate; C_3 acids = alanine, pyruvate or PEP; C_4 acids = malate or aspartate. Alternative decarboxylating enzymes: NADPme = malic enzyme (EC 1.1.1.40), NADme = NAD malic enzyme (EC 1.1.1.39), PEPck = PEP carboxykinase (EC 4.1.1.49).

lysed by phosphoenolpyruvate carboxylase (PEPc; orthophosphate:oxalacetate carboxylase, E.C. 4.1.1.31):

$$\text{phosphoenolpyruvate} + HCO_3^- \rightarrow \text{oxalacetate} + \text{orthophosphate}.$$

Compared with RuBPc, this enzyme has a much greater turnover number and is not inhibited by O_2 [4] and is probably located in the cytoplasm of the mesophyll cells of C4 plants. These attributes mean that the difficulties of providing a high carboxylation activity per unit area of leaf-cell surface, while not introducing too large a diffusion barrier for CO_2 between the intercellular air spaces and the enzyme (see above), are much smaller for PEPc than for RuBPc-o. While the properties of this enzyme fit it for the role of scavenger of any CO_2 which leaks out of the bundle sheath cells as well as for the efficient fixation of exogenous CO_2, the efficient operation of the C4 mechanism requires minimal leakage and recycling of CO_2.[41, 57, 62, 74]

Most analyses of the 'CO_2 pump' and its role in reducing v_o/v_c tacitly assume that the O_2 concentration in the bundle sheath chloroplasts is similar to that in air-equilibrated water. However, the combination of net O_2 evolution in the bundle sheath chloroplasts with the barrier to gas diffusion out of these cells implicit in the maintenance of a high CO_2 concentration in the neighbourhood of RuBPc means that there must be a build-up of O_2 in these cells.[69] The significance of this build-up for the ratio $[O_2]_e/[CO_2]_e$ [equation (4)], and hence v_o/v_c, will be analysed for two types of C4 plants.

In Gramineae with the NADPme pathway of carbon fixation (e.g. *Zea mays, Sorghum bicolor*) only 10% of the whole-leaf capacity for O evolution is found in the bundle sheath chloroplasts.[26, 27, 30, 55, 56] O_2 uptake in the light can occur in the bundle sheath cells by 'dark' respiratory pathways (50% of the whole-leaf capacity), and by RuBPo and other mechanisms of glycolate synthesis. With whole-leaf 'dark' respiration in the light at 5% of the rate of gross photosynthesis, and with a rather high estimate of glycolate synthesis of 10% of the C fixation rate (v_o/v_c of 0.05), it is possible to show that O_2 evolution by photosystem two in the bundle sheath is almost exactly balanced by the O_2 uptake by 'dark' respiratory pathways and by the synthesis and metabolism of glycolate.[3, 9, 30, 38, 39, 52, 61–63] This absence of net O_2 production or consumption in the bundle sheath cells means that the O_2 concentration there is near the air-equilibrium concentration of 250 μM. It has been argued earlier that the *in vivo* value of v_o/v_c [equation (4)] is 0.25 when $[O_2]_e/[CO_2]_e$ is 250/7. Thus the v_o/v_c of 0.05 in *Zea* and *Sorghum* could be achieved with a value of $[CO_2]_e$ of 35 μM.

Assuming that the stomata are wide open, the CO_2 concentration in the intercellular spaces is 10 μM. Thus the diffusion gradient for CO_2 between the bundle sheath chloroplasts and these spaces is (35–10) or 25 μM, i.e. 25 nmoles cm^{-3}. The centrifugally arranged bundle sheath chloroplasts in these plants are relatively close to the intercellular spaces, and the diffusion resistance to CO_2movement from the chloroplasts to the air spaces is about 30 s cm^{-1} (in terms of the area of bundle sheath cell exposed to the air spaces: Ref. 62). The 25 nmole cm^{-3} gradient and the 30 s cm^{-1} resistance yields a CO_2 efflux into the intercellular spaces of about 0.85 nmoles CO_2 (cm^2 bundle sheath exposed to intercellular space)$^{-1}$ s^{-1}; this is some 0.2 of the net rate of CO_2 fixation, and is consistent with the data of Mahon *et al.*[52] on the amount of CO_2 recycled through PEPc in *Zea* in air. This recycling imposes an energetic penalty on the photosynthetic mechanism of these plants of 0.4 ATP per *net* CO_2 fixed,[53] i.e. an extra 0.4 quanta per CO_2. This is consistent with the observed equality of quantum requirements of net CO_2 fixation in C_3 and C_4 plants, taking the energetic requirements of glycolate synthesis and scavenging in C3 plants into account.[2, 19, 29, 53, 63]

Dicotyledonous C4 plants with NADme as the decarboxylating enzyme (e.g. *Amaranthus edulis, Atriplex spongiosa*) contrast with the *Zea* and *Sorghum* discussed above in several ways. They have a much greater percentage (70%) of the whole-leaf capacity for O_2 evolution in bundle sheath chloroplasts; the bundle sheath chloroplasts are disposed centripetally in the cell; and 20% of the net carbon fixed in the bundle sheath cells is routed through glycolate.[3, 9, 26, 27, 30, 38, 39, 52, 53, 56, 61–63] Comparing the O_2 evolved in illuminated bundle sheath cells with that taken up in 'dark' respiration and glycolate synthesis and metabolism in these NADme dicotyledons shows that there is a large net O_2 evolution; this amounts to some 2.3 nmoles O_2 (cm^2 bundle sheath exposed to intracellular spaces)$^{-1}$ s^{-1}. The greater distance of the bundle sheath chloroplasts from the air spaces means that the diffusion resistance for O_2 or CO_2 loss from the chloroplasts to the air spaces is higher (130 s cm^{-1}) in these dicotyledons than in the NADPme gramineae (30 s cm^{-1}). The disposal of 2.3 nmoles O_2 cm^{-2}s^{-1} through a resistance of 130 s cm^{-1} requires a driving force (concentration difference) of 300 nmole O_2 cm^{-3} (300 μM). With 250 μM in the intercellular spaces this gives a total O_2 concentration in the bundle sheath chloroplasts of 550 μM.

With a glycolate formation of 0.2 of the net carbon fixed in these plants, v_o/v_c is 0.1. Since v_o/v_c is 0.25 when $[O_2]_e/[CO_2]_e$ is 250/7 [see above; equation (4)], a v_o/v_c of 0.1 is obtained when $[O_2]_e/[CO_2]_e$ is (250/7)(0.1/0.25) or 100/7. With the bundle sheath chloroplast value of $[O_2]_e$ of 550 μM, a v_o/v_c of 0.1 requires a $[CO_2]_e$ of 550 $\times$ (7/100) or about 40 μM. This CO_2 concentration in the bundle sheath chloroplasts is 30 μM higher than that in the intercellular air spaces, and would lead to an efflux through the diffusive resistance of 130 s cm^{-1} of some 0.25 nmole (cm^2 bundle sheath exposed to intercellular air spaces)$^{-1}$ s^{-1}. This is a smaller fraction of net CO_2 fixation which is recycled through PEPc (about 0.05) than in the NADme Gramineae (0.2): in energetic terms, this smaller recycling must be set against the larger energy requirement for glycolate synthesis and metabolism in the NADme dicotyledons, where glycolate is 0.2 of the net C fixed as opposed to the 0.1 or less in NADPme Gramineae.

This analysis shows that, in these two contrasting types of C4 plants, the properties of RuBPc-o, the O_2 evolution capacity of the bundle sheath chloroplasts, the 'CO_2 pump' and resistances to O_2 and CO_2 diffusion can be used to explain the rate of glycolate synthesis and the extent of CO_2 recycling within the context of the 'classical' interpretation of the structure-function relationships in C4 plants (Ref. 3; cf. Refs. 13, 36, 41).

CONCLUSIONS

Consideration of the diffusive resistance to CO_2 and O_2 movement between the atmosphere and the site of enzyme activity is essential if the *in vivo* activity of RuBPc-o is to be understood. In terrestial C3 plants the enzyme operates *in vivo* under conditions even less conducive to a high rate of carboxylation and a low rate of oxygenation than occurs *in vitro* in air-equilibrated solution. This is because, in the dilute *in vitro* solutions of RuBPc-o, little depletion of CO_2 and accumulation of O_2 occurs. *In vitro* the enzyme is much more concentrated relative to the diffusive supply of CO_2 and removal of O_2. In order to supply the CO_2 and remove the O_2 in steady-state photosynthesis, concentration gradients of CO_2 and O_2 are built up, altering the air-equilibrium concentrations in solution of 250 μM O_2 and 10 μM CO_2 to (say) 253 μM O_2 and 7 μM CO_2, and decreasing the RuBPc activity while increasing that of RuBPo.

In evolutionary terms it is likely that RuBPc-o evolved, and became the cornerstone of autotrophic CO_2 assimilation, when the earth's atmosphere and oceans were essentially anoxic. Under these conditions the RuBPo activity would be an insignificant encumbrance.[73] By the time photosynthetic O_2 production had caused an accumulation of sufficient O_2 for the RuBPo reaction to become significant, RuBPc and the related reactions had become such an important part of photosynthetic metabolism that no alternative primary carboxylation could compete with the increasingly inefficient RuBPc-o. The CO_2/O_2 ratio was probably already low enough at the time of the evolution of the first land plants for RuBPo to be a serious problem for the early land plants in the Silurian era.[64] These C3 plants and their C3 successors could not abolish RuBPo activity and glycolate synthesis, and merely scavenged the glycolate back to more useful compounds. These plants had a further constraint on their rate of CO_2 fixation on a leaf cell area basis due to the large size and low activity of RuBPc-o which restricts the number of enzyme molecules/cell area which can be supplied with CO_2 and have O_2 removed efficiently by diffusion.

Relatively recently in terms of the vascular land plants (less than 100 million years ago out of their history of 400 million years), a wide range of angiosperm families evolved, in parallel, a CO_2 pump which yields a high CO_2 concentration at the site of RuBPc-o activity through an auxiliary carboxylation catalysed by PEPc (which is O_2-insensitive and has a larger turnover number than RuBPc-o). If this mechanism is to overcome the apparent inadequacy of RuBPc-o, the CO_2 pumped into the bundle sheath cells must not leak out to the atmosphere very rapidly. However, the diffusion barriers which prevent this also reduce the rate of O_2 leakage, and O_2 also increases in bundle sheath cells with an active photosystem two. Some C4 plants tolerate this, while others have most of their O_2 production in mesophyll cells rather than bundle sheath cells. Thus CO_2 and O_2 diffusion must be taken into account in analysing the C4 pathway as well as the C3 pathway which, teleologically, the C4 pathway is making more efficient under at least some envrionmental conditions.

Acknowledgement

Dr. S.M. Glidewell has contributed much useful discussion, and Dr. M. Lewandovska commented on an early draft of the manuscript.

REFERENCES

1. BADGER, M.R. and ANDREWS, T.J. Effects of CO_2, O_2 and temperature on a high-affinity form of ribulose diphosphate carboxylase-oxygenase from spinach. *Biochem. biophys. Res. Commun.* **60** (1974) 204-210.

2. BJORKMANN, O. and EHLERINGER, J. Comparison of the quantum yields for CO_2 uptake in C_3 and C_4 plants. *Carnegie Inst. Wash. Yb.* **74** (1975) 760-761.

3. BLACK, C.C. Jr. Photosynthetic carbon assimilation in relation to net carbon fixation. *Annu. Rev. Plant Physiol.* **24** (1973) 233-286.

4. BOWES, G. and OGREN, W.L. Oxygen inhibition and other properties of soybean ribulose diphosphate carboxylase. *J. biol. Chem.* **247** (1972) 2171-2176.

5. BOWLING, D.J.F. Measurement of a gradient of oxygen partial pressure across the instant root. *Planta* **111** (1973) 323-328.

6. BRIGGS, G.E. and ROBERTSON, R.N. Diffusion and absorption in disks of plant tissue. *New Phytol.* **47** (1948) 265-283.

7. BUERK, D.G. and LONGMUIR, I.S. Evidence for nonclassical respiratory activity from oxygen gradient measurements in tissue slices. *Fedn. Proc.* **35** (1976) 527.

8. CHAMPIGNY, M.L. and MOYSE, A. Effects of temperature on spinach chloroplast carboxylation activity. *Biochem. Physiol. Pflanz.* **168** (1975) 575-583.

9. CHAPMAN, E.A. and OSMOND, C.B. The effect of light on the tricarboxylic acid cycle in green leaves — III. A comparison between some C_3 and C_4 plants. *Pl. Physiol.* **53** (1974) 893-898.

10. CHEVILLOTTE, P. Relation between the reaction cytochrome oxidase-oxygen and oxygen uptake in cells *in vivo*. The role of diffusion. *J. theor. Biol.* **39** (1973) 277-295.

11. CLEVER, M.L. and BATTING, R. The solubility of gases in liquids. In DACK, H.R.J. (ed.) *Techniques of Chemistry*, Vol. VIII, *Solutions and solubilities*, Part I. Wiley-Interscience, New York (1975) pp. 379-441.

12. CODD, G.A. and STEWART, W.D.P. Polyhedral bodies and ribulose diphosphate carboxylase of the blue-green alga *Anabaena cylindrica*. *Planta* **130** (1976) 323-326.

13. COOMBS, J.W. Interactions between chloroplasts and cytoplasm in C_4 plants. In BARBER, J. (Ed.) *The Intact Chloroplast.* Elsevier, Amsterdam (1976) pp. 279-313.

14. ERDMAN, W. and CLARK, D.K. Oxygen supply and related neurogenic cell function of *Aplysia californica* neurones. *Fedn. Proc.* **35** (1976) 830.

15. FINDENEGG, G.R. Beziehungen zwischen carboanhydraseaktivitat und aufnahme von HCO_3^- und Cl^- bei den Photosynthese von *Scenedesmus obliquus. Planta* **116** (1974) 123-131.

16. FINDENEGG, G.R. Estimation of bicarbonate fluxes in *Scenedesmus obliquus.* In THELLIER, M., MONNIER, A., DEMARTY, M. and DAINTY, J. (eds.) *Echanges ioniques transmembranaires chez les vegetaux.* C.N.R.S./ Publications de l'Université de Rouen (1977) 275-281.

17. FISCHKOFF, S. and VANDERKOOI, J.M. Oxygen diffusion in biblogical and artificial membranes determined by the fluorochrome pyrene. *J. gen. Physiol.* **65** (1975) 663-676.

18. FORSTER, R.E. The rate of CO_2 equilibration between red cells and plasma. In FORSTER, R.E., EDSALL, J.T., OTIS, A.B. and ROUGHTON, F.J.W. (Eds.) *CO_2: chemical and physiological aspects.* NASA SP-188, Washington D.C. (1969) pp. 275–286.

19. GLIDEWELL, S.M. and RAVEN, J.A. Measurement of simultaneous oxygen evolution and uptake in *Hydrodictyon africanum. J. exp. Bot.* **26** (1975) 479–488.

20. GLIDEWELL, S.M. and RAVEN, J.A. Photorespiration: ribulose diphosphate oxygenase or hydrogen peroxide? *J. exp. Bot.* **27** (1976) 200–204.

21. GOLD, H.J. Comments on the letter of A.N. Stokes on rapid oxygen entry into unperfused tissue slices. *J. Theoret. Biol.* **60** (1976) 477.

22. GROSS, G. and MOLL, W. The diffusion of CO_2 in erythrocytes and haemoglobin solutions. *Pflügers Arch.* **324** (1971) 249–266.

23. GROSS, G. and MOLL, W. Facilitated diffusion of CO_2 across albumin solutions. *J. gen. Physiol.* **64** (1974) 356–371.

24. GROSS, G., MOLL, W., HOPPE, H. and GROSS, H. Proton transport by phosphate diffusion: a mechanism of facilitated CO_2 transfer. *J. gen. Physiol.* **67** (1976) 773–790.

25. GUNNING, B.E.S., STEER, M.W. and COCHRANE, M.P. Occurence, molecular structure and induced formation of the 'stromacentre' in higher plants. *J. cell. Sci.* **3** (1968) 445–456.

26. GUTIERREZ, M., GRACEN, V.E. and EDWARDS, G.E. Biochemical and cytological relationships in C_4 plants. *Planta* **119** 279–300.

27. HATCH, M.D., and KAGAWA, T. Photosynthetic activities of isolated bundle sheath cells in relation to differing mechanisms of C_4 photosynthesis. *Arch. biochem. Biophys.* **175** (1976) 39–53.

28. HATCH, M.D. and OSMOND, C.B. Compartmentation and transport in C_4 photosynthesis. In STOCKING, C.R. and HEBER, U. (eds.) *Encyclopedia of Plant Physiology,* New Series, Vol. III, Springer-Verlag, Berlin (1976) 144–184.

29. HATCH, M.D. and SLACK, C.R. Photosynthetic CO_2-fixation pathways. *A. Rev. Pl. Physiol.* **21** (1970) 141–162.

30. HATCH, M.D., KAGAWA, T. and CRAIG, S. Subdivisions of C_4-pathway species based on differing C_4 acid decarboxylating systems and ultrastructural features. *Austr. J. Pl. Physiol.* **2** (1975) 111–128.

31. HATRICK, A.A. and BOWLING, D.J.F. A study of the relationship between root and shoot metabolism. *J. exp. Bot.* **24** (1973) 607–613.

32. JACOBSEN, B.S., FONG, F. and HEATH, R.L. Carbonic anhydrase of spinach. Studies on its location, inhibition and physiological function. *Pl. Physiol.* **55** (1975) 468–474.

33. KAWASHIMA, N. and WILDMAN, S.G. Fraction 1 protein. *Annu. Plant Physiol.* **21** (1970) 325–358.

34. KAWSHIRO, T. and SCHELD, P. Measurement of Krogh's diffusion constant of CO_2 in respiring muscle at various CO_2 levels: evidence for facilitated diffusion. *Pflügers Archiv.* **362** (1976) 127–134.

35. KAWASHIRO, T., NUSSE, W. and SCHELD, P. Determination of diffusivity of O_2 and CO_2 in respiring tissue: results in rat skeletal muscle. *Pflügers Archiv.* **359** (1975) 231–251.

36. KENNEDY, R.A. Photorespiration in C_3 and C_4 plant tissue cultures. Significance of Kranz anatomy to low photorespiration in C_4 plants. *Pl. Physiol.* **58** (1976) 573–576.

37. KONING, J.De., VAN HAREN, R., HOOFD, L.J.C. and KREUZER, F. Experimental evidence for facilitation of oxygen diffusion by myoglobin in respiring chicken gizzard muscle. *Fedn. Proc.* **35** (1976) 831.

38. KU, S.B. and EDWARDS, G.E. Photosynthesis in mesophyll protoplasts and bundle sheath cells of various types of C_4 plants — IV. Enzymes of respiratory metabolism and energy utilizing enzymes of photosynthetic pathways. *Z. Pflanzenphysiol.* **77** (1975) 16–32.

39. KU, S.B., GUTIERREZ, M., KANAI, R. and EDWARDS, G.E. Photosynthesis in mesophyll protoplasts and bundle sheath cells of C_4 plants — II. Chlorophyll and Hill reaction studies. *Z. Pflanzenphysiol.* **72** 320–337.

40. KUIPER, P.J.C. Temperature dependence of photosynthesis of bean plants as affected by decenylsuccinic acid. *Pl. Physiol.* **40** (1965) 915–918.

41. LAETSCH, W.M. The C_4 syndrome: a structural analysis. *Annu. Rev. Plant Physiol.* **25** (1974) 27–52.

42. LAING, W.A., OGREN, W.L. and HAGEMANN, R.H. Regulation of soybean net photosynthetic CO_2 fixation by the interaction of CO_2, O_2 and ribulose-1, 5-diphosphate carboxylase. *Pl. Physiol.* **54** (1974) 678–685.

43. LILLEY, R.McC. and WALKER, D.A. Carbon dioxide assimilation by leaves, isolated chloroplasts and ribulose bisphosphate carboxylase from spinach. *Pl. Physiol.* **55** (1975) 1087–1092.

44. LONGMUIR, I.S. Reply to a letter to the Editor by A.N. Stokes. *J. Theor. Biol.* **60** (1976) 479–480.

45. LONGMUIR, I.S. and PASHKO, L. The rate of induction of cytochrome P-450 synthesis in hypoxia. *Fedn. Proc.* **35** (1976) 524.

46. LONGMUIR, I.S., SUN, S. and SOUCIE, W. Possible role of cytochrome P450 as a tissue oxygen carrier. In KING, T.E., MASON, H.S. and MORRISON, M. (Eds.) *Oxidases and related redox systems, Proc. Int. Symp., 2nd,* Vol. 2. University Park Press, Baltimore,Md (1973) 451–461.

47. LONGMUIR, I.S., MARTIN, D.C., GOLD, M.J. and SUN, S. Non-classical respiratory activity of tissue slices. *Microvascular Res.* **3** (1971) 125–141.

48. LUCAS, W.J. Photosynthetic fixation of ^{14}Carbon by internodal cells of *Chara corallina. J. exp. Bot.* **26** (1975) 331–346.

49. LUCAS, W.J. The influence of Ca^{2+} and K^+ on $H^{14}CO_3^-$ influx in internodal cells of *Chara corallina. J. exp. Bot.* **27** (1976) 32–42.

50. MAHON, J.D., FOCK, B. and CANVIN, D.T. Changes in specific radioactivity of sunflower leaf metabolites during photosynthesis in $^{14}CO_2$ and $^{12}CO_2$ at normal and low O_2. *Planta* **120** (1974a) 125–134.

51. MAHON, J.D., FOCK, B. and CANVIN, D.T. Changes in specific radioactivity of sunflower leaf metabolites during photosynthesis in $^{14}CO_2$ and $^{12}CO_2$ at three concentrations of CO_2. *Planta* **120** (1974*b*) 245–254.

52. MAHON, J.D., FOCK, M., MAHLER, T. and CANVIN, D.T. Changes in specific activity of sunflower leaf metabolites during photosynthesis in $^{14}CO_2$ and $^{12}CO_2$ at normal and low oxygen. *Planta* **120** (1974*c*) 113–124.

53. MAYNE, B.C., DEE, A.M. and EDWARDS, G.E. Photosynthesis in mesophyll cell protoplasts and bundle sheath cells of various types of C_4 plants – III. Fluorescence emission spectra, delayed light emission and P_{700} content. *Z. Pflanzenphysiol.* **74** (1975) 275–291.

54. NOBEL, P.S. *An introduction to Biophysical Plant Physiology*. W.H. Freeman, San Francisco (1974).

55. OSMOND, C.B. The absence of photorespiration in C_4 plants: real or apparent? In HATCH, M.D., OSMOND, C.B. and SLATYER, R.C. (Eds.) *Photosynthesis and photorespiration.* Wiley, New York (1971) pp. 472–482.

56. OSMOND' C.B. Carbon reduction and photosystem two deficiency in leaves of C_4 plants. *Aust. J. Plant Physiol.* **1** (1974) 41–50.

57. OSMOND, C.B. and SMITH, F.A. Symplastic transport of metabolites during C_4 photosynthesis. In GUNNING, B.E.S. and ROBARDS, A.W. (Eds.) *Intercellular Communication in Plants: Studies in Plasmodesmata.* Springer-Verlag, Berlin (1976) pp. 229–242.

58. POINCELOT, R.P. Uptake of bicarbonate ion in darkness by isolated chloroplast envelope membranes and intact chloroplasts of spinach. *Pl. Physiol.* **54** (1974) 520–526.

59. RACKHAM, O. Radiation, transpiration and growth in a woodland annual. In BAINBRIDGE, R., EVANS, G.C. and RACKHAM, O. (Eds.) *Light as an Ecological Factor.* Blackwell Scientific Publications, Oxford (1968) pp. 167–185.

60. RAVEN, J.A. Exogenous inorganic carbon sources in plant photosynthesis. *Biol. Rev.* **45** (1970) 167–221.

61. RAVEN, J.A. Endogenous inorganic carbon sources in plant photosynthesis — I. Occurrence of the dark respiratory pathways in illuminated green cells. *New Phytol.* **71** (1972*a*) 227–247.

62. RAVEN, J.A. Endogenous inorganic carbon sources in plant photosynthesis — II. Comparison of total CO_2 production in the light with measured CO_2 evolution in the light. *New Phytol.* **71** (1972*b*) 995–1014.

63. RAVEN, J.A. Division of labour between chloroplasts and cytoplasm. In BARBER J. (Ed.) The Intact Chloroplast. Elsevier, Amsterdam (1976) pp. 403–443

64. RAVEN, J.A. The evolution of vascular land plants in relation to supracellular transport processes. *Adv. Bot. Res.* **5** (1977*a*) 153–219.

65. RAVEN, J.A. Regulation of solute transport at the cell level. *Symp. Soc. Exp. Biol.* **31** (1977*b*) 73–99.

66. RAVEN, J.A. and GLIDEWELL, S.M. Photosynthesis, respiration and growth in the shade alga *Hydrodictyon africanum. Photosynthetica* **9** (1975) 361–371.

67. RAVEN, J.A. and GLIDEWELL, S.M. C_4 characteristics of chlorophyte photosynthesis. *Abstr. 4th Int. Cong. Photosynthetic Rsch, Reading, September 1977* (1977) 311.

68. SAMISH, Y.B. The rate of photorespiration as measured by means of O_2 uptake and its respiratory quotient. *Pl. Physiol.* **48** (1971) 345–348.

69. SAMISH, Y.B. Oxygen build-up in photosynthesising leaves and canopies is small. *Photosynthetica* **9** (1975) 372–375.

70. SCHELD, P., KAWASHIRO, T. and PUPER, J. Facilitated diffusion of CO_2 in rat skeletal muscle. *Fedn. Proc.* **35** (1976) 828.

71. SILVERMAN, D.N., TU, C. and WYNNS, G.C. Depletion of ^{18}O from $C^{18}O_2$ in erythrocyte suspensions. The permeability of the erythrocyte membrane to CO_2. *J. biol. Chem.* **251** (1976) 4428–4435.

72. SLABAS, A.R. and WALKER, D.A. Enzymic reconstitution of photosynthetic carbon assimilation. Pentose phosphate-dependent O_2 evolution by illuminated envelop-free chloroplasts from *Spinacia oleracea. Archs. Biochem. Biophys.* **175** (1976) 590–597.

73. SMITH, B.N. Evolution of C_4 photosynthesis in response to changes in carbon and oxygen concentrations in the atmosphere through time. *Biosystems* **8** (1976) 24–32.

74. SMITH, F.A. Transport of solutes during C_4 photosynthesis. In HATCH, M.D., OSMOND, C.B. and SLATYER, R.O. (Eds.) *Photosynthesis and Photorespiration.* Wiley, New York (1971) pp. 302–306.

75. STEEMAN–NIELSEN, E. Photosynthesis of aquatic plants with special reference to carbon sources. *Dansk. Bot. Arch.* **12**, No. 8 (1947) 1–71.

76. STOKES, A.N. Rapid oxygen entry into unperfused tissue slices: an experimental artifact? *J. Theor. Biol.* **60** (1976) 473–475.

77. SUCHDEO, S.R. and SCHULTZ, J.S. Mass transfer of CO_2 across membranes: facilitation by the presence of bicarbonate ion and the enzyme carbonic anhydrase. *Biochim. biophys. Acta* **352** (1974) 412–440.

78. WALKER, D.A. Chloroplast and cell — the movement of certain key substances, etc. across the chloroplast envelope. In NORTHCOTE, D.H. (Ed.) *MTP International Review of Sciences, Biochemistry Series One,* Vol. 11, *Plant Biochemistry.* Butterworths, London (1975) pp. 1–50.

79. WALKER, D.A. CO_2 fixation by intact chloroplasts: photosynthetic induction and its relation to transport phenomena and control mechanisms. In BARBER, J. (Ed.) *The Intact Chloroplast.* Elsevier, Amsterdam (1976) pp. 235–278.

80. WERDAN, K. and HELDT, H.W. Accumulation of bicarbonate in intact chloroplasts following a pH gradient. *Biochim. biophys. Acta* **283** (1972) 430–441.

81. WITTENBERG, B.A., WITTENBERG, J.B. and CALDWELL, P.R.B. Role of myoglobin in the oxygen supply to red skeletal muscle. *J. biol. Chem.* **250** (1975) 9038–9043.

ADDENDUM

The approach used in this commentary has been developed further in a number of review articles.[82, 83, 97, 100, 102] The view that RuBPo is the predominant source of glycolate for the photorespiratory carbon oxidation cycle[94, 95] is strongly supported by recent work involving $^{18}O_2$ feeding to illuminated leaves and chloroplasts[84, 100] and the isolation and characterization of mutants of *Arabidopsis* deficient in phosphoglycolate phosphatase.[103]

An important recent contribution is that of Wong *et al.*[107] who showed that a number of treatments which alter the rate of net photosynthesis in C_3 and C_4 plants via effects on the biochemical capabilities of the leaf induce parallel changes in stomatal conductance; this phenomenon is relevant to hypotheses of optimization of the ratio of carbon gain in net photosynthesis to water loss in transpiration.[87, 88] The phenomenon[107] involves relative constancy of the intercellular CO_2 level in the leaf during photosynthesis under a variety of conditions, and can account quantitatively for the magnitude and constancy of C isotope discrimination in C_3 vascular land plants (Farquhar, unpublished; Raven, unpublished). Farquhar's model takes the $\delta^{13}C$ value of products of RuBPc activity *in vitro* (i.e. free from diffusive constraints) and the differences in gas-phase diffusivity of $^{12}CO_2$ and $^{13}CO_2$ to predict successfully the $\delta^{13}C$ of plants with measured ratios of stomatal conductance to net photosynthesis.

This internally consistent body of data has important repercussions for the analysis of RuBPc-o function. One important conclusion is that the ratio of $[O_2]$ to $[CO_2]$ at the site of RuBPc-o is held nearly constant despite variations in the rate of photosynthesis, which means that the ratio v_o/v_c (equation (4)) is also nearly constant. Another point is that the $\delta^{13}C$ values can be accounted for with the variable (stomatal) fraction of the total diffusion resistance being the major portion of this resistance. This implies that the boundary-layer (gas phase) and mesophyll transport (liquid phase) resistances are small relative to the total transport resistance. Both of these non-stomatal transport resistances are, in part, under the long-term control of the plant. The boundary-layer resistance is a function of the overall plant geometry (which determines the fraction of free air movement which the leaves experience) and leaf size and shape (which determines the effect of air movements on leaf gas exchange). The liquid phase diffusion resistances are related to the balance between photosynthetic capacity on an external leaf area basis and the ratio of internal to external leaf area.[100]

The stomatal resistance in homoiohydric plants can thus be seen to reduce the fraction of the biochemical potential for CO_2 fixation to some 0.7 or so of that which would be achieved with infinite stomatal conductance. This incomplete use of the biochemical potential may be construed as a "Cost of Homoiohydry".[64] The author[98, 99] has attempted to analyse the non-stomatal (poikilohydric) terrestrial plants (free-living and lichenized algae, and the gametophytes of bryophytes and pteridophytes) with respect to photosynthetic conductances. This analysis is summarized below, with the help of a few key references.

All of the non-stomatal land plants seem to use RuBPc as their initial carboxylase.[90, 106] Bryophytes are "good" C_3 plants in terms of net CO_2 exchange in the light as a function of $[CO_2]$ and $[O_2]$, and of their $\delta^{13}C$ (refs. 85, 89, 105, 106). Lichenized algae are less readily categorized as thoroughgoing C_3 plants, and may retain some aspects of the "CO_2 concentrating mechanism" of their algal ancestors.[98]

The $\delta^{13}C$ values for terrestrial bryophytes[105] are at the negative extreme for terrestrial C_3 plants, i.e. are close to the value for RuBPc fixing atmospheric CO_2 *in vitro* with no diffusive limitation. A part of this large, negative $\delta^{13}C$ value may be attributed to the relatively negative $\delta^{13}C$ value for the CO_2 fixed by bryophytes in their natural habitat. Bryophyte photosynthesis in the boundary layer of air may be unable to offset soil respiration, with the result that the air available to the bryophytes is depleted in $^{13}CO_2$ relative to the bulk atmosphere. However, it would appear that the high, negative $\delta^{13}C$ of terrestrial bryophytes does in part reflect a lower sum of diffusive resistances relative to carboxylation resistance in the bryophytes than in the homoiohydric plants (provided comparable bases for expression are used[93]).

In the homoiohydric land plant the diffusion resistances include the gas-phase boundary-layer resistance, the stomatal resistance and the liquid-phase transport resistance. In the gametophytes of bryophytes the stomatal resistance is absent, and the total diffusion resistance is lower, but there are grounds for believing that both the gas-phase boundary-layer resistance and the liquid-phase transport resistance can exceed those in homoiohydric land plants. Boundary-layer resistances are increased due to the small stature of the plants; they are within the boundary layer of the soil, rock or vascular plant on which they are growing.[91] The liquid-phase resistances may be increased relative to those of homoiohydric plants by surface films of water outside the cell walls, and by thicker cuticular layers than are found on the mesophyll cells of homoiohydric plants,[64, 104] which impede the movement of CO_2 and O_2 as well as H_2O. To the extent that both water-repellancy and water-proofing are functions of the extent to which the cell wall is impregnated with cutin or suberin[64] the two contributions to increased liquid-phase transport resistance are mutually exclusive.

These considerations suggest that bryophytes do not benefit as much as one might at first sight expect from their lack of stomatal resistance to gas transport; they probably cannot use more than 0.8 of their biochemical capacity for photosynthesis (based on analysis of the $\delta^{13}C$ values of Troughton[105]). There is the further restriction that a non-homoiohydric plant may spend a greater fraction of its time not using its photosynthetic apparatus at all, a factor which $\delta^{13}C$ measurements do not bear upon. The extent to which diffusive resistances (whose magnitude is essentially invariant over the time scale of stomatal responses in many homoiohydric plants) in various bryophytes represent optimal values for maximizing the carbon fixed per unit of water lost under specific environmental conditions remains to be investigated.

Turning to lichens, Collins and Farrar[86] have attempted a "resistance analysis" of photosynthesis in the lichen *Xanthoria parietina.* Unfortunately, their analysis is flawed by their taking the CO_2 concentration in the intercellular spaces during steady-state photosynthesis as being equal to the concentration at the CO_2 compensation point. The CO_2 compensation point in *Xanthoria*[86] was higher than that found in most terrestrial C_3 plants or in various species of the lichen genus *Umbilicaria.*[92] The $\delta^{13}C$ values obtained for a number of lichens[101] are more positive than those for other terrestrial C_3 plants, and suggest that diffusion resistances are relatively high (although the authors[101] prefer a biochemical explanation for the values of $\delta^{13}C$ and their variation with environmental conditions). The variations in $\delta^{13}C$ with water status are consistent with a high diffusion resistance in *dry* rather than *wet* thalli, i.e. a decreased gas-phase conductance as the thallus dries out rather than an increased conductance due to the loss of surface films of water. We may tentatively conclude that diffusion resistances in lichens means that 0.8 or less of the biochemical potential for photosynthesis by the phycobiont can be used in the lichen.

The general conclusion to be drawn from this analysis of poikilohydric land plants is that they have "Costs" in terms of transport resistances which can equal the "Cost of Homoiohydry" (i.e. the stomatal resistance). The implication for RuBPc-o functioning *in vivo* is that the fraction of the total RuBPc activity which can be used, and the ratio v_o/v_c (equation (4)) may be quite similar in poikilohydric and in homoiohydric C_3 land plants.

A final point on the function of RuBPc and RuBPo in land plants is their possible role in the avoidance of photoinhibition. Photoinhibition occurs when the rate at which the photosynthetic apparatus absorbs light energy exceeds the rate at which the various energy-transforming and detoxifying mechanisms can dissipate the products of chlorophyll excitation.[94-96] In homoiohydric C_3 land plants there is evidence for a role for RuBPo and the photosynthetic carbon oxidation cycle in dissipating absorbed light energy under conditions of high light energy input and (ref. 107 notwithstanding) very restricted exogenous CO_2 supply due to a large decrease in stomatal conductance relative to biochemical capacity.[94-96] In C_4 and CAM plants RuBPo probably plays a less central role in the avoidance of photoinhibition under conditions of high irradiance and relatively low stomatal conductance than is the case for C_3 vascular plants.[94-96]

RuBPo thus seems to be an important part of the mechanism for dissipating potentially harmful excess excitation energy in homoiohydric land plants. However, it is not possible to use this as an argument that RuBPo has always fulfilled this selectively advantageous role in land plants. In extant bryophyte gametophytes, and in the astomatous terrestrial ancestors of homoiohydric plants, there is no obvious way of reducing the $[CO_2]/[O_2]$ ratio perceived by RuBPc-o under natural conditions as to make RuBPo and the photosynthetic carbon oxidation cycle a major protection against photoinhibition. It is thus unlikely that the avoidance of photoinhibition was the selective role for RuBPo during the period between the build-up of substantial ambient O_2 levels and the evolution of homoiohydry; its retention during this early phase of the evolution of homoiohydry in plants living on land, and in extant bryophytes, may be related to the difficulties in modifying so central a photosynthetic enzyme. Thus while a role in avoiding photoinhibition may confer selective advantage on RuBPo in certain plants, this role cannot explain its ubiquity in autotrophs using RuBPo. Poikilohydric plants are of interest in the context of photoinhibition in that many of them retain chlorophyll during desiccation, and may thus be at risk from photoinhibition when desiccation has prevented the activity of RuBPc-o and the photosynthetic carbon reduction and photorespiratory carbon oxidation cycles as energy sinks for excitation energy. RuBPo activity clearly cannot have a protective role here.

ACKNOWLEDGEMENT

This Addendum has benefited greatly from discussions with Dr. G.D. Farquhar, Dr. K. Winter, Professor C.B. Osmond and Dr. S.B. Powles (Australian National University) and Dr. M.C.W. Evans (University College, London).

REFERENCES

82. BENEDICT, C.R. Nature of obligate autotrophy. *Ann. Rev. Pl. Physiol.* **29** (1978) 67–93.

83. BERRY, J.A. and FARQUHAR, G.D. The CO_2 concentrating function of C_4 photosynthesis: a biochemical model. HALL, D.O., COOMBS, J. and GOODWIN, T.W. (eds.) *Photosynthesis* **77**, Biochemical Society, London (1978) pp. 119–132.

84. BERRY, J.A., OSMOND, C.B. and LORIMER, G.H. Fixation of $^{18}O_2$ during photorespiration. Kinetic and steady-state studies of the photorespiratory carbon oxidation cycle with intact leaves and isolated chloroplasts of C_3 plants. *Pl. Physiol.* **62** (1978) 954–967.

85. BJÖRKMAN, O., GAUHL, E., HIESEY, W.M., NICHOLSON, F. and NOBS, M.A. Growth of *Mimulus, Marchantia* and *Zea* under different oxygen and carbon dioxide levels. *Carnegie Institution of Washington Year Book* **67** (1968) 477–478.

86. COLLINS, C.R. and FARRAR, J.F. Structural resistances to mass transfer in the lichen *Xanthoria parietina. New Phytol.* **81** (1978) 71–83.

87. COWAN, I.R. Stomatal behaviour and environment. *Adv. Bot. Res.* **4** (1977) 117–228.

88. COWAN, I.R. and FARGHAR, G.D. Stomatal function in relation to leaf metabolism and environment. *Symp. Soc. Exp. Biol.* **31** (1977) 471–501.

89. FOCK, H., KROTKOV, G. and CANVIN, D.T. Photorespiration in liverworts and leaves. In METZNER, H. (ed.) *Progress in Photosynthesis Research,* Vol. 1, I.U.B.S., Tubingen (1969) pp. 482–487.

90. HILL, D.J. The physiology of lichen symbiosis. In BROWN, D.H., HAWKSWORTH, D.L. and BAILEY, R.H. (eds.) *Lichenology: Progress and Problems.* Academic Press, London (1976) pp. 457–496.

91. HOFFMAN, G.R. and GATES, D.M. An energy budget approach to the study of water relations in cryptograms. *Bull. Torrey Bot. Club* **97** (1970) 361–366.

92. LARSON, D.W. Preliminary studies on the physiological ecology of *Umbilicaria* lichens. *Can. J. Bot.* **57** (1979) 1398–1406.

93. NOBEL, P.S. Internal leaf area and cellular CO_2 resistance: photosynthetic implications of variations with growth conditions and plant species. *Physiol. Pl.* **40** (1977) 137–144.

94. OSMOND, C.B. Integration of photosynthetic carbon metabolism during stress. In RAINS, D.W. and VALENTINE, R.C. (eds.) *Genetic Engineering of Osmoregulation: Impact on Plant Productivity for Food, Chemicals and Energy,* Plenum Publishing Corporation, New York (1980) in press.

95. OSMOND, C.B., WINTER, K. and POWLES, S.B. Adaptive significance of CO_2 cycling during photosynthesis in water stressed C_3, C_4 and CAM plants. In TURNER, N.C. and KRAMER, P.J. (eds.) *Plant Responses to Water and High Temperature Stress.* Wiley Interscience, New York (1980) in press.

96. POWLES, S.B. The role of carbon assimilation and photorespiration in the avoidance of photoinhibition. Ph.D. Thesis, A.N.U., Canberra (1979).

97. RAVEN, J.A. Photosynthesis in cells and tissues. In HALL, D.O., COOMBS, J. and GOODWIN, T.W. (eds.) *Photosynthesis 77*, Biochemical Society, London (1978) pp. 147–155.

98. RAVEN, J.A. Nutrient transport in microalgae. *Adv. Microb. Physiol.*, in press.

99. RAVEN, J.A. The cost of homoiohydry. In preparation.

100. RAVEN, J.A. and GLIDEWELL, S.M. Processes limiting photosynthetic conductance. In JOHNSON, C.D. (ed.) *Processes Limiting Plant Productivity*, Butterworths, London (1980) in press.

101. SHOMER-ILAN, A., NISSENBAUM, A., GALUN, M. and WAISEL, Y. Effect of water regime on carbon isotope composition of lichens. *Pl. Physiol.* **63** (1979) 201–205.

102. SMITH, F.A. and WALKER, N.A. Photosynthesis by aquatic plants: effects of unstirred layers in relation to assimilation of CO_2 and HCO_3^-, and to carbon isotope discrimination. *New Phytol.*, submitted for publication.

103. SOMERVILLE, C.R. and OGREN, W.L. A phosphoglycolate phosphatase-deficient mutant of *Arabidopsis. Nature* **280** (1979) 833–836.

104. TOBIESSEN, P.L., MOTT, K.A. and SLACK, N.G. A comparative study of photosynthesis, respiration and water relations in four species of epiphytic moss in relation to their vertical distribution. *Bryophytum Bibliotheca* **13** (1977) 253–277.

105. TROUGHTON, J.H. Aspects of the evolution of the photosynthetic carboxylation reaction in plants. In HATCH, M.D., OSMOND, C.B. and SLATYER, R.O. (eds.) *Photosynthesis and Photorespiration.* Wiley-Interscience, New York (1971) 124–129.

106. VALANNE, N., ARO, E.-M. and REPO, E. Changes in photosynthetic capacity and activity of RuBPc-ase and glycolate oxidase during the early growth of moss protonemata in continuous and rhythmic light. *Z. PflPhysiol.* **88** (1978) 123–131.

107. WONG, S.C., COWAN, I.R. and FARQHAR, G.D. Stomatal conductance correlates with photosynthetic capacity. *Nature* (1979) in press.

ACTINOMYCETE GENETICS AND ANTIBIOTICS

DAVID A. HOPWOOD

John Innes Institute, Norwich NR4 7UH

TOPICS COVERED:

THE NATURE OF THE ACTINOMYCETES: COMPLEX BACTERIA

The actinomycetes, one of the most important groups of microorganisms, are often misunderstood. Confusion arises because of their mycelial growth habit and because members of several genera differentiate into chains of dry, powdery spores ("conidia").[14] Thus the colonies of some actinomycetes resemble, superficially, those of certain fungi. Although some early microbiologists went so far as to group the actinomycetes with the fungi on these grounds, most recognised that the individual hyphal units were much smaller than those of fungi. They also realised that actinomycetes forming complex colonies consisting of coherent systems of branching hyphae were linked by a rather complete series of forms with more rudimentary, or more transient, mycelial growth to organisms like those now classified as mycobacteria, corynebacteria and arthrobacteria; in the latter organisms, the only tendency to hyphal development was the production of somewhat pleomorphic, occasionally branched, cells. The actinomycetes were therefore often thought to be in some sense "intermediate" between bacteria and fungi.

It is now clear that the correct taxonomic position for the actinomycetes is firmly amongst the bacteria. Any resemblance to the fungi represents a parallel evolution, presumably resulting from comparable selective pressures: the need to remain firmly attached to a food source which requires to be solubilised by the action of extracellular hydrolytic enzymes.[12] As for many fungi, the habitat is typically a soil environment where decomposing organisms are the food source. Certain genera of actinomycetes, like some fungi, invade animal tissues to cause diseases[14], such as tuberculosis (*Mycobacterium tuberculosis*) and leprosy (*M. leprae*) and a few others of lesser importance. In contrast to fungi, few actinomycetes cause plant diseases – *Streptomyces scabies*, the Potato Scab organism, is one – but some have nitrogen-fixing symbiotic relationships with the roots of non-leguminous plants.[2]

The fundamental resemblances of actinomycetes to other bacteria are many. They include cytological features, such as the absence of nuclear membranes (making the organisms prokaryotic by definition), the possession of ribosomes of the bacterial type, characteristic Gram-positive mucopeptide cell walls, and mesosomes but not mitochondria.[55] The organisms are hosts to viruses of typical bacteriophage morphology and other properties.[25] In their genetics, too, the actinomycetes resemble other bacteria. Like them, they have single circular chromosomes with a considerable degree of clustering of genes involved in the same sequence of biochemical reactions. They also possess plasmids which act as sex factors and carry genes concerned with various dispensible metabolic systems, just as do the plasmids of other bacteria.[10,40,46]

Although all the actinomycetes are prokaryotes they show strong indications of being polyphyletic: the mycelial growth habit which defines them has probably evolved more than once. This is suggested by the properties of a genus of thermophilic organisms, *Thermoactinomyces*, which includes some of those causing the allergic disease Farmer's Lung.[8] In other groups of actinomycetes the various asexual spore structures develop from the parent hyphae by processes of segmentation in which all the contents of the hyphae end up in the spores. In *Thermoactinomyces*, in contrast, true endospores develop inside mother cells just as they do in the eubacterial genera *Bacillus* and *Clostridium*; much of the cytoplasm of the mother cells then degenerates. The endospores of *Thermoactinomyces* are morphologically extremely similar to these eubacterial spores and, like them, contain dipicolinic acid and are extremely heat resistant, differing in these respects from other actinomycete spores. The DNA base composition of *Thermoactinomyces* (about 50% guanine plus cytosine) is fairly similar to that of bacilli (typically 40–50% G + C) and very different from that in most other actinomycetes (72–74% for *Streptomyces* spp.).

Moreover, mycelia of *Thermoactinomyces vulgaris* exchange genes as naked DNA in a typical transformation process[28], just as many bacilli do[59] whereas, as discussed below, the other groups of actinomycetes which have been studied recombine through a plasmid-mediated mating process comparable with conjugation in Gram-negative groups such as the enterobacteria and pseudomonads.

THE IMPORTANCE OF THE ACTINOMYCETES: ANTIBIOTICS AND OTHER USEFUL PRODUCTS

The actinomycetes are undoubtedly "important" in recycling carbon and nitrogen in the soil; as always in microbial ecology, however, their role is hard to quantify. It is presumably considerable since large numbers of actinomycetes are found in soil and composts.[36]

The actinomycetes are of direct importance to man because they are uniquely prolific as antibiotic producers, and are significant also in other areas of the fermentation industry, notably in the production of amino acids and some enzymes.

Of about 3000 chemically different antibiotics described by 1974, the actinomycetes accounted for over 2000, the remainder being split about 2:1 between the fungi and the eubacteria.[3] Thus, although half the world sales of antibiotics (estimated at about $\$4.2 \times 10^9$ per annum in 1974) originate from fungi, these are made up of 30 or so artificial derivatives of just two chemically related antibiotics, penicillin G and cephalosporin C. In contrast, the 70 or so commercially produced actinomycete antibiotics[44], which make up most of the other half of world sales, include a great variety of chemical structures. Examples are such clinically important compounds as the various tetracyclines, erythromycin, the aminoglycosides (streptomycin, neomycin, kanamycin and several others), chloramphenicol, lincomycin (all derived from *Streptomyces* spp.), rifamycins (from *Nocardia mediterranei*) and gentamicin (another aminoglycoside, from *Micromonospora purpurea*). Some further compounds which are invaluable research tools in molecular biology and biochemistry, because they perturb biosynthetic processes in specific ways, can also be added to the list: mitomycin C (DNA synthesis), actinomycin D and streptolydigin (different steps in transcription, along with rifamycin), puromycin (translation, along with chloramphenicol and the aminoglycosides) and cerulenin (fatty acid synthesis and related processes).

A further significant outlet for actinomycete antibiotics is as feed additives for livestock. For reasons which are not entirely clear, antibiotics lead to faster and more efficient weight gain. The practise of supplementing animal feeds with clinically useful antibiotics received a severe and justified setback, particularly in the United Kingdom, with the publication in 1969 of the Swann Report[53] which highlighted the role of feed additive antibiotics in the build-up of plasmid-borne and therefore transmissible drug resistance in human pathogenic bacteria. The correct response by the antibiotics industry was to develop, specifically as feed additives, antibiotics which are not useful clinically and which do not select for organisms cross-resistant to clinically valuable antibiotics, and here the very diversity of actinomycete products has been valuable; monensin is a notable example.[51] However, irresponsible promotion and use of certain antibiotics as feed additives continues.

Actinomycete antibiotics are also used in the treatment of some plant diseases. An interesting case is the injection of palms in Florida to control Lethal Yellowing Disease caused by a mycoplasma infection in the phloem.[37] The trees are not completely cured, but trunk injections a few times a year with oxytetracycline keep the palms reasonably healthy. Another mycoplasma, causing Western X-Disease of stone fruits in California, is similarly controlled with annual injections.[42] Economically more important, in Japan, is the control of the bacterial pathogen which causes Rice Blast by spraying with kasugamycin; and of *Alternaria kikuchiana* causing Black Spot of pear, and other fungal diseases, with polyoxin.[31]

As producers of amino acids, the genera *Corynebacterium, Brevibacterium* and some of their relatives are the most important in a huge area of industrial activity.[58] These "coryneform organisms" are grouped as relatives of the actinomycetes in Bergey's Manual[4] although how close they are phylogenetically to the typical mycelial actinomycetes is of course uncertain. In another sector of the fermentation industry, *Arthrobacter* (also in the coryneform group) and *Streptomyces* itself are important as producers of glucose isomerase. This is a large-volume product used in the corn syrup industry; it isomerises some of the glucose – produced from maize starch – to fructose, greatly increasing the sweetness of the syrup.

ANTIBIOTIC RESISTANCE: THE DESTRUCTION OF ANTIBIOTICS

In this section we explore the possibility that the actinomycetes have provided, not only many antibiotics which are valuable in fighting bacterial diseases, but also the means whereby the pathogenic bacteria fight back.

Not long after antibiotics came into widespread use, resistant pathogenic bacteria appeared and spread, sometimes rendering a previously valuable antibiotic useless in treating particular infections. At first, new mutations to antibiotic resistance were involved. However, from the late 1950's onwards, first in Japan and later in other parts of the world, a new mechanism for the acquisition of drug resistance was demonstrated: the "infectious" transfer of resistance genes from harmless bacteria to pathogens. The natural vectors of such genes were shown to be plasmids, some very similar to the *F*-factor which controls mating in *Escherichia coli* and its relatives. These new plasmids, called *R*-factors[19], could acquire different resistance genes successively in response to the selective pressures created by antibiotic use (and misuse) so that a single *R*-factor could confer resistance to six or more chemically unrelated antibiotics or synthetic antibacterial agents such as sulphonamides. How did this happen and where did these various resistance genes come from?

The mechanism which links different resistance genes together on the same plasmid has been clarified by recent exciting advances in bacterial genetics: many of these genes form parts of DNA segments called transposons[13, 34, 52] which can move from one site to another, on a bacterial chromosome, a bacteriophage or a plasmid, by mechanisms which do not depend on crossing-over between homologous DNA sequences. Thus an *R*-factor can pick up (or lose), successively, a series of resistance genes which may have originated quite separately: the more so because some *R*-factors can promote conjugation, and so transfer themselves, between bacteria belonging to different genera. The most promiscuous *R*-factors – plasmids of the so-called P1 incompatibility group – can be transferred into virtually any genus of Gram-negative bacteria. Such plasmids could have collected transposable DNA sequences coding for antibiotic resistances from a wide variety of different sources.

The ancestral home of the resistance genes now found in pathogenic bacteria has been the subject of much speculation. A particularly stimulating hypothesis[17] is that genes conferring resistance to actinomycete antibiotics arose in the actinomycetes themselves and were transferred on plasmids into human pathogens, probably via a series of other saprophytic bacteria. An organism producing an antibiotic must be resistant to it. The way in which antibiotic-producing actinomycetes tolerate their antibiotics is not established, and there may well be a multiplicity of mechanisms. However, it is suggestive that many actinomycetes produce enzymes which inactivate antibiotics of the aminoglycoside family by transferring phosphate, acetate or adenylate groups to them. They do not destroy the bulk of the antibiotic in the culture medium but probably act locally in the outer layers of the cell.[17] Enzymes with similar properties are determined by the plasmid-borne resistance genes of pathogenic bacteria and it was this correspondence which led to the theory of their origin. Later, chloramphenicol acetylating enzymes, like those coded by *R*-factors, were also found in several *Streptomyces* spp.[41, 49]; these differ from those acting on the aminoglycosides by rapidly inactivating the antibiotic in the culture medium. In the case of the aminoglycoside antibiotics, there was some correspondence between the antibiotic produced by a streptomycete and the specificity of an inactivating enzyme it might possess, although the correspondence was not perfect.[17] However, *S. venezuelae*, the species noted for chloramphenicol production, although very resistant to it, lacks the chloramphenicol acetylating enzyme; on the other hand, certain chloramphenicol non-producers possess it, without being markedly resistant.

Perhaps protection against one's own antibiotic is not the primary role of these enzymes. Instead, if the enzymes are indeed protective in actinomycetes, as they certainly are in clinical isolates, their target could often be the antibiotics produced by other actinomycetes. The genus *Streptomyces* contains hundreds of "species" of very similar growth form and properties, and producing many different antibiotics. The most serious competitors of a particular streptomycete are probably other streptomycetes. This may explain the widespread occurrence of the antibiotic-inactivating enzymes amongst the actinomycetes and it does not of course invalidate the theory that the corresponding genes may have become transferred to other groups of bacteria.

Very little is known about the genetic determination of antibiotic inactivating enzymes in actinomycetes. In the single published case, chloramphenicol acetylation is determined by a chromosomal, not a plasmid gene, in *Streptomyces acrimycini*.[56] However, this does not mean that such a gene could not have been picked up by a plasmid.

Not all antibiotic resistance genes code for inactivating enzymes. Tetracycline resistance is due to a protein which reduces the net uptake of tetracycline into the cell, thereby keeping the internal concentration below a toxic level[19], but this function might also operate in actinomycetes. Whatever the origin of antibiotic resistance genes, resistance to synthetic antibacterial agents, like sulphonamides or trimethoprim, presumably has a different explanation. However, β-lactamases which destroy the characteristic *fungal* antibiotics of the penicillin and cephalosporin families, could be added to the actinomycete list, since β-lactam antibiotics (cephamycins) and β-lactamases which destroy them are now being found in streptomycetes.[43, 50]

STREPTOMYCES GENETICS: CONJUGATION MEDIATED BY SEX FACTORS

Just as a particular isolate of *E. coli*, the K-12 strain, has been the subject of the vast majority of work on enterobacterial genetics, so the A3(2) strain of *Streptomyces coelicolor* has been the focus of most of the work in actinomycete genetics. However, because such a minute amount of effort has been applied to *S. coelicolor*, in comparison to *E. coli*, really detailed knowledge of most aspects of the molecular genetics of actinomycetes is lacking.

S. coelicolor has a single circular chromosome. It was the deduction of circularity from the results of genetical experiments[21], following its original unexpected discovery in *E. coli*[32] and in its relative *Salmonella typhimurium*, which suggested that chromosome circularity might be general amongst bacteria. More recently, circularity has been shown for *Bacillus subtilis*[60] and for species of *Rhizobium* and *Acinetobacter*[5, 35, 38, 54], tending to confirm the original prediction.

A special feature of the *S. coelicolor* linkage map is its symmetry: genes controlling different steps in the same biosynthetic pathway often lie diametrically opposite one another. A possible explanation[22] for this state of affairs (which has recently been seen also in *E. coli*[61]), is that crossing-over between two identical circular chromosomes occupying the same cellular compartment — and this is likely to be a common situation in *Streptomyces* because the space between adjacent septa in the mycelium almost certainly contains several genomes' equivalent of DNA[26] — would have led to a double sized (diploid) chromosome with every gene opposite its duplicate. The situation would have been subject to reversal by a further, internal, crossover and so some kind of equilibrium would have existed between haploid and diploid circles. However, as time went on and recessive lethal mutations tended to accumulate in heterozygous condition, the double circle would have become stabilised, since most classes of haploid would have contained a lethal mutation. Successive deletions of redundant DNA would then have occurred, and the genome would have approached its original size and haploid information content, though retaining some initially redundant DNA which could have evolved to code for new functions. Symmetry of the map would have remained because of the tendency, which is widespread in bacteria, for genes involved in the same sequence of events to be clustered on the chromosome. Thus, deletion of some genes of a cluster at one point on the map and others from the duplicate cluster would have left the type of symmetry that we observe. The theory has remained unmodified and untested in the ten years since it was proposed, but the prospect of gene cloning may allow an experimental test since some degree of residual base-sequence homology of opposite segments of the chromosome might be expected to remain.

We describe the process leading to genetic recombination in *S. coelicolor* as a "conjugation" because it requires contact between the prospective parent organisms, as in the well known process in *E. coli.* Moreover, plasmid sex factors are involved in mating in both organisms. In *E. coli*, a mating appendage, the sex pilus, is the major identified product of genes borne on the *F* plasmid and is essential for mating.[47] We do not know whether or not anything comparable is required for gene exchange in *S. coelicolor.*

In *S. coelicolor* two sex factors (SCP1 and SCP2) have been found which promote recombination.[7, 25] SCP1 has so far defied all attempts at physical isolation but the genetic evidence for its existence and behaviour is very considerable. SCP2, on the other hand, is defined both genetically and physically and is a plasmid of about 19×10^6 daltons.[7] SCP2 is recognised phenotypically purely in terms of its behaviour as a sex factor, whereas SCP1, in addition, carries genes controlling the biosynthesis of an antibiotic called methylenomycin A, and resistance to it.[57]

From the A3(2) "wild-type" containing both SCP1 and SCP2 autonomously (termed $SCP1^+$ $SCP2^+$), it has been possible, by recognising the various fertility levels of strains carrying the plasmids and their derivatives in different combinations, to isolate strains from it which have lost either or both plasmids. $SCP1^-$ $SCP2^-$ strains are not completely sterile amongst themselves:

recombinants appear in such "matings" at a frequency of 10^{-7}–10^{-8}. If SCP1 is present autonomously in one strain in a mixture, the frequency is about 10^{-5}, and slightly *higher* if it is in both parents. On the other hand, SCP2 autonomously in one parent gives a recombination frequency of about 10^{-5}, but this is *reduced* to about 10^{-6} if SCP2 is in both parental strains. Variants of SCP2, called SCP2*, promote an increased recombination frequency: about 10^{-3} when SCP2* is present in one or both parents. The fertility levels due to combinations of SCP1 with SCP2 or SCP2* are essentially additive, rather than multiplicative.[6]

There is so far no evidence of a direct physical interaction of SCP2 (or SCP2*) with the chromosome, but nor is there evidence against it. (At least a transient interaction may be implied by the comparatively high fertility of matings involving SCP2*.) On the other hand SCP1 certainly undergoes various interactions of this kind. Its integration into the chromosome at various positions gives strains which donate, at very high frequencies in matings with SCP1⁻ strains, chromosomal markers on one or both sides of the point of integration. It can also incorporate various chromosomal segments to become one of a series of SCP1′ factors. When they are transferred to SCP1⁻ strains, such SCP1′ factors render the cell functionally diploid for genes on the inserted segment.[29]

Almost 100% of a parental SCP1⁻ strain in a mating mixture with SCP1⁺, SCP1′ or integrated-SCP1 acquires the plasmid during the course of the mixed culture – a period of three or four days of growth. An unknown proportion of these transfers are from newly infected recipients to further recipients rather than direct from the original donor strain. The same applies in SCP2⁺ X SCP2⁻ or SCP2* X SCP2⁻ matings. When integrated-SCP1 strains are mated with SCP1⁻, pieces of the donor chromosome transferred with SCP1 have a high probability of recombining with the resident SCP1⁻ chromosome, so that nearly 100% of the progeny in such matings are recombinants.[25]

Although there is a wide taxonomic gulf between *E. coli* and *S. coelicolor*, and a difference in composition of the outermost layers of the cell wall which might be expected to have an important bearing on mating processes (*E. coli*, as a Gram-negative organism, has a lipoprotein "outer membrane" external to the rigid mucopeptide layer which all walled bacteria possess), the presence of plasmid sex factors in both organisms may imply a close resemblance in conjugation mechanisms. What are the main *apparent* resemblances and differences?

The behaviour of SCP1′ and F′ strains are closely similar[29], but it could be argued that this is simply a consequence of having, in the same cell, a complete chromosome and a plasmid carrying a chromosomal segment inserted into it; it need not imply a similar means of plasmid transfer *between* cells.

In *E. coli*, many sex factors repress their own transfer functions so that an established plasmid-carrying culture has poor donor ability, whereas cells that have newly received a sex factor by mating are transiently de-repressed and highly transfer proficient. Constitutively de-repressed mutants can be isolated, and the naturally de-repressed *F*-factor may be a wild example of such a mutant.[39] Although it is more difficult to analyse the situation in *S. coelicolor* because the mycelial growth habit rules out simple kinetic experiments, there is evidence of self-repression of the sex factor activity of SCP2; and SCP2* may represent a de-repressed mutant.[6]

The various donor types of *S. coelicolor*, involving integrations of SCP1, certainly show differences in behaviour from the Hfr state which arises in *E. coli* when *F* integrates into the chromosome, but it may be possible to reconcile at least some of the differences.

It is now known that the *F*-factor carries a small number of specific DNA "insertion sequences", about 1400 or about 5000 base pairs long, in a particular short segment of the circular plasmid molecule. These sequences can undergo reciprocal recombination with identical sequences on the host chromosome, resulting in co-integration of the two circular DNA molecules, chromosome and plasmid, to give the Hfr state.[16,52] When mating with an F^- cell occurs, DNA is transferred, starting at a special *origin* within the *F*-factor and proceeding in a particular direction from this origin. The origin is not adjacent to the region of the *F*-factor involved in integration into the host chromosome. Therefore, as conjugation proceeds, first a leading segment of *F* DNA is transferred, then chromosomal genes progressing around the circular chromosome from the point of plasmid integration, and ending with the remaining segment of the *F*-factor back to the origin. Since the unions between mating cells are fragile, DNA transfer is usually interrupted long before the whole chromosome is transferred, and so there is a gradient of transmission of donor genes to the progeny, falling from high values at one side of the point of integration of the *F*-factor in any particular Hfr strain to very low values at the tail end of the chromosome. Moreover, since only those very rare progeny which inherit the whole donor chromosome and the "tail" of *F* as well as the "head" can be donors, nearly all the recombinant progeny are found to be F^- recipients.

In sharp contrast, the integrated-SCP1 donors of *S. coelicolor*[29] yield recombinant progeny which are all donors like the parent, even when they inherit only donor markers near to the point of SCP1 integration. On the face of it, this seems an obvious difference from *E. coli*. However, the difference could be reconciled by postulating that SCP1 carries a transfer origin *adjacent* to the point of interaction with the chromosome in such an orientation that the whole of SCP1 enters the recipient as a leading sequence before the first chromosomal genes, which are then inherited by recombinants with a frequency that falls with their distance from the site of SCP1 integration.

A second difference applies to some, only, of the integrated-SCP1 donors. For these (of which the donor called NF is the type specimen) the gradient of inheritance of donor markers by the progeny is "bidirectional": that is, donor markers from *both* sides of the site of plasmid integration have high frequencies in the progeny and the gradient of inheritance falls with increasing distance *in both directions* from this point.[25] The explanation for this behaviour is by no means obvious, although models involving integration of more than one SCP1 at the same site might be entertained. Other classes of donors, referred to as "unidirectional", are more comparable with *E. coli* Hfr strains in generating a single gradient of inheritance of donor genes, on one side of the point of integration.[29]

We see, then, that some at least of the apparent differences in sex factor behaviour between *E. coli* and *S. coelicolor* may be real, but we shall probably not know for certain until physical studies can be made of the DNA of SCP1 in its autonomous and integrated forms, or until donor states involving SCP2, whose DNA can already be studied, are isolated. It was only by DNA studies that a fairly complete picture of the role of *F* in conjugation emerged, and even now, after much dedicated work, aspects of the process remain obscure; in particular the role of the sex pilus which may, or perhaps may not, be the actual organ through which DNA passes to the female cell.[15]

RECOMBINATION WITHOUT SEX FACTORS: PROTOPLAST FUSION AND HETEROKARYOSIS

Recently, fusion of many kinds of cells and protoplasts has been brought about through the agency of "fusogens", the simplest and most generally applicable being polyethylene glycol (PEG). This material causes the non-specific fusion of biological membranes so that animal cells, organelles, and plant and bacterial protoplasts can be fused in many combinations, sometimes with interesting biological consequences: for example in the production of hybrid lines between human and rodent cells[45], of somatic hybrids between different plant species[18], or of heterokaryons between different fungal species.[1,33] In streptomycetes, PEG-induced fusion of artificially prepared protoplasts leads to extremely high frequencies of recombination, in the absence of known sex factors. Thus, in a mixture of two $SCP1^-$ $SCP2^-$ strains of *S. coelicolor* A3(2), in which recombinants are very rare (10^{-7} or less) in a mating, recombination frequencies as high as 20% can be achieved by protoplast fusion[30]: a millionfold increase!

Streptomyces protoplasts are prepared by growing the mycelium in shaken liquid culture in a medium containing a critical concentration of glycine, which renders the cell walls susceptible to digestion by lysozyme. Digestion is done in a medium containing an osmotically stabilising concentration of sucrose, and particular concentrations of calcium, magnesium and other salts. The protoplasts from two genetically marked strains are then washed, mixed, treated briefly with PEG, and washed again. They are then plated on a medium, also osmotically stabilised, on which they regenerate into mycelium which grows normally and sporulates. When samples of these spores are characterised for the genetic markers of the parents, high proportions of recombinants are found. When the parents differ for several markers, numerous multiple crossover recombinant colonies are produced. Fusion of protoplasts derived from the two strains probably gives rise to units in which complete chromosomes of each parent are initially present, but the units do not remain effectively diploid (D. A. Hopwood and H. M. Wright, unpublished results). Probably at least one, and sometimes both, parental chromosomes fragment, with loss of some pieces, so that individual colonies arising from fusions never contain both parental genotypes, nor complementary recombinant classes. Instead, one parental genotype and a group of recombinant classes is often found, the genotypes varying in respect of some, but not all, of the markers which distinguished the two parents. A fragment of one parental genome seems to be available for a number of recombination events with the other, complete, parental genome; in other cases, genome fragments from *both* parents apparently interact.

Efficient intra-strain recombination through protoplast fusion occurs in every streptomycete so far tested, provided that care is taken to minimise contamination of protoplast preparations by viable cells which failed to be converted to protoplasts. Such units, which will not fuse because their plasma membranes are not exposed to the action of the PEG, sometimes develop precociously, and therefore into larger colonies, on the regeneration plates, diluting out the recombinants arising from the regenerating fused protoplasts.[30] Certain inter-specific mixtures of protoplasts have yielded recombinants, but the frequencies have been very variable: insights have still to be gained into the requirements for successful inter-specific fusions.

In fungi, a frequent occurrence in a mixed culture of two genetically marked derivatives of the same species is heterokaryosis: the formation of a mycelium containing nuclei of different genotypes. This also occurs when two streptomycete derivatives carrying complementary nutritional markers grow together. Individual hyphae apparently really do contain both parental types of chromosome since mycelial fragments isolated by means of a micromanipulator give rise to strains of each parental genotype.[9] In most fungal heterokaryons, nuclear fusion to give diploid nuclei occurs only occasionally and mitotic crossovers in such nuclei are also rare, so that recombinants arise only very rarely in heterokaryotic growth.[48] In *Streptomyces* heterokaryons, recombinants are also very rarely produced, in spite of the fact that there are no nuclear membranes to separate the different genomes. This suggests that some feature of the cellular organisation of the *Streptomyces* mycelium normally keeps to a low frequency the occurrence of recombinational events between chromosomes in the same cytoplasm. In contrast, when genome fragments are introduced by a sex factor in a mating, or as a result of protoplast fusion, recombination between them and the resident chromosomes is very frequent.

Probably, heterokaryons form by a mechanism that does not involve sex factor activity. The occasional breakdown of the barriers to recombination between chromosomes in such heterokaryons could well account for the very low level of recombination in $SCP1^-$ $SCP2^-$ matings of *S. coelicolor* A3(2).

GENETICS OF OTHER ACTINOMYCETES: VARIATIONS ON A THEME

Genetic recombination has been reported in some 16 other species of *Streptomyces* besides *S. coelicolor* A3(2), and heterokaryons without recombination in a further four species (see ref. 27). In most of the cases in which recombination was detected, heterokaryons were also reported, and it seems probable that heterokaryon formation is universal, or nearly so, in the genus. In *S. coelicolor* A3(2), heterokaryons are not selected under the conditions in which crosses are normally analysed for recombination, whereas in several other species they are; however, heterokaryons can readily be studied in *S. coelicolor* A3(2) under other conditions, for example when two parental strains are grown together for 24 hr on cellophane over a fully supplemented medium and then transferred to a selective medium for further heterokaryotic growth to occur.[48] Such a procedure might well reveal heterokaryons in the other species in which only recombinants were found. What of the studies in which only heterokaryons were reported? It seems likely that such strains lacked sex factors, so that any recombination that could have occurred would have taken place only within heterokaryons by the route discussed earlier. Very rare recombinants would probably not have been seen against the background of frequent heterokaryons that arose. A genuine case of absolute sterility would be interesting, but is not established by the published data.

In some other streptomycetes, a circular linkage map, much less abundantly marked than that of *S. coelicolor* A3(2)[25], has been deduced, as in strains of *S. rimosus, S. glaucescens, S. olivaceus, S. achromogenes* var. *rubradiris, S. bikiniensis* var. *zorbonensis, S. venezuelae* and *S. acrimycini* (see ref. 27). There seem to be considerable resemblances between the gene sequences on these various maps, insofar as the functions of genes recognised in the different species can be deduced to correspond. Evidently, as in the enterobacteria, in which gene arrangements in *E. coli* K-12 and *S. typhimurium* LT-2 are extremely similar, there is considerable conservation of genome architecture in the genus *Streptomyces.* This is probably a feature of the naked DNA chromosomes of prokaryotes, distinguishing them from eukaryotes in which chromosomal re-arrangements occur much more commonly, disrupting gene sequences over short periods of evolution.

Recombination through conjugation has also been detected in species of *Micromonospora, Nocardia* and *Mycobacterium*, the former two being producers of valuable antibiotics (gentamicin, sisomicin, rifamycins, ristocetin[27]).

Only in *S. rimosus,* apart from *S. coelicolor* A3(2), have plasmid sex factors so far been recognised. In an isolate of this species, a single sex factor, SRP1, has been found[20]; $SRP1^+ \times SRP1^+$ matings give recombinants at a frequency of about 10^{-6}, $SRP1^+ \times SRP1^-$ at about 10^{-4} and $SRP1^- \times SRP1^-$ matings, like $SCP1^-\ SCP2^- \times SCP1^-\ SCP2^-$ crosses of *S. coelicolor*, have a residual fertility of about 10^{-8}. Since recombination is equally frequent in several of the other species in which it has been studied, plasmids are doubtless at work and could be recognised by a search for strains of altered fertility. In other cases, as suggested above, a low level of recombination may be occurring without sex factor activity; perhaps recombination in such strains could be enhanced by transferring sex factors to them.

GENETICS OF ANTIBIOTIC PRODUCTION: PLASMIDS AND CHROMOSOMAL GENES

An enormous body of information on the genetics of antibiotic production might be expected since the development of actinomycete strains yielding high concentrations of antibiotics has been a major concern in the fermentation industry, and because natural recombination systems suitable for gene mapping are widespread in the actinomycetes. For certain important antibiotics, the strains now used on an industrial scale produce very much higher yields (several hundredfold or more) than the strains originally isolated from Nature. However, these improvements have been achieved by the repeated selection, on *empirical* criteria, of higher yielding mutants. Only very recently has information been sought on the arrangement of the genes involved in antibiotic synthesis. The most complete knowledge exists for two non-commercial antibiotics made by *S. coelicolor* A3(2). The biosynthetic pathway of one of them, actinorhodin, is controlled by a cluster of closely-linked chromosomal genes; for a second, methylenomycin A, the genes are borne on the SCP1 plasmid. In several other species, some genes involved in antibiotic production are found to be plasmid-linked. However, in the best studied examples, oxytetracycline in *S. rimosus* and chloramphenicol in *S. venezuelae*, in contrast to the situation for methylenomycin, the plasmid appears to play some kind of regulatory role and the structural genes for biosynthetic pathway enzymes are chromosomal. These cases are discussed in more detail elsewhere.[24]

It is unlikely, *a priori*, that the case of methylenomycin is unique and we can expect other interesting examples of plasmid-borne structural genes for antibiotic pathways to be discovered as this field of research now expands after a slow start. A detailed understanding of the part played by plasmid-borne regulatory genes in controlling antibiotic synthesis will also be interesting and important.

POSSIBILITIES FOR GENETIC MANIPULATION: ENORMOUS POTENTIAL TO EXPLOIT

Although a great deal could be done to harness the natural genetic recombination systems available in the actinomycetes to produce improved genotypes, the prospect of supplementing these systems with artificial procedures is even more attractive. Two such procedures are, or will shortly be, available.

For the *random* recombination of genome segments from two (or more) strains, protoplast fusion, with its high frequency of multiple crossover recombinants, appears to be very suitable.

Such recombination would be useful in the construction of complex genotypes and in breeding for a quantitative character such as increased yield of an antibiotic. For the moment, protoplast crosses could be made empirically between divergent lines in a mutation and selection programme. However, if the full potential of protoplast fusion is to be realised, a model biometrical study would be very informative in order to gain insight into the best strategy of intercrosses, back-crosses, mutagenic treatments and selection lines.[23] Such a quantitative genetical analysis has been made of penicillin production in the fungus *Aspergillus nidulans*[11], using its sexual cycle to generate recombinants, but no comparable analysis exists for a prokaryote. Protoplast fusion in a strepto-mycete appears to offer the ideal recombinational tool for such a study.

For the manipulation of particular genes, and for the construction of strains containing segments of DNA from other species, recombinant DNA techniques are likely to be invaluable. These will require the development of suitable vectors for the actinomycetes themselves, since the cloning of actinomycete genes into *E. coli* on currently available vectors is of more limited interest. No actinomycete vectors are yet available, but some temperate bacteriophages and plasmids are currently under development for this purpose. In particular the SCP2 plasmid of *S. coelicolor* A3(2), with single cleavage sites for the useful restriction enzymes *Eco*RI and *Hin*dIII and small numbers of sites for *Bam*HI and *Sal*P1 (≡ *Pst*I)[7], seems promising, once suitable conditions for trans-forming the plasmid DNA back into the host are found.

The cloning of DNA segments into such vectors will have many applications in funda-mental studies of genome organisation in *Streptomyces*: for example to probe the possible redundancy of sequences on opposite sides of the linkage map discussed earlier; to investigate the transcriptional specificity of genes acting at different stages in the complex cycle of differentiation of the substrate mycelium, aerial mycelium and spores — a uniquely complex cycle for a prokaryote[12]; and to analyse the fine structure of gene clusters coding for the multi-enzyme complexes which are likely to be involved in the synthesis of many antibiotics, particularly those produced by the "polyketide" route of biosynthesis, which is a variant of fatty acid synthesis.[27] Actinorhodin is an academic example of a polyketide compound, but many valuable antibiotics such as erythromycin and other macrolides, tetracyclines and rifamycin, fall into the same class.

Gene cloning within actinomycetes will also have industrial applications. Some of the most obvious have been discussed elsewhere.[23,24] They include the possibility of transferring the genes for synthesising a new valuable antibiotic to an industrially more amenable host, perhaps one which had already been selected for increased production of an antibiotic of the same chemical family; or constructing a hybrid containing parts of two different antibiotic pathways in order to generate a "hybrid" antibiotic. Numerous other possibilities will doubtless occur once the work gets under way.

CONCLUSION

The actinomycetes were a Cinderella group for many decades after the description of the first "ray fungus" in the latter part of the nineteenth century. Selman Waksman, more than any other microbiologist, deserves the credit for his extensive studies, between the wars, of what was then a relatively unfashionable group of microbes. His dedication was rewarded with the discovery, in 1943, of streptomycin, the first in a long list of important actinomycete antibiotics which is still growing: some 150 new actinomycete antibiotics are currently described in supplements to the *Journal of Antibiotics* each year. But the interest of the actinomycetes extends far beyond their importance as antibiotic producers. This article has discussed some special features of their genetics; other interesting aspects of their biology have not been dealt with. Notable amongst these is their position as prokaryotes with one of the most complex cycles of colonial differentiation, a topic reviewed elsewhere by Chater and Merrick.[12] Analysis of this differentiation could provide important insights into some aspects of morphogenesis in general.

REFERENCES

1. ANNÉ, J., EYSSEN, H. and SOMER, DE P. Somatic hybridisation of *Penicillium roquefortii* with *P. chrysogenum* after protoplast fusion. *Nature* **262** (1976) 719–721.

2. BECKING, J.H. Endophyte and association establishment in non-leguminous nitrogen-fixing plants. In W. NEWTON, J.R. POSTGATE and C. RODRIGUEZ-BARRUECO (Eds) *Recent Developments in Nitrogen Fixation.* Academic Press, London (1977), pp. 551-567.

3. BÉRDY, J. Recent developments of antibiotic research and classification of antibiotics according to chemical structure. *Adv. Appl. Microbiol.* **18** (1974) 309–406.

4. Bergey's Manual of Determinative Bacteriology, 8th Edition. R.E. BUCHANAN and N.E. GIBBONS (Eds) Williams and Wilkins, Baltimore, (1974).

5. BERINGER, J.E., HOGGAN, S.A. and JOHNSTON, A.W.B. Linkage mapping in *Rhizobium leguminosarum* by means of R plasmid-mediated recombination. *J. gen. Microbiol.* **104** (1978) 201–207.

6. BIBB, M.J. Genetical and physical studies of a *Streptomyces coelicolor* plasmid. Ph.D. Thesis, University of East Anglia, Norwich (1978).

7. BIBB, M.J., FREEMAN, R.F. and HOPWOOD, D.A. Physical and genetical characterisation of a second sex factor, SCP2, for *Streptomyces coelicolor. Molec. gen. Genet.* **154** (1977) 155–166.

8. BLYTH, W. Farmer's lung disease. In G.SYKES and F.A. SKINNER (Eds) *Actinomycetales: Characteristics and Practical Importance.* Academic Press, London (1974) pp. 261-276.

9. BRADLEY, S.G. and LEDERBERG, J. Heterokaryosis in *Streptomyces. J. Bact.* **72** (1956) 219–225.

10. BRODA, P. *Plasmids.* W.H. Freeman, Reading (In press).

11. CATEN, C.E. and JINKS, J.L. Quantitative genetics. In K.D. MACDONALD (Ed) *Second International Symposium on the Genetics of Industrial Microorganisms.* Academic Press, London (1976) pp. 93-111.

12. CHATER, K.F. and MERRICK, M.J. Streptomycetes. In J.H. PARISH (Ed) *Development and morphogenesis in prokaryotes.* Blackwell Scientific Publications, Oxford (In press).

13. COHEN, S.N. Transposable genetic elements and plasmid evolution. *Nature* **263** (1976) 731–738.

14. CROSS, T. and GOODFELLOW, M. Taxonomy and classification of the actinomycetes. In G. SYKES and F.A. SKINNER (Eds) *Actinomycetales: Characteristics and Practical Importance.* Academic Press, London (1974) pp. 11–112.

15. CURTISS, R. Bacterial conjugation. *A. Rev. Microbiol.* **23** (1969) 69–136.

16. DAVIDSON, N., DEONIER, R.C., HU, S. and OHTSUBO, E. The DNA sequence organization of F and F-primes and the sequences involved in Hfr formation. *Microbiology* **1** (1975) 56–65.

17. DAVIES, J. and BENVENISTE, R.E. Enzymes that inactivate antibiotics in transit to their targets. *Ann. N.Y. Acad. Sci.* **235** (1974) 130–136.

18. EVANS, P.K. and COCKING, E.C. In H.E. STREET (Ed) *Plant Tissue and Cell Culture.* 2nd Edition. Blackwell, Oxford (1977) pp. 103–135.

19. FALKOW, S. *Infectious Multiple Drug Resistance*, Pion, London (1975).

20. FRIEND, E.J., WARREN, M. and HOPWOOD, D.A. Genetic evidence for a plasmid controlling fertility in an industrial strain of *Streptomyces rimosus. J. gen. Microbiol.* (In press).

21. HOPWOOD, D.A. Lack of constant genome ends in *Streptomyces coelicolor. Genetics* **54** (1966) 1177–1184.

22. HOPWOOD, D.A. Genetic analysis and genome structure in *Streptomyces coelicolor. Bacteriol. Rev.* **31** (1967) 373–403.

23. HOPWOOD, D.A. Genetic recombination and strain improvement. *Dev. Ind. Microbiol.* **18** (1977) 111–123.

24. HOPWOOD, D.A. Extrachromosomal control of antibiotic production. *A. Rev. Microbiol.* **32** (In press)

25. HOPWOOD, D.A., CHATER, K.F., DOWDING, J.E. and VIVIAN, A. Advances in *Streptomyces coelicolor* genetics. *Bacteriol. Rev.* **37** (1973) 371–405.

26. HOPWOOD, D.A. and GLAUERT, A.M. The fine structure of *Streptomyces coelicolor* II. The nuclear material. *J. Biophys. Biochem. Cytol.* **8** (1961) 267–278.

27. HOPWOOD, D.A. and MERRICK, M.J. Genetics of antibiotic production. *Bacteriol. Rev.* **41** (1977) 595–635.

28. HOPWOOD, D.A. and WRIGHT, H.M. Transformation in *Thermoactinomyces vulgaris. J. gen. Microbiol.* **71** (1972) 383–398.

29. HOPWOOD, D.A. and WRIGHT, H.M. Interactions of the plasmid SCP1 with the chromosome of *Streptomyces coelicolor* A3(2). In K.D. MACDONALD (Ed) *Second International Symposium on the Genetics of Industrial Microorganisms.* Academic Press, London (1976) pp. 607–619.

30. HOPWOOD, D.A., WRIGHT, H.M., BIBB, M.J. and COHEN, S.N. Genetic recombination through protoplast fusion in *Streptomyces. Nature* **268** (1977) 171–174.

31. HORI, M., EGUCHI, J., KAKIKI, K. and MISATO, T. Studies on the mode of action of polyoxins. VI. Polyoxin-sensitive and resistant strains of *Alternaria kikuchiana. J. Antibiot.* **27** (1974) 260–266.

32. JACOB, F. and WOLLMAN, E.L. Genetic and physical determinations of chromosomal segments in *Escherichia coli. Symp. Soc. Exp. Biol.* **12** (1958) 75–92.

33. KEVEI, F. and PEBERDY, J.F. Interspecific hybridization between *Aspergillus nidulans* and *Aspergillus rugulosus* by fusion of somatic protoplasts. *J. gen. Microbiol.* **102** (1977) 255–262.

34. KLECKNER, N. Translocatable elements in procaryotes. *Cell* **11** (1977) 11–23.

35. KONDOROSI, A., KISS, G.B., SORRAI, T., VINCZE, E. and DANSALVI, Z. Circular linkage map of the *Rhizobium meliloti* chromosome. *Nature* **268** (1977) 525–527.

36. LACEY, J. Actinomycetes in soils, composts and fodders. In G. SYKES and F.A. SKINNER (Eds) *Actinomycetales: Characteristics and Practical Importance.* Academic Press, London (1974) pp. 231–251.

37. MCCOY, R.E., CARROLL, V.J., POUCHER, C.P. and GWIN, G.H. Field control of coconut lethal yellowing with oxytetracycline hydrochloride. *Phytopath.* **66** (1976) 1148–1150.

38. MEADE, H.M. and SIGNER, E.R. Genetic mapping of *Rhizobium meliloti. Proc. natn. Acad. Sci. USA* **74** (1977) 2076–2078.

39. MEYNELL, E., MEYNELL, G.G. and DATTA, N. Phylogenetic relationships of drug-resistance factors and other transmissible bacterial plasmids. *Bacteriol. Rev.* **32** (1968) 55–83.

40. MEYNELL, G.G. *Bacterial Plasmids*, Macmillan, London, (1972).

41. NAKANO, H., MATSUHASHI, Y., TAKEUCHI, T. and UMEZAWA, H. Distribution of chloramphenicol acetyltransferase and chloramphenicol-3-acetate esterase among *Streptomyces* and *Corynebacterium. J. Antibiot.* **30** (1977) 76–82.

42. NYLAND, G. and SACHS, R.M. Control aspects of plant mycoplasma diseases chemotherapy in the field. *INSERM*, **33** (1974) 235–242.

43. OGAWARA, H. Production and property of beta-lactamases in *Streptomyces. Antimicrob. Agents Chemother.* **8** (1975) 402–408.

44. PERLMAN, D. The fermentation industries – 1977. *ASM News* **43** (1977) 82–89.

45. PONTECORVO, G. Alternatives to sex: genetics by means of somatic cells. In R. MARKHAM, D.R. DAVIES, D.A. HOPWOOD and R.W. HORNE (Eds) *Modification of the Information Content of Plant Cells.* North Holland/American Elsevier, Amsterdam (1975) pp. 1–14.

46. RICHMOND, M.H. Plasmids and chromosomes in prokaryotic cells. *Symp. Soc. gen. Microbiol.* **20** (1970) 249–277.

47. ROWBURY, R.J. Bacterial plasmids with particular reference to their replication and transfer properties. *Prog. Biophys. molec. Biol.* **31** (1977) 271–317.

48. SERMONTI, G. *Genetics of Antibiotic-Producing Microorganisms.* Wiley-Interscience, London (1969).

49. SHAW, W.B. and HOPWOOD, D.A. Chloramphenicol acetylation in *Streptomyces. J. gen. Microbiol.* **94** (1976) 159–166.

50. STAPLEY, E.O., JACKSON, M., HERNANDEZ, S., ZIMMERMAN, S.B., CURRIE, S.A., MOCHALES, S., MATA, J.M., WOODRUFF, H.B. and HENDLIN, D. Cephamycins, a new family of β-lactam antibiotics. I. Production by actinomycetes, including *Streptomyces lactamdurans* sp. n. *Antimicrob. Agents Chemother.* **2** (1972) 122–131.

51. STARK, W.M. Lilly contributions to fermentation products. *Dev. Ind. Microbiol.* **18** (1977) 27–37.

52. STARLINGER, P. and SAEDLER, H. IS-elements in microorganisms. *Curr. Topics in Microbiol. and Immunol.* **75** (1976) 111.

53. Swann Report. Joint Committee on the use of Antibiotics in Animal Husbandry and Veterinary Medicine. *Cmnd.* **4190**. HMSO. (1969).

54. TOWNER, K.J. and VIVIAN, A. RP4-mediated conjugation in *Acinetobacter calcoaceticus. J. gen. Microbiol.* **93** (1976) 355–360.

55. WILLIAMS, S.T., SHARPLES, G.P. and BRADSHAW, R.M. The fine structure of the actinomycetales. In G. SYKES and F.A. SKINNER (Eds) *Actinomycetales: Characteristics and Practical Importance.* Academic Press, London (1974) pp. 261–276.

56. WRIGHT, H.M. and HOPWOOD, D.A. A chromosomal gene for chloramphenicol acetyltransferase in *Streptomyces acrimycini. J. gen. Microbiol.* **102** (1977) 417–421.

57. WRIGHT, L.F. and HOPWOOD, D.A. Identification of the antibiotic determined by the SCP1 plasmid of *S. coelicolor* A3(2). *J. gen. Microbiol.* **95** (1976) 96–106.

58. YAMADA, K., KINOSHITA, S., TSUMODA, T. and AIDA, K. (Eds) *The Microbial Production of Amino Acids.* Kodansha, Tokyo; Wiley, London (1972).

59. YOUNG, F.E. and WILSON, G.A. Genetics of *Bacillus subtilis* and other gram-positive sporulating bacilli. In H.O. HALVORSEN, R. HANSON and L.L. CAMPBELL (Eds) *Spores V.* American Society for Microbiology, Washington, D.C. (1972) pp. 77–106.

60. YOUNG, F.E. and WILSON, G.A. *Bacillus subtilis.* In R.C. KING (Ed) *Handbook of Genetics*, Vol. 1. Plenum Press, New York (1974) pp. 69–114.

61. ZIPKAS, D. and RILEY, M. Proposal concerning mechanism of evolution of the genome of *Escherichia coli. Proc. natn. Acad. Sci. USA* **72** (1975) 1354–1358.

ADDENDUM

Perhaps the most important new development in actinomycete genetics has been the discovery of conditions for the efficient transformation of *Streptomyces* by plasmid DNA (see p. 138). The procedure[62] is to expose protoplasts of the organism to plasmid DNA in the presence of PEG (at a lower concentration than that required for optimal protoplast fusion[66]) and then to allow the protoplasts to regenerate into a mycelial culture. By this technique, a very high proportion of the regenerating protoplasts (up to 20%) can be transformed with the SCP2* plasmid, and this has allowed the first recombinant DNA experiments in *Streptomyces* to be performed (M.J. Bibb *et al.*, unpublished results). The technique works also for transfection of protoplasts by the DNA of various *Streptomyces* phages,[67] and phage vectors for the introduction of foreign DNA into Streptomycetes are now under development.[64] Plasmid and phage vectors may be expected to have complementary advantages in the genetic manipulation of the organisms.[65]

An interesting extension of the protoplast fusion technique (see p. 135) is to irradiate with ultraviolet light, immediately before fusion, the protoplast preparations of the two strains to be recombined. This kills many of the protoplasts which are not involved in fusion events, while fusing protoplasts can give rise to viable progeny by recombining out the UV-induced lethal hits in their DNA, which tend to occur at different positions in the chromosomes of different members of the population. Thus there is a very significant increase in the proportion of recombinant progeny from the fusion.[66]

The idea that conjugation in *Streptomyces* may be a simpler process than in *E. coli* (see pp. 133-4) has recently received support from the discovery of a new plasmid (SLP1) in a relative of *S. coelicolor* called *S. lividans*.[63] This plasmid is much too small (it can be as small as 6.25×10^6 daltons in molecular size) to code for a conjugation apparatus as complex as the sex pilus of *E. coli*, yet it apparently promotes its own transfer by conjugation.

REFERENCES

62. BIBB, M.J., WARD, J.M. and HOPWOOD, D.A. Transformation of plasmid DNA into *Streptomyces* at high frequency. *Nature* **274** (1978) 398–400.

63. BIBB, M.J., WARD, J.M. and HOPWOOD, D.A. The development of a cloning system for *Streptomyces. Dev. Ind. Microbiol.* (In press).

64. CHATER, K.F. Actinophage DNA. *Dev. Ind. Microbiol.* (In press).

65. HOPWOOD, D.A. and CHATER, K.F. Fresh approaches to antibiotic production. *Proc. Roy. Soc.* B (In press).

66. HOPWOOD, D.A. and WRIGHT, H.M. Factors affecting recombinant frequency in protoplast fusions of *Streptomyces coelicolor. J. gen. Microbiol.* **111** (1979) 137–143.

67. SUAREZ, J.E. and CHATER, K.F. Polyethyleneglycol-assisted transfection of *Streptomyces* protoplasts. *J. Bacteriol.* (In press).

GENETIC IMPROVEMENT OF PROTEIN AND YIELD IN FOOD LEGUMES*

ALICE M. EVANS and H.E. GRIDLEY

Department of Applied Biology, University of Cambridge

(1) INTRODUCTION

(a) *Food legumes as protein sources*

The food legumes or pulse crops are important sources of protein and calories in human diets, particularly in the developing world. In the U.K. these grain legumes are more important in animal feed than in human diets but with the increased cost of animal proteins this situation may well change.

Although food legumes are rich sources of protein they make a smaller contribution to world protein production than do the cereals. Of the protein consumed by the human population of the world, 70% comes from plants and only 30% from animal sources. Using crop production estimates for 1968, Jalil and Tahir[37] calculated that cereals contributed about 70% and legumes about 18% to the supply of plant protein in the world. So, although legumes are considerably higher in protein than cereals they are grown over a smaller hectarage and are generally lower yielding. Whether cereals are intrinsically higher yielding than legumes (because in legumes there is greater competition for carbohydrate and protein production) is difficult to determine. It is true however, that more breeding effort has been applied to improving cereal yields. The task for plant breeders engaged in improving food legumes must be firstly to increase their yield and secondly to improve their protein content and quality.

(b) *Food legumes of the world: their production and productivity*

Food legumes are distributed world-wide and include several botanical genera. The most important species are shown in Table 1, and it can be seen that only two food legumes are currently important in Britain, namely the field bean and the pea. They occupy 40,000 hectares (ha) and 31,000 ha respectively and are produced to the extent of 110,000 and 82,000 metric tonnes (m.t.).[22] We have indicated that the common bean the runner bean and lupins may possibly be grown as pulse crops in Britain in the future but this will only be achieved if considerable breeding effort is devoted to these species. We have also speculated that the soya bean, the lentil and the chick pea may also find a place in this country.

*Originally published under the title: "Prospects for the improvement of protein and yield in food legumes".

TABLE 1. THE IMPORTANT FOOD LEGUMES

	Status in the U.K.	Botanical species	Common names	Region of origin	Region of major production
(i)	Currently important in the U.K.	*Vicia faba* (L.)	Field bean Broad bean	Mediterranean	China Europe
		Pisum sativum (L.)	Field peas	Mediterranean	USSR China
(ii)	Potentially important in the U.K.	*Phaseolus vulgaris* (L.)	Dry bean Common bean	Latin America	Latin America
		P. coccineus (L.)	Runner bean	Latin America	Latin America
		Lupinus spp.	Lupins	Eurasia	Australia Europe
(iii)	Other 'possibilities' for the U.K.	*Glycine max.* Merr. (L.)	Soya bean	Far east	N. America China
		Lens esculenta (Moench)	Lentil	Eurasia	India Near East
		Cicer arietinum (L.)	Chick pea	Eurasia	India
(iv)	Tropical pulses (unlikely to find a place in the U.K.)	*Arachis hypogaea* (L.)	Groundnut Peanut	S. America	India Africa
		Cajanus cajan (L.) Millsp.	Pigeon pea	India	India
		Phaseolus lunatus (L.)	Lima bean	Latin America	Latin America
		Psophocarpus tetragonolobus (L.)	Winged bean	S.E. Asia	S.E. Asia
		Vigna unguiculata (L.) Walp.	Cow pea	Africa	Africa
		V. radiata (L.) Wilcz.	Green gram	Asia	Asia
		V. mungo (L.) Hepper.	Black gram	Asia	Asia

On a world-wide scale the food legumes are grown to the extent of 70 million ha but their total production is only 50 m.m.t. The hectarage, production and productivity of 10 important grain legumes is given in Fig. 1. By far the greatest world production comes from the soyabean and the groundnut which are mostly grown as oil crops and only secondarily as protein crops. The most productive food legumes are the soyabean, the field bean and the pea with average yields exceeding 1 t/ha. The remaining crops are all rather low yielding with considerable potential for improvement by breeding and cultivation.

(c) *Protein content of the important food legumes*

Many workers have reported on the range of protein content found in the food legumes. With the development of the International Crop Research Centres, germplasm collections of the various legumes are being screened for many characters including protein content. A few selected reports on the protein content of some of our food legumes is given in Table 2.

FIGURE I. WORLD AREA, PRODUCTION AND PRODUCTIVITY OF PULSES, 1976

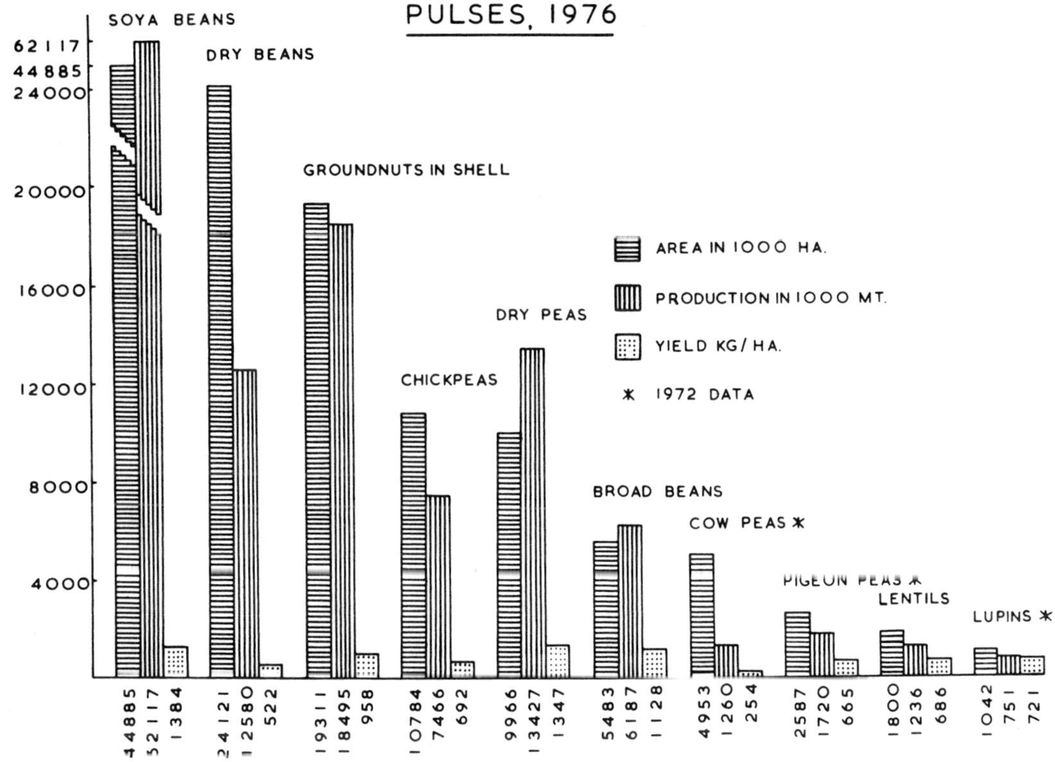

Although there is a great range in protein content in these legumes which at first sight would suggest that there is considerable scope for increasing the protein content by selection, it is apparent that much of the range is due to environmental effects. In *Phaseolus vulgaris* for example, the environmental effects were three times as great as the effects due to genotype.[48, 70]

The environment affects both the yield and the protein content of crops and improvement programmes must therefore take account of the relationships between these factors.

(2) THE GENETIC CONTROL OF YIELD, PROTEIN CONTENT AND QUALITY

A knowledge of the relationship between yield and protein and their genetic control is essential to maximise progress in breeding and selection programmes. More information is available for the soyabean crop than for other grain legumes because of the greater economic importance of soyabeans.

TABLE 2. PROTEIN CONTENT IN FOOD LEGUMES

Species	No. of varieties	% Protein range	% Protein mean	Authority
Field beans	511	22–37		Hawtin *et al.*[34]
		24–33	27.0	Bond[9]
Peas	108	17–31	24.9	Kaul[49]
Common beans	4524	17–35		Kelly[43]
	336	20–34	26.8	Woolfe and Hamblin[86]
Soyabeans		37–50	40.5	Hartwig[33]
Lupins	51	20–44	34.5	Gardiner *et al.*[24]
Lentils	1688	23–36		Hawtin *et al.*[34]
	100	20–31	24.8	Kaul[40]
Chick peas	2677	19–28		Hawtin *et al.*[34]
	192	12–28	19.5	Kaul[40]
Pigeon peas	2262	18–29		Hawtin *et al.*[34]
	55	18–26	21.5	Kaul[40]
Cow peas	224	18–32		Tobin[82]
	118	20–28	28.0	Hawtin *et al.*[34]
Black gram	20	22–30	25.0	Tobin[82]
Green gram	45	18–34	28.0	Tobin[82]

TABLE 3. PHENOTYPIC (P) AND GENOTYPIC (G) CORRELATIONS BETWEEN YIELD AND PROTEIN IN SOME LEGUME CROPS

Species	Correlations		Material	No. of environments	Authority
Common bean	−0.12 to −0.26	P	Single plants	1	Cheah and Evans[14]
	−0.23	P	343 lines	1	Rutger *et al.*[70]
	−0.30	P	Single plants	1	Kelly[41]
	−0.45**	P	Single plants	1	Leleji *et al.*[49]
	−0.48**	P	32 lines	3	Gridley and Evans[29]
	−0.64**	P	25 varieties	2	Tandon *et al.*[80]
Peas	0.02 to −0.34**	P	99 F_3 progenies	1	Pandey and Gritton[61]
	0.22 to −0.49	G	from 4 crosses	1	
	0.14 to −0.35**	P	99 F_4 progenies	1	
	0.19 to −0.57	G	from 4 crosses	1	
Cow pea	−0.14	P	11 varieties	3	Bliss *et al.*[3]
	−0.38	G			
Field bean	0.23 to −0.29	P	10 to 56 varieties	7	Bond[9]

(a) *Yield and protein relations*

In a recent review Brim[12] gave details from a series of studies on soyabeans in which phenotypic and genotypic correlations between yield and protein ranged from 0.35 ($P = >0.05$) to -0.58 ($P = <0.05$) and 0.22 ($P = >0.05$) to -0.42 ($P = <0.01$) respectively. Negative correlations predominated for both the phenotypic and genotypic correlations. These studies dealt with F_3 or more advanced generations replicated over locations and years, and are therefore meaningful.

Table 3 gives correlations between yield and protein for other food legumes. The data were collected from material which included varieties, breeding lines and single plants in early generations and only in 4 cases was there replication over locations and years. Although negative associations predominate the correlations tend to be small and not always statistically significant indicating that selection for high yield may not necessarily be accompanied by a decrease in protein content.

The relationships between seed weight and protein content were investigated in 60 lines of mung beans[51] and positive correlation was found. But in 33 varieties of chick peas the correlation was negative.[71] However, no correlation was found between seed weight and protein content in 220 lines of pigeon peas[17] and it was suggested that as seed size is an important yield component, it should be possible to develop high yielding lines with high protein content in pigeon peas.

The greatest successes in selecting high yielding lines with high protein content have been achieved in the soyabean[12, 33] and in the field bean[9] suggesting that similar results may possibly be obtained in other food legumes.

(b) *Genetic control of yield and protein*

The effects of gene action for yield and protein have been described in terms of additive and non-additive genetic variation, general and specific combining ability effects and broad and narrow sense heritabilities.

Non-additive gene action for yield has been demonstrated in soyabeans,[32] field beans[0, 50] and pigeon peas.[18, 73] Both additive and non-additive effects for yield have been shown to occur in soyabeans,[84] peas,[20, 39] mung beans[78, 79] and chick peas.[45]

Protein content has been shown to exhibit only additive genetic variation in soyabeans[32] and common beans[65] but both additive and non-additive gene action in peas,[20] mung beans[75] and chick peas.[71] Heritability values for yield and protein in the food legumes are shown in Table 4. In general the narrow sense heritabilities for both yield and protein content in the food legumes are low. Both these characters will therefore present a considerable challenge to the breeder concerned with their improvement.

Even though protein yield per hectare must be an important objective for the breeder of temperate food legumes, one cannot ignore the importance of a high protein percentage in developing countries. This means a high protein content in the food bowl, the individual meal. Sheldrake[74] has argued that when the effective income for the undernourished increases, they eat more food so protein per cent may not be so important.

While increased protein content is a primary objective, protein quality must also be con-

TABLE 4. HERITABILITY VALUES FOR YIELD AND PROTEIN IN SOME LEGUME CROPS

	Yield		Protein	
Species	h^2_b*	h^2_n†	h^2_b	h^2_n
Common bean		0.09–0.11[15]	0.82[65] 0.32–0.75[41]	0.49[65] 0.32–0.79[41]
Cow pea	0.12[3]		0.29[3]	
Groundnut	0.81[46]			
Lupins	0.57[27]		0.83[27]	
Peas		0–0.41[61]		0.17–0.56[61]
Pigeon pea		0.11[18]		
Soya bean	0.03–0.58[12]		0.57–0.90[12]	

* h^2_b = broad sense heritability.

† h^2_n = narrow sense heritability.

sidered in terms of higher levels of the limiting essential amino acids and also improved protein digestibility.

Ultimately it is the nutritional value of food legumes in mixed diets or animal feed which must be taken into account rather than their value as single foods. Furthermore legumes cause flatulence, and they contain anti-nutritional factors like trypsin inhibitors and haemagglutinins. Some reports have suggested that breeders should aim to overcome the trypsin inhibitor problem genetically. However Putztai[66] has shown that trypsin inhibitors in *Phaseolus vulgaris* have exceptionally high sulphur amino acid contents. Since these trypsin inhibitors are completely inactivated during normal cooking procedures, the correct strategy might be to breed for increased rather than decreased amounts.

These aspects will not be considered further in this review since no adequate screening methods which can be easily used by plant breeders have yet been developed.

(c) *Variation in protein quality*

In order to improve the protein quality of legumes it is necessary to change the amino acid composition since legumes are deficient in the sulphur amino acids, methionine and cystine. The main achievements in improving the amino acid composition of crops has been made in the cereals, particularly in improving the lysine content of maize.[57] The superior protein quality of opaque-2-maize,[52] Hiproly and Risø 1508 barley[56] and the high lysine sorghums[77] result from changes in the relative amounts of the protein fractions. To date there have been no reports of major genes that either control protein synthesis or alter the amino acid composition of legume seeds in a manner similar to that found in cereals.

In legumes about 80% of the total protein consists of globulins which are storage proteins. However the albumins or metabolic proteins are richer in methionine per unit of nitrogen. It has been suggested that breeding for reduced globulin to albumin ratio might be an appropriate method for increasing methionine content of legume proteins, particularly in the common bean.[4] Differences in the sulphur amino acid content of the globulin fraction have also been reported for peas,[19] groundnuts,[13] soyabeans,[69] lupins,[36] the cowpea and field beans.[10] So there may be opportunity for improving the content of S-amino acids by selecting for that fraction of the globulins which is higher in these amino acids.[53]

Other workers have explored the possibilities of altering the proportions of the component parts of the seed of legumes. Otoul[60] found that the embryo was richer in arginine and methionine than the cotyledons but since the embryo comprised little more than 1% of the weight of the seed, he concluded that selection for larger embryo size in legumes is unlikely to be a productive approach.

A wide range of values for methionine and cystine have been reported in legumes[59] and in the common bean, where 20,000 lines were screened for available methionine,[42] 63 lines showed an increase of 33% above the standard variety. Similarly Hackler and Dickson[30] found in their survey of some common bean lines that the highest in methionine was 25% above the mean, the highest in lysine 30% and the highest in threonine 12% above the mean. Furthermore in the Kiev mutant of lupins the lysine content was shown to be 6.8% which compares with normal varieties which have a lysine content of 4–5%.[25]

These latter workers have paid attention to the lysine content which is generally high in legumes. It has been argued that in mixed diets with cereals, lysine is likely to be in short supply since cereals are deficient in lysine and they constitute the higher proportion of the diet. There is therefore some merit in increasing the lysine content of legumes as well as the methionine and cystine. However, even if the levels of the S-amino acids in legumes are adequately increased to keep them from being limiting, their tryptophan would become limiting. It has been estimated by nutritionists that beans, for example, need about twice the tryptophan content and three times the present levels of S-amino acids to make all 24% of the protein available as a dietary protein.

A further problem is the apparent negative correlation between the protein percentage and the methionine content as a percentage of protein,[1] although this correlation is usually rather low.[4] This negative correlation also applies to lysine (g per 16 g N) and per cent protein.[23] Thus selection for one particular trait should not be conducted without paying attention to the other affected traits.

Although the level of total protein and thus of amino acids in legumes can vary greatly according to the environment in which they are grown, the proportion of each amino acid varies to a smaller extent[48] and low to medium heritabilities for the S-amino acids have been reported.[4]

It thus appears that there is some scope for selection for improved amino acid composition, although it must be added that amino acid analysis is subject to such experimental error that it is difficult to know how far differences in reported amino acid values are true and how far an artefact arising from analytical difficulties.[83]

(3) GENETIC IMPROVEMENT OF YIELD AND PROTEIN

Improvement of any crop character depends on the identification and selection of superior genotypes in a genetically variable population. But the rate of progress from selection will depend on the accuracy, speed and hence amount of genetic material that can be screened for the desired characters. Experimental designs are available which can accurately assess the yield of large numbers of genotypes in replicated field trials. Total protein can also be determined on seed from these trials. Present techniques are relatively slow although an extremely rapid infra-red method, which requires no weighing of chemical reagents is presently under test.[85] The same technique may also be used to determine specific amino acids which presently depend on complicated and slow ion-exchange chromatography.

Aside from problems of determination the improvement of yield and protein is hindered by both characters being quantitatively inherited and having relatively low heritabilities. Difficulties in simultaneous improvement stem from the negative correlation between the two characters. To date there is considerable experience on breeding for yield but little on selection for improved protein and even less on simultaneous selection for both yield and protein.

In this section breeding and selection for improvement of yield and protein will be considered covering progress from pure line, mass and pedigree selection through modified pedigree methods to population breeding and recurrent selection.

(a) *Pure line, mass and pedigree selection*

The variation in existing varieties or land races of self-pollinating crops, which includes most of the food legumes, will largely be accounted for by differences between homozygous genotypes. Such variation has been exploited through pure line and mass selection. The former was used for protein improvement in eight Brazilian varieties of common bean.[16] In three of these there was sufficient genetic variability for single plant selection to be effective in isolating progenies higher in protein than the parent varieties. New varieties produced by this system usually trace back to a single pure line.

In contrast, new varieties produced by mass selection can be traced back to a number of pure lines. In Nigeria two cycles of mass selection applied to local varieties of cowpea achieved yield increases of 18–30%.[58] Mass selection was also effective in field beans for increasing and decreasing protein content in three populations over three years;[55] seed size and other yield parameters were not affected in this study. In a related study in groundnuts there was an indication of increased yield after one generation of selection for increased seed size in a local Zambian variety.[81]

Ultimately both these systems become ineffective as the genetic variation in the basic population is exhausted and a breeder must then attempt to create new genetic variation. A system often adopted is enforced crossing of inbreeding varieties followed by pedigree breeding in which selection for desired characters is practised in the segregating generations.

Selection is usually started at the F_2 between single plants to produce progenies for testing in later generations. Shannon *et al.*[72] took a random sample of single plants in six crosses of soyabeans between heavy yielding and high protein lines. A number of resulting F_4 progenies significantly surpassed the best parent for yield and protein. A similar improvement was obtained in lupins[27] where, although F_2 plants were selected for low alkaloid content in 38 crosses, F_3 progenies could be obtained which were significantly better for both characters than the best parent. Pandey and Gritton[62] took a random sample of 99 single plants in four crosses of peas and showed that the observed changes in the values of the F_3 progenies from selecting for high protein and yield ranged from −0.84 to 5.85% and 3.22 to 12.53% of the population means respectively. In work at Cambridge,[29] although selection was delayed till the F_3 in two crosses of common bean, there were F_5 progenies which showed heavier yield and higher protein than the best parent.

Although single plants can be selected in the F_2 the practice is questionable for characters which show low heritability and are subject to genotype x environment interactions. Such an example is cited by Coyne[15] who could obtain no yield increases in the F_3 of one cross of common bean by selecting the top 5% of F_2 plants for seed yield or for each of the three yield components.

The evidence indicates that single plant progenies can be isolated from genetically variable populations that are superior to the original parents in yield and protein. Owing to lack of seed most early generation trials are conducted in a single environment and a proper assessment of the success of breeding and selection methods can only come from trials replicated over sites and years.

Success in pure line, mass and pedigree selection is dependent on retention of single plants or genotypes that give superior progeny. This may often be a matter of chance with characters of low heritability.[31] Although random sampling may help there is a limit to the number of genotypes that can be selected and tested. In pedigree selection, especially, this may only represent a small portion of the genetic variation. Further, as a result of self pollination in subsequent generations there is a rapid approach to homozygosity which limits segregation and recombination.

In an attempt to overcome these problems breeders have used modified forms of pedigree selection and bulk population breeding methods.

(b) *Modified pedigree and bulk breeding methods*

These methods aim to preserve genetic variation by maintaining F_2 single plants or individual crosses as populations which can be selected in later generations when the majority of loci are homozygous. The former is initially similar to a standard pedigree system except that reselection is only practised after one or more generations of testing, often conducted in more than one environment. This system was used in four crosses of soyabeans.[5, 6, 7] In three crosses from F_2 single plants, whose progeny were tested at two sites in the F_3 and F_4, progenies at F_5 could be obtained which were significantly heavier yielding than the best parent. In one cross reselection in the F_3 and F_4 isolated F_5 progeny that were heavier in yield than the earlier F_5 progeny.

Empig and Fehr[21] evaluated three bulk population breeding methods in three crosses of soyabeans. No differences were found in the mean yield of 45 F_6 progenies derived from F_5 single plant selection for the three methods, but differences in variance were found depending on the type of bulk method used.

A further modification of the pedigree method is single seed descent (SSD)[11] which consists of advancing each F_2 plant in the population through subsequent generations by harvesting a single seed from each plant. Assuming additive genetic variance and obligate inbreeding, the genetic variance at homozygosity between progenies is twice that of F_2 progenies and within progeny variance is zero. Also by this method genetic variance is easily maintained for characters of low heritability.

In both the above studies the SSD technique was started at the F_2 in all crosses. In the first study[5, 7] SSD was as effective as reselection in F_3 and F_4 in isolating high yielding F_5 lines. In the second study[21] SSD compared favourably with the best bulk population breeding methods and also maintained high genetic variability.

Reports for other legumes although not comparing different breeding methods have advocated the use of bulk breeding. Yield increases of 23–30% were obtained from applying a bulk pedigree system to three early generations of a diallel cross in cowpeas.[58] From two studies in the common bean[31, 15] it was suggested that single plant selection should be delayed until later generations following bulk breeding. And in pigeon peas it was considered that as yield and yield components showed low heritability, bulk population breeding methods should be applied in early segregating generations.[18]

The limited evidence above, which has dealt solely with yield, does not critically compare pure line, mass and pedigree selection with SSD and bulk population breeding. The soyabean work does show that SSD compares favourably with bulk methods and that pedigree selection can be effective in later generations.

In all breeding methods considered, selection or crossing is followed by strict self pollination with a rapid approach to homozygosity, permitting little further chance for recombination and reassortment of linked genes. This is particularly relevant to the improvement of yield and protein which are quantitatively inherited and negatively correlated. Also, only a few parental lines can be used in the initial crossing which further restricts the potential for obtaining new recombinant types.

(c) *Recurrent selection and population improvement*

Selection procedures that maximise recombination, by recurrent crossing between selected genotypes, have received little attention in legume breeding. These crops are predominantly self pollinated (with the exception of the field bean and pigeon pea) and in the absence of male sterility they require emasculation and hand pollination which is labour intensive. This also limits seed supplies.

The results available on the use of recurrent selection methods indicate the potential of such an approach. A scheme was proposed in soyabeans for selecting and intermating the top 20 lines which would give an expected gain of 235 kg per ha.[32] Brim[12] applied the system to protein improvement in soyabeans and found significant increases for four cycles of recurrent selection. Although no selection was practised for yield a concurrent although non-significant increase was obtained for the unselected trait.

In the common bean a recurrent selection programme has been carried out at Cambridge[29] between a heavy yielding and a high protein F_3 population. In the first cycle, hybrid populations, derived from intercrosses between F_3 single plants of the two populations, could be selected that were higher in yield and protein than the original parents. These populations also compared favourably with F_5 progenies selected as F_3 single plants within each population.

To overcome the restriction on the number of parents used in the initial crossing and to retain the advantages of recurrent selection Jensen[38] proposed a diallel selective mating system. This system uses more than two parents and a combination of a selfing series with selective mating cycles and has been primarily used in cereals. However in a programme to improve the yield of chick pea, Singh and Auckland,[76] although using the more classical approach of limited crossing and pedigree selection, have also initiated a programme using this form of recurrent selection.

Other workers with legumes are also considering this form of selection because of lack of progress with traditional methods. It was found that mass selection and bulk pedigree breeding quickly exhausted variation in a cowpea programme and that further progress would only stem from an increase in variation from hybridisation or mutation.[58] More recently breeding procedures have been described in detail using male sterility or natural crossing in legumes for multiple crossing in population improvement programmes for yield.[67, 34] A more restricted crossing programme was suggested in the common bean for simultaneous improvement of yield and protein, involving intercrossing of high protein plants within high yielding families.[44] In the pigeon pea, which is one of the few grain legumes exhibiting a degree of natural crossing, recurrent selection and population improvement programmes could be easily applied and should yield promising results.[45]

(d) *Inducing new variability by mutation*

An alternative means of creating genetic variation is the use of ionising radiations or chemical agents which cause changes in and rearrangement of chromosomal material. This variation can then be manipulated by standard selection and breeding procedures.

X-ray and neutron radiation were used in two populations of soyabeans and gave a five fold increase in genetic variability for yield.[68] Selection in the M_3 of these populations for yield and protein gave increases in the range of 48–176 kg per ha and 0.86–0.91% protein.[63] In another study[35] on soyabeans, larger increases in protein varying from 0.11 to 3.42% were obtained in three populations after irradiation with gamma rays.

Larger improvements have been reported in the common bean and peas. Gamma radiation in red and pinto populations of beans gave mutant lines with protein levels up to 6.67 and 10.27% higher respectively.[2] Pea mutants were obtained from X-ray and neutron radiation and ranged in yield increase of crude protein per plant from 80 to 96% of the parent variety.[26]

X-rays have also been successfully employed in groundnuts for yield improvement. In one study[28] a yield advance of 15% was obtained in the F_4 after a diallel cross between five deleterious macromutants. In a study in India[64] several generations of selection in a population derived from a cross between two mutants, from the same variety, isolated progenies which gave yield increases of 27–46% of the original variety.

The technique has also been shown to be particularly useful in the common bean in altering a single qualitative character such as seed coat colour to meet specific consumer preferences.[54]

Many breeders feel that there is presently sufficient variability for yield and protein, within the relatively unexploited food legumes, to obtain adequate improvement by the use of the more conventional breeding and selection methods. However, inevitably as variation is exhausted, breeders must consider new methods of creating variation and studies on mutation breeding may have considerable benefit for the future.

(4) CONCLUSIONS: IMPLICATIONS AND PROSPECTS

There are evidently considerable prospects for the improvement of seed yield in the food legumes, both by improved cultivation practices and by breeding. There is therefore the possibility for improving the protein yield per hectare.

There is also the opportunity for improving the per cent protein in these crops but because of the negative relationship between per cent protein and seed yield, strict screening will have to be carried out in the segregating generations to identify and breed for the desirable recombinant types which combine high yield and high protein content. The negative correlation between these characters is usually rather low and not always statistically significant indicating that genetic advance is possible.

Traditional methods of breeding by pedigree selection in the self pollinating legumes have shown lack of significant progress in breeding for yield and protein. Both characters tend to show low heritabilities although the narrow sense heritability for protein content does not appear to be as low as that for seed yield. The use of single seed descent methods would appear to have merit in preserving genetic variation and in accelerating the breeding for these two characters. However, recurrent selection methods are likely to offer the most promising methods since they include selfing and intercrossing of genotypes and thereby achieve the breakage of linkages and allow the accumulation of desirable genes for high yield and high protein and an opportunity to select efficient genotypes.

These methods will however demand a considerable amount of breeding effort in carrying out hand pollinations in crops which are predominantly self pollinating. The search for male sterility becomes imperative in these crops since it would alleviate much of the labour involved in developing recurrent selection methods. Progress in the pigeon pea and the field bean which show a degree of cross pollination should be more easily achieved.

Already high yielding lines with high protein content have been isolated in the soyabean crop[12, 73] and the field bean[9] suggesting that this may also be possible in the other food legumes.

There is some scope for selection for improved protein quality in the food legumes but the plant breeder will need to be provided with a more reliable and easier screening technique for the limiting sulphur amino acids. At present the techniques available show nearly as much error in determination of methionine and cystine as there is genotypic variation in the levels of these amino acids. The screening and improvement of the lysine content in legumes may be easier and may be more relevant to the improvement of protein quality in mixed diets with cereals.

Ultimately it is the nutritional value of food legumes for human consumption in developing countries or for animal feed in the developed world which is the main objective. We still require a greater understanding of the reasons for the low digestibility of legume proteins.

The future objective for plant breeders will be to concentrate firstly on breeding for seed yield in the food legumes and then for protein improvement. However research must proceed on the development of methods for the simultaneous improvement of these two characters which are of such economic importance.

Acknowledgement
The work was carried out within a research project sponsored by the Ministry of Overseas Development whose support is gratefully acknowledged.

REFERENCES

1. ADAMS, M.W. On the quest for quality in the field bean. In M. MILNER (Ed.) *Nutritional Improvement of Food Legumes by Breeding.* P.A.G. Symposium. FAO, Rome, Italy, Wiley (1975) pp. 143–149.

2. AMIRSHAHI, M.C. and TAVAKOLI, M. Protein content of five different species of pulse crops. In *Improving Plant Proteins by Nuclear Techniques.* IAEA, Vienna (1970) pp. 331–335.

3. BLISS, F.A., BARKER, L.N., FRANCHOWIAK, J.D. and HALL, T.C. Genetic and environmental variation of seed yield, yield components and seed protein quantity and quality of cowpea. *Crop Sci.* **13** (1973) 656–660.

4. BLISS, F.A. and HALL, T.C. Food legumes – compositional and nutritional changes induced by breeding. *Cereal Fds. Wld.* **22** (1977) 106–113.

5. BOERMA, H.R. and COOPER, R.L. Comparison of three selection procedures for yield in soybeans. *Crop Sci.* **15** (1975) 225–229.

6. BOERMA, H.R. and COOPER, R.L. Performance of pure lines obtained from superior yielding heterogeneous lines in soybeans. *Crop Sci.* **15** (1975) 300–302.

7. BOERMA, H.R. and COOPER, R.L. Effectiveness of early generation yield selection of heterogeneous lines in soybeans. *Crop Sci.* **15** (1975) 313–315.

8. BOND, D.A. Yield and components of yield in diallel crosses between inbred lines of winter beans (*Vicia faba*). *J. agric. Sci.* **67** (1966) 325–336.

9. BOND, D.A. The development of field beans as a crop in Britain. *Proc. Nutr. Soc.* **29** (1970) 74–79.

10. BOULTER, D., EVANS, I.M. and DERBYSHIRE, E. Proteins of some legumes with reference to environmental factors and nutritional value. *Qualitas P1.* **23** (1973) 239–249.

11. BRIM, C.A. A modified pedigree method of selection in soybeans. *Crop Sci.* **6** (1966) 220.

12. BRIM, C.A. Quantitative genetics and breeding. In B.E. CALDWELL (Ed) *Soybeans: Improvement, Production and Uses.* Am. Soc. Agron. Inc., Madison, U.S.A. (1973) pp. 155–186.

13. BROWN, W.L. The threonine, serine, cystine and methionine content of peanut proteins. *J. biol. Chem.* **142** (1943) 299-301.

14. CHEAH, C.H. and EVANS, A.M. Correlating crude protein percentage with seed yield in segregating generations of *P. vulgaris* crosses. *Bean Improvement Cooperative Ann. Rep.* (1973) 29–31.

15. COYNE, D.P. Correlation, heritability and selection of yield components in field beans, *Phaseolus vulgaris* L. *Am. Soc. Hort. Sci.* 93 (1968) 388–396.

16. CROCOMO, O.J., TULMANN, N.A., BLIXT, S. and MIKAELSEN, K. Breeding for protein in the bean (*Phaseolus vulgaris* L.) I. Inventory of some Brazilian varieties and a number of lines of differing origin. In *Evolution of Seed Protein Alterations by Mutation Breeding.* IAEA, Vienna (1976) pp. 197–208.

17. DAHIYA, B.S. and BRAR, J.S. The relationship between seed size and protein content in pigeon pea (*Cajanus cajan* (L.) Millsp.) *Trop. Grain Legume Bull.* **3** (1976) 18–19.

18. DAHIYA, B.S. and BRAR, J.S. Diallel analysis of genetic variation in pigeon pea (*Cajanus cajan*). *Expl. Agric.* **13** (1977) 193–200.

19. DANIELSON, C.D. An electrophetic investigation of vicilin and leguminin from seeds of peas.

Acta Chem. Scand. **4** (1950) 762–771.

20. DAS, K. and KUMAR, H. Genetic analysis of total protein content in pea. In S. RAMANUJAN and R.D. IYER (Eds) *Proceedings of the Second General Congress of the Society for the Advancement of Breeding Researches in Asia and Oceania.* Session XII. Improvement of Grain Legumes (1974) pp. 777–783.

21. EMPIG, L.T. and FEHR, W.R. Evaluation of methods for generation advance in bulk hybrid soybean populations. *Crop Sci.* **11** (1971) 51–54.

22. FAO. Crops. In *FAO Production Year Book.* Statistics Division of the Economic and Social Policy Department. FAO, Rome, Italy. **30** (1976) pp. 89–186.

23. GAMLIN, L.J. Evaluation of a new method of lysine determination and a study of lysine and protein content and relationship with seed yield in *Phaseolus vulgaris.* Postgraduate Diploma thesis, Dept. Appl. Biol. Library, University of Cambridge (1974) (unpublished).

24. GARDINER, M.L. and GORMAN, R.C. Feed analysis and mineral content of lysine associated with lupinosis of sheep in Western Australia. *Aust. J. exp. agric. Anim. Husbandry,* **5** (1965) 410–416.

25. GOLOVCHENTO, V.I., SOLONENKO, L.P., CHERNYI, I.V., KLVOSTOBA, V.V. and TROFIMOVA, O.S. Amino acid composition of protein in radiation-induced varieties of lupin for fodder and in economically valuable mutants of spring wheat. In *Improving Plant Protein by Nuclear Techniques.* IAEA, Vienna (1970) pp. 149–162.

26. GOTTSCHALK, L. and MULLER, H. Monogenic alteration of seed protein content and protein pattern in X-ray-induced *Pisum* mutants. In *Improving Plant Proteins by Nuclear Techniques.* IAEA, Vienna (1970) pp. 201–212.

27. GREEN, A.G., ORAM, R.W. and READ, B.J. Genetic variation for seed yield, protein content, oil content and seed weight in *Lupinus albus. Aust. J. agric. Res.* (1977) 785–793.

28. GREGORY, W.C. Mutation breeding. In K.J. FREY (Ed). *Plant Breeding.* University Press, Ames (1966) pp. 189–218.

29. GRIDLEY, H.E. and EVANS, A.M. Yield/protein relations in *P. vulgaris.* In *University of Cambridge Grain Legumes Research Project.* Report to the Ministry of Overseas Development (1977) pp. 6–12. (unpublished).

30. HACKLER, L.R. and DICKSON, M.H. A comparison of the amino acid and nitrogen content of pods and seeds of beans (*Phaseolus vulgaris* L.) *Search (agric.) N.Y. State Agric. exp. Stat., Geneva* **3** (1973) 1–6.

31. HAMBLIN, J. Plant breeding interpretations of the effects of bulk breeding on four populations of beans (*Phaseolus vulgaris* L.) *Euphytica* **26** (1977) 157–168.

32. HANSON, W.D., PROBST, A.M. and CALDWELL, B.E. Evaluation of a population of soybean genotypes with implications for improving self-pollinated crops. *Crop Sci.* **7** (1967) 99–103.

33. HARTWIG, E.E. Breeding soybeans for high protein and quality. In *New Approaches to Breeding for Improved Plant Protein.* IAEA, Vienna (1969) pp. 67–70.

34. HAWTIN, G.C., RACHIE, K.O. and GREEN, J.M. Breeding strategy for the nutritional improvement of pulses. In J.H. HULSE, K.O. RACHIE and L.W. BILLINGSLEY (Eds) *Nutritional Standards and Methods of Evaluation for Food Legume Breeders.* IDRC, Ottawa (1977) pp. 48–51.

35. HIRAIWA, S., TANAKA, S. and NAKAMURA, S. Induction of mutants with higher protein content in soybeans. In *Evaluation of Seed Protein Alterations by Mutation Breeding.* IAEA, Vienna (1976) pp. 185–196.

36. HOVE, E.L. Composition and protein quality of sweet lupin seed. *J. Sci. Fd. Agric.* **25** (1974) 851–859.

37. JALIL, M.E. and TAHIR, W.M. World supplies of plant proteins. In J.W.G. PORTER and B.A. ROLLS (Eds). *Proteins in Human Nutrition.* Academic Press, London (1973) pp. 35–46.

38. JENSEN, N.F. A diallel selective mating system for cereal breeding. *Crop. Sci.* **10** (1970) 629–635.

39. KARUP, A. and DAVIS, D.W. Inheritance of seed yield and its components in a six parent diallel cross in peas. *J. Am. Soc. Hort. Sci.* **95** (1970) 795–797.

40. KAUL, A.K. Research in combining nutritive quality with high yield. In *Recent Research in the Improvement of Protein and Nutritive Properties of Food and Food Plants.* IARI Research Service 6 (1971) pp. 7–32.

41. KELLY, J.D. Genetic modifications of protein quantity and quality in beans *Phaseolus vulgaris* L. *Diss. Abstr. Int.* **B36** (1975) 86B.

42. KELLY, J.F. Genetic variation in the methionine levels of mature seeds of common bean (*Phaseolus vulgaris* L.) *J. Am. Soc. Hort. Sci.* **96** (1971) 561–563.

43. KELLY, J.F. Increasing protein quantity and quality. In M. MILNER (Ed) *Nutritional Improvement of Food Legumes by Breeding* P.A.G. Symposium. FAO, Rome. Wiley, New York (1975) pp. 174–184.

44. KELLY, J.D. and BLISS, F.A. Heritability estimates of percentage seed protein and available methionine and correlation with yield in dry beans. *Crop Sci.* **15** (1975) 753–757.

45. KHAN, T.N. A new approach to the brreding of pigeon pea (*Cajanus cajan* Millsp.): formation of composites. *Euphytica* **22** (1973) 373–377.

46. KUSHAWA, J.S. and TAWAR, M. Estimates of genotypic and phenotypic variability in groundnut (*Arachis hypogea* L). *Indian J. Agric. Sci.* **43** (1973) 1049–1054.

47. LAL, B. Diallel analysis of grain yield and some other quantitative traits in Bengal gram (*Cicer arietanum* L.). Unpublished M.Sc. thesis, Punjab Agricultural University, Ludhiana (India) Library, p. 95. (Quoted in Singh, K.B. and Auckland, A.K.[76]).

48. LANTZ, E.M., GRUGH, H.W. and CAMPBELL, A.M. Effect of variety, location and years on the protein and amino acid content of dried beans. *J. Agric. Food Chem.* **6** (1958) 58–60.

49. LELEJI, O.I., DICKSON, M.H., CROWDER, L.V. and BOURKE, J.B. Inheritance of crude protein percentage and its correlation with seed yield in beans, *Phaseolus vulgaris. Crop Sci.* **12** (1972) 168–171.

50. MAHMOUD, S.A. Heterosis and combining ability in some broad beans (*Vicia faba*) diallel crosses. *Pl. Breed. Abstr.* **47** (1977) 12177.

51. MALHOTRA, V.V. and SINGH, K.B. Genetic variability and path coefficient analysis for protein in green gram (*Phaseolus aureus* Roxb.) *Egypt. J. gen. Cytol.* **5** (1976) 170–173.

52. MERTZ, E.T., BATES, L.S. and NELSON, O.E. Mutant gene that changes protein composition and increases lysine content of maize endosperm. *Science* **145** (1964) 279–280.

53. MILLERD, A. Biochemistry of legume seed proteins. *Ann. Rev. Plant Physiol.* **26** (1975) 53–72.

54. MOH, C.C. Mutation breeding in seed-coat colors in beans (*Phaseolus vulgaris* L.). *Euphytica* **20** (1971) 119–125.

55. MUNCK, L., HOLMBERG, E., KARLSSON, K.E., LOFQUIST, B. and SJODIN, J. Breeding for protein quantity and quality in the field bean (*Vicia faba*) and barley (*Hordeum vulgare*). In M. MILNER (Ed) *Nutritional Improvement of Food Legumes by Breeding.* P.A.G. Symposium. FAO, Rome. Wiley, New York (1975) pp. 197–215.

56. MUNCK, L., KARLSSON, K.E. and HAGBERG, A. Selection and characterization of a high-protein, high-lysine variety from the world barley collection. In R.A. NILAN (Ed) *Proceedings of Second International Barley Genetics Symposium.* Washington State University Press, Washington (1971) pp. 544–558.

57. NELSON, O.E. Genetic modifications of protein quality in plants. *Adv. Agron.* **21** (1971) 171–194.

58. OJOMO, O.A. Yield potential of cowpeas, *Vigna unguiculata* (L.) Walp: results of mass and bulk pedigree selection methods in Western Nigeria. *Nigerian Agric. J.* **11** (1974) 150–156.

59. ORR, M.L. and WATT, B.K. Amino acid content of foods. In *U.S.D.A., Home Economics Research Report No. 4* (1957) pp. 16–54.

60. OTOUL, E. Répartition des principaux acides aminés dans les differentes parties de la graine d'un cultivar de *Phaseolus vulgaris* L. *Bull. Rech. Argron. Gembloux* IV (1969) 287–301.

61. PANDEY, S. and GRITTON, E.T. Gentotypic and phenotypic variances and correlations in

peas. *Crop Sci.* **15** (1975) 353–356.

62. PANDEY, S. and GRITTON, E.T. Observed and predicted response to selection for protein and yield in peas. *Crop Sci.* **16** (1976) 289–292.

63. PAPA, K.E., WILLIAMS, J.H. and HANWAY, O.G. Effectiveness of selection for quantitative characters in the third generation following irradiation of soybean seeds with X-rays and thermal neutrons. *Crop Sci.* **1** (1961) 87–90.

64. PATIL, S.H. Trombay groundnut selection for increased oil content and yield. *Indian J. Agric. Sci.* **43** (1970) 370–376.

65. PORTER, W.M. Genetic control of protein and sulphur contents in dry beans. *Phaseolus vulgaris.* Ph.D. Thesis. Lafayette, Indiana. Purdue University (1972).

66. PUTZTAI, A. The isolation of two proteins. Glycoprotein I and trypsin inhibitor, from seeds of kidney bean (*Phaseolus vulgaris* L.). *Biochem. J.* **101** (1966) 379–384.

67. RACHIE, K.O. and GARDNER, C.O. Increasing efficiency in breeding partially outcrossing grain legumes. In *International Workshop on Grain Legumes.* International Crops Research Institute for the Semi-Arid Tropics. Begumpet, India (1975) pp. 285–300.

68. RAWLINGS, J.O., HANWAY, O.G. and GARDNER, C.O. Variation in quantitative characters of soybeans after seed irradiation. *Agron. J.* **50** (1960) 524–528.

69. ROBERTS, L.C. and BRIGGS, D.R. Isolation and characterization of the 7S component of soybean globulins. *Cereal Chem.* **42** (1965) 71–85.

70. RUTGER, T.N. Variation in protein content and its relation to other characters in beans (*Phaseolus vulgaris* L.) In *Report of the Tenth Dry Bean Research Conference, Davis, Calif., U.S.A.* **10** (1970) pp. 59–69.

71. SANDHU, S.S., KEIM, W.F., HODGES, H.F. and NYQUIST, W.E. Inheritance of protein and sulfur content in seeds of chick peas. *Crop Sci.* **14** (1974) 649–652.

72. SHANNON, J.G., WILCOX, J.R. and PROBST, H.H. Estimated gains from selection for protein and yield in the F_4 generation of six soybean populations. *Crop Sci.* **12** (1972) 824–826.

73. SHARMA, O. and GREEN, J.M. Perspective of pigeon pea and ICRISAT'S breeding program. In *International Workshop on Grain Legumes.* International Crops Research Institute for the Semi-Arid Tropics. Begumpet, India (1975) pp. 19–29.

74. SHELDRAKE, A.R. Session Review. In *International Workshop on Grain Legumes.* International Crops Research Institute for the Semi-Arid Tropic, Begumpet, India (1975) pp. 225–226.

75. SINGH, T.P. Epistatic bias and gene action for protein content in green gram (*Phaseolus aureus* Roxb.) *Euphytica* **23** (1974) 459–465.

76. SINGH, K.B. and AUCKLAND, A.K. Chickpea breeding at ICRISAT. In *International Workshop on Grain Legumes.* International Crops Research Institute for the Semi-Arid Tropics. Begumpet, India (1975) pp. 3–17.

77. SINGH, R. and AXTELL, J.D. High lysine mutant gene (hl) that improves protein quality and biological value of grain sorghum. *Crop Sci.* **13** (1973) 535–539.

78. SINGH, K.B. and JAIN, R.P. Analysis of diallel cross in *Phaseolus aureus.* Roxb. *Theor. appl. Genet.* **42** (1971) 279–281.

79. SINGH, T.P. and SINGH, K.B. Heterosis and combining ability in *Phaseolus aureus* Roxb. *Theor. appl. Genet.* **44** (1974) 12–16.

80. TANDON, O.B., BRESSANI, R., SCRIMSHAW, N.S. and LE BEAU, F. Nutritive value of beans. Nutrients in Central American beans. *J. agric. Food Chem.* **5** (1957) 137–142.

81. THOMAS, R.L., PRIOR, A.J. and GRAFIUS, J.E. Improving the quality of a groundnut population in Zambia by bulk selection of seed. *Expl. Agric.* **10** (1974) 185–192.

82. TOBIN, G. Unpublished data. Dept. Appl. Biol. University of Cambridge (1973).

83. TOBIN, G. and CARPENTER, K.J. The nutritional value of the dry bean (*Phaseolus vulgaris*) — a literature review. *Nutr. Abstr. Rev.* (1978) in press.

84. WEBER, C.R., EMPIG, L.T. and THORNE, J.C.

Heterotic performance and combining ability of two-way F_1 soybean hybrids. *Crop Sci.* **10** (1970) 159–160.

85. WILLIAMS, P.C. Application of near infrared reflectance spectroscopy to the analysis of cereals and oilseeds. *Cereal Chem.* **55** (1975) 561–576.

86. WOOLFE, J.A. and HAMBLIN, J. Within and between genotypes variation in crude protein content of *Phaseolus vulgaris* L. *Euphytica* **23** (1974) 28–30.

CURRENT RESEARCH INTO TRACHEARY ELEMENT FORMATION

J.R. BARNETT

Plant Science Laboratories, University of Reading,
Whiteknights, Reading, Berks. RG6 2AS

There is a growing appreciation of the importance of wood as a major source of renewable energy, especially for developing nations. The need to halt and, if possible, reverse the continuing depletion of the world's forest resources has been the subject of an increasing number of reports and publications.[10, 19, 40] If these reports are heeded, and management of plantations and existing forests for energy supply is to be as efficient as possible, the research effort into all aspects of this type of forestry will have to be intensified. It will also be important to understand as fully as possible the basic process of wood formation, since it is ultimately this process which stores the energy of photosynthesis in a form which man can use.

Despite increasing interest in xylogenesis in recent years no single aspect of the process is completely understood. The differentiation of primary[20, 39, 55, 56] and secondary[55, 56] xylem has been the subject of earlier reviews.

The initiation and mechanism of xylem differentiation have been studied both *in vivo* and *in vitro*. *In vivo* investigations have involved mainly the removal of differentiating tissues from a stem, sometimes after treatments aimed at modifying the differentiation process, for examination by light or electron microscopy. These treatments have been physical, such as manipulation of light intensity and temperature,[15] and chemical, involving applications of hormones.[16, 17, 29] Such of this work as has been carried out on secondary xylem has been mostly by researchers who are directly or indirectly connected with the forest industry. By far the greater part of the work on xylogenesis has, however, been carried out on primary tissues. These are easier or more convenient to handle than the hard woody tissues of a tree stem which, in any case, are rather difficult to remove from the stem without damage.

The difficulty of observing what is occurring in a living stem, and of measuring hormone and nutrient levels *in situ*, has led many workers to an alternative approach; namely, the use of tissue cultures. These have been manipulated to induce xylogenesis in previously parenchymatous callus, although the number of ways in which this has been effected almost equals the number of species examined. Consequently there is disagreement about the factors responsible for stimulation of xylogenesis. It is also uncertain to what extent results obtained *in vitro* may be extrapolated to the living stem in which the physical and chemical environment of the meristem is almost certainly very different from that of the callus.

In this commentary it is proposed to consider aspects of xylogenesis which are becoming better understood, and to discuss some of the problems which remain to be explained. For a comprehensive review of the topic the reader is referred to Roberts.[55, 56]

THE ROLE OF CELL DIVISION

All the cells in the differentiated plant body arise by division of a meristem or its derivatives. Whether division plays a part in the differentiation process itself or merely provides more cells to undergo differentiation has been a subject of interest to many workers. Roberts[55, 56] has described attempts made to resolve this problem using tissue culture techniques and hormones known to stimulate or inhibit mitosis.

Among significant observations in recent years Torrey and Foskett[65] found that cytokinin was essential for the differentiation of xylem in cultured pea root segments. Pea segments in a medium without kinetin underwent only a proliferation of the pericycle, which caused the cortex to break up forming a friable callus in which the cells were polyploid. Addition of kinetin was followed by division in cortical cells and formation of tracheary elements with reticulate secondary walls. Wright and Northcote,[73, 74] observed that tetraploid cells in sycamore callus did not differentiate. It was found necessary to add NAA and kinetin at particular levels to induce differentiation. The effect of cell division was thought to involve reduction of the state of ploidy in the cell to a level at which differentiation was permitted to occur.

However, *in vivo* studies on differentiation of metaxylem in *Allium cepa* root tips[24] supported earlier observations[36, 62] that there is an increase in nuclear DNA in differentiating cells, and it has recently been shown that somatic polyploidy is common in xylem precursor cells.[33] Recent studies[18, 44] have led to the conclusion that the ploidy level of differentiating cells is not an important factor in differentiation as similar levels were found in both differentiating tracheary elements and undifferentiated cells of cultured explants of Jerusalem artichoke tuber. The observations did suggest, however, that any precursor cells synthesising DNA must divide prior to differentiation. As in the observations of Wright and Northcote[73, 74] auxin and cytokinin were found to be necessary for cell division and differentiation.

Some doubt has, however, been thrown on the view that cell division must always precede differentiation.[64] Tracheary element formation was induced in single isolated cells in culture, which had not been observed to divide since culturing. The possibility exists, of course, that the cells which differentiated had undergone a division prior to isolation which had predisposed them to differentiation.

Cell division is clearly an essential part of xylem differentiation in an intact stem, in that the arrangement of cells in the vascular bundle of primary tissue reflects the pattern of division which has followed that of the primary meristem. In the secondary body, division of the cambium is necessary to maintain growth in the stem, periclinal divisions increasing the radius of the stem and anticlinal divisions maintaining continuity in the cambium as the circumference increases. The pattern of these divisions and the fate of the cells produced has been the subject of several painstaking studies.[2, 3, 4, 13, 14]

Spiral grain in conifers has been attributed to a tendency for pseudotransverse divisions to favour a particular direction. In fact the orientation of the mitotic spindle at late metaphase does show some tendency to favour division in such a way that it would reinforce spirality in a particular direction. Harris[22] reviewed the evidence for this and concludes as a result of his own work that spiral grain arises mainly from differential growth of the fusiform initials, possibly reflecting a spirality in auxin flow down the stem, which produces curvature at the cell tips. It was suggested that the skewing of the mitotic spindle reflected rather than caused the cytoplasmic stress leading to spirality.

It seems certain therefore, that cell division has an important part to play in xylem differentiation, probably triggering the change from a parenchymatous behaviour pattern by creating the appropriate nuclear DNA conditions. It seems certain also, that the pattern of cell division is the single main factor in determining the stem anatomy in any species. Further studies using callus to examine the effects of division on differentiation and nuclear DNA levels, coupled with examination of DNA levels in cambium and differentiating xylem, should lead to a better understanding of this part of differentiation.

THE ROLE OF NUTRIENTS AND HORMONES

The effect of various hormones on xylem differentiation has been extensively investigated *in vivo* and *in vitro*.[55, 56] Gibberellic acid, auxins, abscisic acid and cytokinins have all been shown to affect xylem differentiation in a variety of ways depending on the species studies.

Maintenance of growth in tissue cultures invariably requires the presence of an auxin, although this alone is not usually sufficient to induce xylogenesis.[56] It is probable, though, that a supply of auxin is necessary for differentiation, notably during the phase of cell enlargement. The origin of this auxin *in vivo* is still uncertain. While there is a large body of evidence to suggest that auxin which originates in the leaves travels in the phloem to the sites of differentiation,[56] there is also evidence that hormones are formed in the autolysing cytoplasm of xylem cells at maturation.[59, 60] The work of several groups has indicated that two or more hormones acting in unison may be essential for xylogenesis, although these workers have concentrated on the role of hormones in the initiation of the process.[56]

The effect of hormone applications on the quality of xylem produced, e.g. cell dimensions, wall thickness etc., has been examined mainly by forestry oriented scientists working on forest tree species.[16, 29] Jenkins[29] found that IAA applied to seedling stems of *Pinus radiata* D. Don had no effect on tracheid diameter, but led to a marked increase in wall thickness. Abscisic acid reduced cell radial diameter and rate of cell production. Denne and Wilson[16] found that IAA had no effect on tracheid dimensions and wall thickness in disbudded shoots of *Picea abies* Karst and *Picea sitchensis* (Bong.) Carr. However, lateral application of IAA to intact shoots of these species did increase both tracheid diameter and wall thickness. Denne and Wilson[16] suggested from this that IAA acts synergistically with another growth regulator normally supplied by the buds, controlling cell size and wall thickness by affecting the time spent by the differentiating cell in these respective stages of development.

It has been apparent for some years that factors other than hormones are involved in xylogenesis in callus cultures. Sucrose has been shown to play an important role in the control of differentiation,[28, 72] and it has been demonstrated that the concentration of sucrose in the nutrient medium may be used to manipulate the direction of differentiation. At a particular auxin level, sucrose at between one and two per cent led to xylem formation; sucrose at four per cent led to phloem formation, while concentrations in between these gave a mixture of xylem and phloem. The resemblance of these concentrations to the concentration gradient that might exist between the differentiating xylem and phloem through the cambium in a living stem led the authors to suggest that such a sucrose gradient might play an important role in the differentiation of cambium *in vivo.* Recent work[69] has re-emphasised the importance of nutrient concentrations in differentiation. Differentiation of simple-pitted parenchyma cells with some secondary wall was induced in callus from *Pinus radiata* by reducing either sucrose or major inorganic nutrient concentrations to half the level required for maintenance of callus growth. By reducing both sucrose and major inorganic nutrient concentrations, however, nodules which contained a stem-like structure were induced to form at the surface of the callus. In both experiments, auxin was in the media at maintenance level. A striking feature of the differentiated nodules was that their stem-like structure was of a secondary type, with phloem, cambium and xylem cells readily identifiable.[7] In the xylem, tracheids had uniform secondary thickening and bordered pits. A second important observation was that the cells of this stem-like structure were oriented only locally, showing that the factor which produces the vertical orientation of tracheary elements in a stem was missing. In this, the nodules resembled the callus nodules formed *in vivo* in resin pockets of *Pinus ponderosa.*[23] In both cases this absence of uniform orientation could be explained by the absence of a polar auxin flow down through the differentiating tissue. In the case of the callus the source was below the tissue in the medium, while the resin pocket nodules only contact with the rest of the stem was via the few ray parenchyma cells from which the nodule had formed on the xylem side of the pocket.

It is noteworthy that differentiation was induced in *P. radiata* callus without the addition of any hormone other than indolebutyric acid. It is possible, however, that the callus may itself produce kinetin or other hormones it requires, given a suitable stimulus. The stimulus in *P. radiata* callus may be starvation in this case. Torrey[64] noted without comment that cells maintained for long periods on his nutrient medium differentiated into tracheary elements. Here again, fall in nutrient levels may have been responsible.

Unfortunately for our understanding of xylogenesis uncertainty remains about the extent to which callus or a stem produces its own differentiation factors additional to those applied by the investigator. Applications of hormones to stems or calluses may in fact be excessive in many cases, so that it is an overdose response which is being studied. It has been suggested that hormones might be produced in the differentiating cells themselves, although this view was recently contested.[16]

It is apparent from studies on the effects of hormones and nutrients on differentiation, that although cell division may be a trigger for differentiation, other factors in the stems' internal environment must be involved in ensuring that the potential for differentiation released by the trigger is realised. It is also clear that no one factor can be held responsible, but that many separate factors must combine in suitable proportions if the process is to go to completion.

CELL ENLARGEMENT

The extent to which tracheary elements enlarge following division of the meristem is of great interest to the forest industry and has led to studies on the effect of hormones and environment on tracheid dimensions in gymnosperms.[15, 16, 29, 34, 75]

The amount of growth which takes place in a cambial derivative prior to the onset of secondary wall formation is high variable. Bailey[1] measured the cambial initials and mature tracheary elements in a variety of gymnosperms and angiosperms and found that growth in length varied from zero in *Gingko biloba* to between 10 and 15% in gymnosperms such as *Thuja occidentalis.* In angiosperms, however, increases in length of several hundred per cent were common for fibre-tracheids while vessel segments in these species were seldom longer than the initials from which they arose. Growth in cell diameter is similarly variable, growth being limited to the radial direction in the tracheids of gymnosperms, while the vessels of angiosperms expand radially and tangentially to become many times larger than their initials.

In a woody stem, any growth must take place against considerable restraints. The mature xylem to the interior of the zone of differentiating cells, and the often thick bark to the outside are both rigid structures. That the thin walled cambial and enlarging cells are not crushed and are able to expand must mean that they possess a high turgor pressure. It is, of course, the older phloem which gives under this pressure, and eventually becomes crushed and non-functional. Ultimately the bark also gives under the stress and takes on its characteristic fissured appearance.

Turgor alone is insufficient to account for cell growth, and many authors have investigated the effect of growth hormones on the cell wall in attempts to explain the role of these substances in growth.[56] That auxin is a major causative agent in cell wall loosening is now beyond doubt, although its mechanism of action is still a matter of dispute. The possibility that it acts in the cell by stimulating a proton pump mechanism which liberates hydrogen ions into the wall has been investigated by a number of workers in recent years.[11, 25, 42, 45, 46, 51, 52, 66] The hydrogen ions are thought to act by breaking hydrogen bonds between wall components. Pope[45] however, concluded that IAA and hydrogen ions promote growth by different mechanisms; IAA breaking covalent linkages in non-cellulosic wall components while protons break the physical linkages. A full understanding of the mechanism of cell wall loosening will probably depend on our acquiring a better model of the structure of the middle lamella/primary wall system than we possess at present.

Whatever the mechanisms of auxin action on cell walls and growth may be, any growth takes place within a continuous, enclosed system. Intercellular spaces are not usually found between fusiform cambium cells, and each cell tip can only grow by somehow forcing its way between its neighbours. Several hypotheses have been developed to explain how this occurs, including the symplastic growth hypothesis[50] and the intrusive growth hypothesis.[61] This early work has been reviewed by Wenham and Cusick,[70] who studied the growth of secondary wood fibres in *Salix viminalis* L. These authors concluded that the growing fibre tip secretes an enzyme which weakens the middle lamella between the cells ahead of it. These cells then round off under the influence of their own internal turgor leaving a space into which the fibre tip may grow.

Studies on expanding tracheids in *Pinus radiata* have shown that it is during the phase of cell enlargement that the sites of bordered pits are determined.[8] This is contrary to the long held view that the pit sites are predetermined in the cambial initials are primary pit fields in the radial cell walls.[31] It was suggested that these pit fields divide with the rest of the cell and are enlarged and restored to their original size with *de novo* formation of more plasmodesmata between cell divisions. The demonstrated absence of plasmodesmata from the radial walls of *Pinus radiata* cambium[8] rules out the possibility that these structures are involved in the determination of pit sites. It has been proposed instead,[8] that a localised thinning of the cell wall, brought about by loosening and rearrangement of the wall structure, gives rise to circular thinner areas around which a thick rim is formed which protrudes into both cells involved in formation of the particular pit pair. This rim correctly locates the pit site for each cell without the need for any other communication, and may act as the template required by the hypothesis of Young and Service[77] for formation of the pit border during the phase of secondary wall formation.

This process of pit site determination may be related to cell enlargement in that the factor required to cause localised loosening of the wall may be the same as that involved in loosening of the wall for growth in the cell tip regions. For example a short-term flux of hydrogen ions released into the wall from a vesicle could initiate the necessary localised wall rearrangement. The presence of vesicles which might perform this task in enlarging cell walls has been demonstrated.[8]

The presence of plasmodesmata, or plasmodesmata-like structures, which are occasionally seen in the torus of a bordered pit, remains to be explained. It is felt that in cases where these pits lie on a tangential wall they are probably true plasmodesmata which remain from the original cell plate formed following division of the initial. Such plasmodesmata persist in pit membranes in the tangential walls of ray cells in *Aesculus hippocastanum* (Barnett-unpublished observation), and in the tangential walls between ray parenchyma cells in gymnosperms.[5, 37] Plasmodesmata-like structures in the bordered pit tori in the radial walls of tracheids of *Pinus radiata* are found, on careful examination, to pass only halfway through the torus, and frequently end in a median nodule at the middle lamella. These are similar to the ectoplasmic stands described by Roland,[58] and are almost certainly produced *de novo* during differentiation.[6] It is possible that they have a role in lignification as they often appear to be connected with the osmiophilic vesicles produced by dictyosomes at the time when lignification of the wall is occurring.[6] It is also possible that these ectoplasmic strands have on occasion been misinterpreted as plasmodesmata, their tendency to end at the middle lamella being thought a consequence of breakage during cell enlargement, or a reflection of their failure to be within the plane of section throughout their length.

SECONDARY THICKENING

Secondary walls are a characteristic feature of tracheary elements. In primary tissue cellulose is laid down in characteristic patterns which allow the cell to retain a certain amount of flexibility at a time when the plant body requires freedom to extend longitudinally in the region of growth. The annular and helical patterns of thickening found in first formed primary tracheary elements which allow this flexibility give way to the reticulate, less flexible patterns as the region of growth ages.[39]

In the secondary xylem on the other hand, the cellulose microfibrils are laid down in the well-known lamellar pattern which gives maximum strength to the wall. At the time this is occurring, the stem has ceased any significant longitudinal growth in the zone where secondary tracheary elements are being formed, and there is no requirement for the cells to be able to elongate following thickening. The cell wall and its synthesis have been the subjects of a recent review.[48] The mechanism of synthesis and orientation of cellulose microfibrils has long been the subject of disagreement among workers, who in recent years became polarised towards two hypotheses. According to one, microtubules, first described by Ledbetter and Porter,[35] have been invoked variously as the synthetic and orienting structures of cellulose microfibrils. The alternative view, proposed by Preston,[47] was that an ordered system of enzyme granules at the plasmalemma surface was responsible for synthesis and orientation. Evidence for the microtubule theory was at first readily obtained, while the granule hypothesis suffered from the difficulty of observing such a system with electron microscopy. The advent of freeze-etching, however, has permitted granules to be visualised at the plasmalemma surface in cellulose synthesising organisms.[9, 57] Strong evidence for Preston's hypothesis has recently come from work by Peng and Jaffe,[43] who were able to visualise, by freeze-etching, granules associated with microfibrils at the plasmalemma surface of *Pelvetia* embryos. Convincing evidence against the microtubule hypothesis was presented by Nelmes *et al.*[38] who demonstrated the absence of these structures from the differentiating xylem cells of apple stems suffering from the condition known as rubbery wood. In this condition the cell walls lack lignin and the wood becomes very pliable. The differentiating xylem cells of normal trees contained microtubules. This supports the contention that microtubules appear to be involved in lignification.[6, 37] Barnett[6] found that in differentiating tracheids of *Pinus radiata*, microtubules are present during cell enlargement, possibly as remnants of the mitotic spindle; and towards the end of secondary wall formation in especially large numbers. Wardrop,[68] has shown that it is at this time that lignification of the secondary wall begins.

Those who adhere to the view that microtubules are involved as orienting agents in cellulose biosynthesis point to the destruction of microtubules and the random orientation of microfibrils which occurs after treatment of cells with colchicine. The danger with the use of this and similar agents is, of course, that they probably affect many cell processes, and the destruction of microtubules may be merely coincidental with the formation of a random wall structure. Further support for the granule hypothesis in higher plants has come from electron microscopy of differentiating fibres and tracheids in which granular material may occasionally be seen between the plasmalemma and developing wall.[6, 37, 54]

LIGNIFICATION

The process of lignification follows that of cellulose deposition in the cell wall.[32, 68] The possible role of dictyosomes and ectoplasmic strands in the process of lignification[C] has already been mentioned. It is not proposed to consider here the complex chemistry of the lignification process, but rather its significance for the cell. Its structural role is clearly illustrated by the effect that its absence has in the case of rubbery wood in apples.[38] In addition to a structural role it now appears probably that lignin protects the wall from digestion by the autolytic enzymes released on the death of the cell. Certainly the effect of these enzymes is very marked in primary tracheary elements,[39] where the lignified annular, spiral and reticulate thickenings remain unaffected while the regions of wall between the thickenings are attacked and become porous, permitting the passage of water. Johnson[30] has suggested that the relative rigidity of the thickenings and the flexibility of the intervening lignin-free wall areas may be involved in the control of water flow through the wall; the more flexible unlignified wall regions flexing inwards or outwards according to the state of turgor or tension within the cell; and acting as a valve in consequence of resultant opening and closing of pores in the wall.

In secondary tracheary elements it is the non-lignified margo which suffers attack by the autolytic enzymes at cell death, while the main part of the cell wall and the lignified torus of the pit membrane remain intact. The mechanism by which selected areas of wall remain unlignified in this way is unknown.

REACTION WOOD

An aspect of xylem differentiation in which lignification is an important factor is the formation of reaction wood. The result of deviation from the vertical in a tree stem is the formation of compression wood in gymnosperms and tension wood in angiosperms as the tree attempts to restore its vertical alignment. The structure and properties of reaction wood have been reviewed by Jane[27] and the formation of compression wood has been discussed by Timell.[63] As it is generally considered to be a defect in timber the mechanism of its formation merits attentions. The classical theory is that compression wood is formed in response to high levels of auxin. Evidence for this comes from experiments in which compression wood has been induced in actively growing stems forming normal wood by addition of excess auxin,[71] and from the observation that in fast growing specimens of some conifers a ring of compression wood may form even in vertical trees. Presumably this reflects the higher auxin levels in these faster growing specimens.

During the formation of compression wood cells, the tracheids develop walls which usually lack an S_3 layer, have a steeper microfibril angle in the S_2 layer than normal wood, have a higher lignin content than normal wood, and possess large intercellular spaces between the adjacent tracheids, which thus have a rounded appearance in transverse section. In addition the compression wood tracheids are shorter on average than the normal tracheids in the same tree. This presumably reflects the more rapid growth rate on the compression wood side of the stem, requiring more pseudotransverse divisions to maintain cambial continuity around the rapidly expanding circumference.

How these compression wood features act to restore the alignment of the stem is not understood. Jane[27] has suggested that differential shrinkage is a factor. He proposes that since, during differentiation, tracheary elements pass from a state of turgor to a state of tension when they become functional conducting elements, shrinkage will occur at this time. The presence of extra lignin in the compression wood tracheids[41] would tend to reduce shrinkage in these cells to a lower value than is found in normal tracheids, so that eventually the stem reassumes a vertical orientation. Indeed, the extra shrinkage of the opposite wood tracheids would tend to put the compression wood tracheid under extra tension, giving rise to separation along the middle lamellae between these cells, and thus the characteristic intercellular spaces.

Tension wood fibres form on the upper side of a leaning stem or a branch of angiosperms. Low levels of auxin are thought to be responsible for their formation. They are known as gelatinous fibres owing to the presence of a gelatinous or G layer interior to the S layers. Wardrop[67] has reviewed work on the structure and formation of tension wood. Of the S layers the S_2 and S_3 may or may not be present.[12, 53] The amount of lignification in the S_2 and S_3 layers is variable, but the G layer is almost pure crystalline cellulose with its microfibrils oriented at about 5° to the long axis of the cell.[49] Gelatinous tracheids have recently been described in gymnosperms,[26] and the formation of these cells, as the formation of gelatinous fibres in angiosperms, has been attributed to low levels of auxin. As with compression wood, we have as yet no real idea of the mode of action of tension wood gelatinous fibres in straightening the stem.

The formation of these 'abnormal' cell types illustrates again the important role of auxin in controlling cell differentiation where slight changes in concentration result in major structural changes.

SENESCENCE

The mature tracheary element is, of course, a hollow tube whose walls with their pits represent the end result of the differentiation process. Thus during differentiation the cytoplasm of the cambial derivative is following a programme which leads to its own destruction. The final stages of this process have been observed microscopically by Wodzicki and Humphreys.[76] The constructive processes of cell enlargement and cell wall formation can hardly be thought of as senescent, but once these processes are complete the cell dies and becomes a functional part of the sap-stream system. Xylem differentiation has been studied as a model for cell death by Gahan[21] who has examined the activities of various hydrolytic enzymes in maturing primary xylem cells. That such enzymes do play an essential role in the formation of tracheary elements is indicated by their action on unlignified regions of the wall in primary tracheary elements, and on the margo of bordered pit membranes in secondary elements. Their action in these cells at the end of wall formation is essential if the cells are to carry out their function of water conduction. The possibility that their action is limited to particular areas of the wall by lignification of areas to be protected has already been mentioned.

The soup of enzymes formed by autolysis of the cytoplasm must also lead to the formation and release of many useful products. The possibility exists that hormones required in the differentiation of further tracheary elements are among these.[59, 60] As it is unlikely that each individual tracheary element immediately joins the water conducting system on maturation, it is possible that many of these breakdown products are absorbed from the tracheary element by adjacent parenchyma cells. In forest trees for example, some of the materials may be absorbed into the rays as heartwood precursors, and then transported to the centre of the tree for their incorporation into heartwood. Alternatively they may be passed to the resin canals in gymnosperms.

Clearly the events occurring at the end of the differentiation process, which result in the removal of most of the cytoplasmic material from the cell lumen, have important positive roles in the formation of xylem. They cannot be considered as merely leading to the removal of materials which are no longer required. Further study of these terminal events may well provide important information about the control of differentiation, and the formation of heartwood and resins.

CONCLUSION

From the above discussion of a few of the aspects of xylem differentiation currently under examination it is apparent that we are still far from understanding the process fully. The problems are, of course, not simple ones. The extent to which each cell can control its own destiny, and to which it relies on external stimuli is not clear. Indeed, we do not yet know what sort of environment the meristematic cells are in within a stem, or what the internal environment of a piece of callus is like. The use of hormones and nutrients to induce differentiation can tell us at best only a part of the story.

It is probably safe to conclude, however, that the process of differentiation is induced in the stem by stimuli related to those applied by investigators, and described briefly above.

In order to understand the full story, it will almost certainly be necessary to improve techniques for *in vivo* examination of xylem formation, with a view to observing the process itself, or even obtaining a true picture of the environment of the differentiating cells. Steady progress is being made, however, and it is likely that this will accelerate as awareness of the importance of efficient xylem production from economic and social viewpoints increases.

REFERENCES

1. BAILEY, I.W. The cambium and its derivative tissues — II. Size variations of cambial initials in gymnosperms and angiosperms. *Am. J. Bot.* **7** (1920) 355–367.

2. BANNAN, M.W. The frequency of anticlinal divisions in fusiform cambial cells of *Chamaecyparis. Am. J. Bot.* **37** (1950) 511–519.

3. BANNAN, M.W. Some aspects of the elongation of fusiform cambial cells in *Thuja occidentalis. Can. J. Bot.* **34** (1956) 175–196.

4. BANNAN, M.W. Ontogenetic trends in conifer cambium with respect to frequency of anticlinal divisions and cell length. *Can. J. Bot.* **38** (1960) 795–802.

5. BARNETT, J.R. Seasonal variation in the ultrastructure of the cambium in New Zealand grown *Pinus radiata* D. Don. *Ann. Bot.* **37** (1973) 1005–1011.

6. BARNETT, J.R. Tracheid differentiation in *Pinus radiata. Wood Sci. Tech.* **11** (1977) 83–92.

7. BARNETT, J.R. Fine structure of parenchymatous and differentiated *Pinus radiata* callus. *Ann. Bot.* **42** (1977) 367–373.

8. BARNETT, J.R. and HARRIS, J.M. Early stages of bordered pit formation in radiata pine. *Wood Sci. Tech.* **9** (1975) 233–241.

9. BARNETT, J.R. and PRESTON, R.D. Arrays of granules associated with the plasmalemma in swarmers of *Cladophora. Ann. Bot.* **34** (1970) 1011–1017.

10. BURLEY, J. Selection of species for fuelwood plantations. *8th World Forestry Congress,* Jakarta, Indonesia. (1978).

11. CLELAND, R. Auxin induced hydrogen ion secretion from *Avena* coleoptiles. *Proc. Nat. Acad. Sci. U.S.A.* **70** (9173) 3092–3093.

12. CÔTÉ, W.A., DAY, A.C. and TIMELL, T.E. A contribution to the ultrastructure of tension wood fibres. *Wood Sci. Tech.* **3** (1969) 257–271.

13. CUMBIE, B.G. Developmental changes in the vascular cambium in *Leitneria floridana. Am. J. Bot.* **54** (1967) 414–424.

14. CUMBIE, B.G. Developmental changes in the vascular cambium of *Polygonum lapathifolium. Am. J. Bot.* **56** (1969) 139–146.

15. DENNE, M.P. Effects of environmental change on wood production and wood structure in *Picea sitchensis* seedlings. *Ann. Bot.* **40** (1976) 1017–1028.

16. DENNE, M.P. and WILSON, J.E. Some quantitative effects of indoleacetic acid on the wood production and tracheid dimensions of *Picea. Planta* **134** (1977) 223–228.

17. DIGBY, J. and WAREING, P.F. The relationship between endogenous hormone levels in the plant and seasonal aspects of cambium activity. *Ann. Bot.* **30** (1966) 607–622.

18. DODDS, J.H. and PHILLIPS, R. DNA and histone content of immature tracheary elements from cultured artichoke explants. *Planta* **135** (1977) 213–216.

19. EARL, D.E. *Forest Energy and Economic Development.* Clarendon Press, Oxford. (1975).

20. ESAU, K. *Vascular Differentiation in Plants.* Holt, Rinehart and Winston, New York. (1965).

21. GAHAN, P.B. A reinterpretation of the cytochemical evidence for acid phosphatase activity during cell death in xylem differentiation. *Ann. Bot.* **42** (1978) 755–758.

22. HARRIS, J.M. Spiral grain and xylem polarity in radiata pine: microscopy of cambial reorientation. *N.Z.J. For. Sci.* **3** (1973) 363–378.

23. HARRIS, J.M. and BARNETT, J.R. Differentiated callus nodules in resin pockets of *Pinus ponderosa* Laws. *N.Z.J. For. Sci.* **5** (1975) 226–229.

24. INNOCENTI, A.M. and AVANZI, S. Some cytological aspects of the differentiation of metaxylem in the root of *Allium cepa. Caryologia* **24** (1971) 283–292.

25. JACOBS, M. and RAY, P.M. Rapid auxin-induced decrease in free space pH and its relationship to auxin-induced growth in maize and pea. *Pl. Physiol.* **58** (1976) 203–209.

26. JACQUIOT, C. and TRENARD, Y. Note sur la présence de trachéides à parois gélatineuses dans des bois résineux. *Holzforsch.* **28** (1974) 73–76.

27. JANE, F.W. *The Structure of Wood.* A. and C. Black, London (1970).

28. JEFFS, R.A. and NORTHCOTE, D.H. The influence of indol-3-yl-acetic acid and sugar on the pattern of induced differentiation in plant tissue cultures. *J. Cell. Sci.* **2** (1967) 77–88.

29. JENKINS, P.A. Influence of applied indoleacetic acid and abscisic acid on xylem cell dimensions in *Pinus radiata* D. Don. In R.L. BIELESKI, A.R. FERGUSON and M.M. CRESSWELL (Eds) *Mechanisms of Regulation of Plant Growth.* Bulletin 12, The Royal Society of New Zealand, Wellington (1974). pp. 737–742. (New Zealand Forest Service Reprint No. 687).

30. JOHNSON, R.P.C. Can cell walls bending round xylem vessels control water flow? *Planta* **136** (1977) 187–194.

31. KERR, T. and BAILEY, I.W. The cambium and its derivative tissues — X. Structure, optical properties and chemical composition of the so-called middle lamella. *J. Arnold Arbor.* **15** (1934) 327–349.

32. KUTSCHA, N.P. and SCHWARZMANN, J.M. The lignification sequence in normal wood of balsam fir (*Abies balsamea*). *Holzforsch.* **29** (1975) 79–84.

33. LAI, V. and SRIVASTAVA, L.M. Nuclear changes during differentiation of xylem vessel elements. *Cytobiol.* **12** (1976) 220–243.

34. LARSON, P.R. Some indirect effects of environment on wood formation. In M.H. ZIMMERMAN (Ed) *The Formation of Wood in Forest Trees.* Academic Press, New York (1964) pp. 345–365.

35. LEDBETTER, M.C. and PORTER, K.R. A microtubule in plant cell fine structure. *J. Cell. Biol.* **19** (1963) 239–250.

36. LIST, A. Some observations of DNA content and cell nuclear volume growth in the developing xylem cells of higher plants. *Am. J. Bot.* **50** (1963) 320–329.

37. MURMANIS, L. Particles and microtubules in vascular cells of *Pinus strobus* L. during wall formation. *New Phytol.* **70** (1971) 1089–1093.

38. NELMES, B.J., PRESTON, R.D. and ASHWORTH, D. A possible function of microtubules suggested by their abnormal distribution in rubbery wood. *J. Cell. Sci.* **13** (1973) 741–751.

39. O'BRIEN, T.P. Primary vascular tissues, In A.W. ROBARDS (Ed) *Dynamic Aspects of Plant Ultrastructure.* McGraw-Hill, London (1974) pp. 414 440.

40. OPENSHAW, K. Wood fuels the developing world. *New Scientist* **61** (1974) 271–272.

41. PARHAM, R.A. and COTE, W.A. Distribution of lignin in normal and compression wood of *Pinus taeda* L. *Wood Sci. Tech.* **5** (1971) 49–62.

42. PARISH, D.J. and DAVIES, P.J. On the relationship between extracellular pH and the growth of excised pea stem segments. *Pl. Physiol.* **59** (1977) 574–578.

43. PENG, H.B. and JAFFE, L.F. Cell-wall formation in *Pelvetia* embryos. A freeze-fracture study. *Planta* **133** (1976) 57–71.

44. PHILLIPS, R. and DODDS, J.H. Rapid differentiation of tracheary elements in cultured explants of Jerusalem artichoke. *Planta* **135** (1977) 207–212.

45. POPE, D.G. Separation of Indol-3-ylacetic acid induced growth from acid induced growth in *Avena* coleoptiles. *Ann. Bot.* **41** (1977) 1069–1071.

46. POPE, D.G. Does indoleacetic acid promote growth via cell wall acidification? *Planta* **140** (1978) 137–142.

47. PRESTON, R.D. Structural and mechanical aspects of plant cell walls with particular reference to synthesis and growth. In M.H. ZIMMERMANN (Ed) *The Formation of Wood in Forest Trees.* Academic Press, New York (1964) pp. 169–188.

48. PRESTON, R.D. *The Physical Biology of Plant Cell Walls.* Chapman & Hall, London (1974).

49. PRESTON, R.D. and RANGANATHAM, V. The fine structure of the fibres of normal and tension wood in Beech (*Fagus sylvatica* L.) as revealed by X-rays. *Forestry* **21** (1947) 92–98.

50. PRIESTLEY, J.H. Studies in the physiology of cambial activity – II. The concept of sliding growth. *New Phytol.* **29** (1930) 96–140.

51. RAYLE, D.L. Auxin induced hydrogen ion secretion in *Avena* coleoptiles and its implications. *Planta* **114** (1973) 63–73.

52. RAYLE, D.L. and CLELAND, R. Enhancement of wall-loosening and elongation by acid solution. *Pl. Physiol.* **46** (1970) 250–253.

53. ROBARDS, A.W. The application of the modified sine rule to tension wood production in the stem of crack willow (*Salix fragilis* L.). *Ann. Bot.* **30** (1966) 513–523.

54. ROBARDS, A.W. Particles associated with developing plant cell walls. *Planta* **88** (1969) 376–379.

55. ROBERTS, L.W. The initiation of xylem differentiation. *Bot. Rev.* **35** (1969) 201–250.

56. ROBERTS, L.W. *Cytodifferentiation in Plants: Xylogenesis as a Model System.* Cambridge University Press, Cambridge (1976).

57. ROBINSON, D.G. and PRESTON, R.D. Plasmalemma structure in relation to microfibril biosynthesis in *Oocystis. Planta* **104** (1972) 234–246.

58. ROLAND, J.C. Aspects infrastructuraux des rélations éxistant entre le protoplasme et la paroi des cellules du collenchyme. *J. Microsc.* **6** (1967) 399–412.

59. SHELDRAKE, A.R. The production of hormones in higher plants. *Biol. Rev.* **48** (1973) 509–559.

60. SHELDRAKE, A.R. and NORTHCOTE, D.H. Some constituents of xylem sap and their possible relationship to xylem differentiation. *J. exp. Bot.* **19** (1968) 681–689.

61. SINNOTT, E.W. and BLOCH, R. Changes in intercellular relations during the growth and differentiation of living plant tissues. *Am. J. Bot.* **26** (1939) 625–634.

62. SWIFT, H. The constancy of deoxyribose nucleic acid in plant nuclei. *Proc. Nat. Acad. Sci. U.S.A.* **36** (1950) 643–654.

63. TIMELL, T.E. Ultrastructure of the dormant and active cambial zones and the dormant phloem associated with formation of normal and compression woods in *Picea abies* (L.) Karst. *Technical Publication No. 96. State University of New York, College of Environmental Science and Forestry* (1973).

64. TORREY, J.G. Tracheary element formation from single isolated cells in culture. *Physiol. Plant.* **35** (1975) 158–165.

65. TORREY, J.G. and FOSKETT, D.E. Cell division in relation to cytodifferentiation in cultured pea root segments. *Am. J. Bot.* **57** (1970) 1072–1080.

66. VANDERHOEF, L.N., FINDLEY, J.S., BURKE, J.J. and BLIZZARD, W.E. Auxin has no effect on modification of external pH by soybean hypocotyl cells. *Pl. Physiol.* **59** (1977) 1000–1003.

67. WARDROP, A.B. The reaction anatomy of arborescent angiosperms. In M.H. ZIMMERMANN (Ed) *The Formation of Wood in Forest Trees.* Academic Press, New York (1964) pp. 405–456.

68. WARDROP, A.B. Occurrence and Formation in Plants. In K.W. SARKANEN and C.H. LUDWIG (Eds) *Lignins.* Wiley, New York. (1971) pp. 19–38.

69. WASHER, J., REILLY, K. and BARNETT, J.R. Differentiation in *Pinus radiata* callus culture. The effect of nutrients. *N.Z.J. For. Sci.* 7 (1978) (in press).

70. WENHAM, M.W. and CUSICK, F. The growth of secondary wood fibres. *New Phytol.* **74** (1975) 247–261.

71. WERSHING, H.F. and BAILEY, I.W. Seedlings as experimental material in the study of "redwood" in conifers. *J. For.* **40** (1942) 411–414.

72. WETMORE, R.H. and RIER, J.P. Experimental induction of vascular tissues in callus of angiosperms. *Am. J. Bot.* **50** (1963) 418–430.

73. WRIGHT, K. and NORTHCOTE, D.H. Induced root differentiation in sycamore callus. *J. Cell Sci.* **11** (1972) 319–337.

74. WRIGHT, K. and NORTHCOTE, D.H. Differences in ploidy and degree of intercellular contact in differentiating and non-differentiating sycamore calluses. *J. Cell Sci.* **12** (1973) 37–53.

75. WODZICKI, T.J. Mechanisms of xylem differentiation in *Pinus sylvestris* L. *J. exp. Bot.* **22** (1971) 670–687.

76. WODZICKI, T.J. and HUMPHREYS, W.J. Cytodifferentiation of maturing pine tracheids: the final stage. *Tissue & Cell* **4** (1972) 525–528.

77. YOUNG, O.A. and SERVICE, M.J. Slow viscous flow and the organisation of the cell wall in conifer tracheids. *Wood Sci. Tech.* **5** (1971) 1–5.

ADDENDUM

The factors responsible for the initiation of xylem differentiation have been the subject of a review by Phillips.[12] Recently, Harrison and Klein[8] have examined the roles of auxin, gibberellin and cytokinins in the control of the direction of differentiation, i.e. towards xylem or phloem production. They conclude that this may be manipulated by varying the combination and relative concentrations of the hormones.

Longman *et al.*,[10] using clonal material, have shown that size and wall-thickness of secondary xylem elements are determined by a combination of genetic and environmental factors, with the degree of response to a particular environment being under genetic control.

Evidence has been obtained by Cline[5] for the presence in elongating cell walls of an enzyme responsible for acid-induced wall-loosening. Cycloheximide treatments of 6 hours were found to be necessary to inhibit acid-induced growth in *Avena* coleoptile segments.

Opinion remains divided on the mechanism of cellulose microfibril synthesis. Robinson[13] has reviewed progress in this field up until 1977. Some evidence has been produced supporting a role for microtubules in microfibril orientation[11] (again based on colchicine treatments), while Chafe has proposed that the enzyme granules now generally accepted to be responsible for the biosynthetic process are responsible for orientation.[4] He suggests that the granules move along the plasmalemma by slow viscous flow, following or forming geodesic lines. On reaching the cell tip a granule would travel around it and back down the cell following a line of opposite spiral to the original. Colvin and Leppard have questioned the idea that the compact, crystalline microfibrils seen with the electron microscope in fact represent the microfibrils as synthesised.[6] They suggest, on the basis of studies of pellicles of *Acetobacter* species, that initially the microfibrils are highly hydrated structures up to 100 nm wide, and that they achieve their more familiar appearance only after removal of the hydrating water molecules. They also suggest that the microfibrils are not synthesised by addition of glucose units to an existing microfibril end as has been generally accepted, but rather that they are formed by aggregation of polyglucan chains secreted by the cell. If this is indeed the case in *Acetobacter*, there seems to be little evidence for this system in other, higher plant species which have been examined.

The biosynthesis of lignin has been reviewed recently by Gross.[7]

Boyd[1] has re-examined the evidence in favour of the gravity/auxin theory of compression wood formation. He suggests that this evidence may be interpreted to show that this type of tissue may be formed as a response to stresses within the stem, without the need to invoke a gravity-mediated redistribution of auxin. This reinterpretation of the classical observations of early workers in this field, and of results obtained by more recent authors is highly controversial, and is unlikely to find ready support among proponents of the auxin theory. On the other hand, Boyd's[2] examination of the evidence for a role for the β-D-1, 3-linked glucan "laricinan" in the action of compression wood in reorientating woody tissue is likely to find more acceptance. This substance, first described by Hoffmann and Timell,[9] has been proposed as the driving force for reorientation.[3, 14] Boyd presents a strong case against a primary role for laricinan, however, suggesting that it may be symptomatic of, rather than a functional agent of, compression wood.

REFERENCES

1. BOYD, J.D. Basic cause of differentiation of tension wood and compression wood. *Aust. For. Res.* **7** (1977) 121–143.
2. BOYD, J.D. Significance of laricinan in compression wood tracheids. *Wood Sci. Tech.* **12** (1978) 25–35.
3. BRODZKI, P. Callose in compression wood tracheids. *Acta Soc. Bot. Polon.* **41** (1972) 321–327.
4. CHAFE, S.C. On the mechanisms of cell wall microfibrillar orientation. *Wood Sci. Tech.* **12** (1978) 205–217.
5. CLINE, M.G. The stability of the postulated wall-loosening enzyme in acid induced growth. *Planta* **145** (1979) 389–391.
6. COLVIN, J.R. and LEPPARD, G.G. The biosynthesis of cellulose by *Acetobacter xylinum* and *Acetobacter acetigenus. Can. J. Microbiol.* **23** (1977) 701–709.
7. GROSS, G.G. Biosynthesis of lignin and related monomers. In LOEWUS, F.A. and RUNECKLES, V.C. (Eds.) *The Structure, Biosynthesis and Degradation of Wood.* Plenum Press, New York (1977) pp. 141–184.
8. HARRISON, Maria A. and KLEIN, R.M. Role of growth regulators in initiation of secondary xylem and phloem cells. *Bot. Gaz.* **140** (1979) 20–24.
9. HOFFMANN, G.C. and TIMELL, T.E. Isolation of a β-1,3-glucan (laricinan) from compression wood of *Larix laricina. Wood Sci. Tech.* **4** (1970) 159–162.
10. LONGMAN, K.A., LEAKEY, R.R.B. and DENNE, M.P. Genetic and environmental effects on shoot growth and xylem formation in a tropical tree. *Ann. Bot.* **44** (1979) 377–380.
11. MARCHANT, H.J. and HINES, E.R. The role of microtubules and cell wall deposition in elongation of regenerating protoplasts of *Mougeotia. Planta* **146** (1979) 41–48.
12. PHILLIPS, R. Cytodifferentiation. In VASIL, I.K. and MURPHY, D.G. (Eds.) *Recent Advances in Plant Tissue Culture* (1980) (In press).
13. ROBINSON, D.G. Plant cell wall synthesis. *Adv. in Bot. Res.* **5** (1977) 89–153.
14. WIOCH, W. Longitudinal shrinkage of compression wood in dependence on water content and cell wall structure. *Acta Soc. Bot. Polon.* **44** (1975) 217–229.

MOLECULAR ASPECTS OF SEED PROTEIN BIOSYNTHESIS

DONALD SPENCER and T.J.V. HIGGINS

CSIRO, Division of Plant Industry, Canberra 2601, Australia

INTRODUCTION

Although approximately 70% of the edible protein produced in the world comes from seeds it is only in the last 10 years that cell biologists in any number have taken an interest in the process of seed protein synthesis and deposition. This increased activity was probably due to greater awareness of the importance of seeds in world food supply and to fears expressed in the late 1960's of a possible world shortage of protein. Reassessment of minimum daily protein requirements for humans have since allayed these fears for the moment, but this phase did serve to highlight the inadequacies of seed proteins where they contributed a major part of the diet of man or his monogastric animals such as poultry and pigs.

Seed proteins in general are not well balanced with respect to the 10 amino acids that are essential in the diet of monogastric animals. Proteins of cereal seeds are typically deficient in lysine and often low in other essential amino acids such as tryptophan, valine and threonine. Legumes on the other hand are usually deficient in methionine and cysteine. Most current research bears on the specific question of whether the protein composition of the seed can be modified to a sufficient extent to improve the dietary balance of amino acids without loss of their biological function in supporting the growth of the germinating seed. It is also likely that in future there will be a demand for seed proteins with particular physical properties that are suited to processes such as texturizing or spinning of protein fibres.

We now need a basic knowledge of seed proteins — their characteristics, their synthesis and deposition, and the controls operating at the genetic, biochemical and physiological levels that regulate the size of the storage protein fraction and the proportion of the individual proteins in this fraction. This knowledge will help define the limits within which seed proteins can be modified and could indicate new selection criteria for plant breeding programmes relating to seed protein improvement.

This Commentary will consider some of the more recent developments in this field which contribute to our understanding of the nature of seed storage proteins and of the cellular processes involved in their synthesis and deposition particularly in cereals and legumes. In general, we have tried to cite the most recent relevant references in the hope that the interested reader will be able to work back from these to earlier reports.

COMPLEXITY OF STORAGE PROTEINS

Seed proteins fall into two broad categories, namely, storage proteins which are unique to the seed and metabolic proteins which have their counterparts elsewhere in the plant. The storage proteins have claimed the most attention primarily because they constitute the major protein fraction in most seeds. If we are to study molecular aspects of storage protein synthesis and deposition it is essential that the individual storage proteins themselves should be well characterized. This has proved no easy task. Even the definition of a storage protein is not straightforward. In legumes, storage proteins could probably be defined as those proteins contained in protein bodies of mature seeds. A less specific definition, however, is needed for cereals where, at least in the starchy endosperm of some species, probably only one fraction of the storage proteins (the prolamins) is localized in protein bodies, and these latter structures may not retain their identity through to maturity. A more general definition along the lines suggested by Derbyshire *et al.*[24] is therefore probably more appropriate, namely, any protein fraction which is unique to the seed and which constitutes more than 5% of the seed protein.

The early work on characterization of seed proteins indicated a great deal of similarity among the storage proteins of cereals on the one hand and those of the legumes on the other. However, as might be expected, as more sophisticated analytical techniques became available, dissimilarities within these groups became obvious. The picture that is emerging today is one of increasing complexity and microheterogeneity of the major storage proteins at both the holoprotein and polypeptide subunit level. This will be illustrated with a limited number of examples from legumes and cereals.

The pioneering work of Osborne around the turn of this century resulted in the characterization of the proteins of many seeds into different classes dependent on their solubility successively in water (albumins), dilute salt (globulins), 70% ethanol (prolamins) and dilute alkali (glutelins). On this basis, the major storage proteins of legumes are globulins whereas those of cereals are largely prolamins and glutelins.

Legumes

Osborne[57] recognized that the globulins of many legume seeds could be fractionated into two distinct classes on the basis of their differential precipitation with ammonium sulphate and differential heat coaguability. These two fractions from *Pisum sativum* and *Vicia faba* were named legumin and vicilin. Danielsson[22] further fractionated the proteins by differential isoelectric precipitation and by analytical ultracentrifugation. He examined 34 species in the Leguminosae and found that 30 contained two major globulin fractions, one in the range 7-9S and the other 11-13S. The few exceptions lacked the 11-13S fraction. This remarkable consistency tended to confirm Osborne's conclusion some 60 years before that "closely related plants contain the same seed globulins". However Danielsson later pointed out[21] that in spite of their apparent homogeneity on ultracentrifugation and moving boundary electrophoresis, the pea globulins, legumin and vicilin, may each consist of complex mixtures of proteins.

Danielsson's speculations have proved well founded. The advent of more rigorous fractionation techniques (particularly the use of gel electrophoresis under both dissociating and non-dissociating conditions, column chromatography and a range of immunological procedures), has brought a realization that, in spite of an overall similarity in many properties, not only are there distinct differences between many of the legumin-like and vicilin-like proteins of different legume species[24] but also considerable heterogeneity exists within these fractions for a particular species, at both the holoprotein and polypeptide level.

Intergeneric and interspecific heterogeneity was clearly shown by immunological studies on the family Leguminosae.[27] While some degree of relatedness to the legumin and vicilin fractions of *P. sativum* and *V. faba* was found widely in the genera of the Fabeae and Trifolieae it was fairly rare in all other tribes examined, including Phaseoleae.

Intraspecific heterogeneity can be illustrated with the storage proteins of *P. sativum*. Using a variety of electrophoretic techniques, differential solubility, isoelectric precipitation, crossed immunoelectrophoresis and molecular weight determination, Thomson and colleagues have shown that vicilin and legumin fractions are each composed of multiple holoproteins and that the number and type of holoproteins are genotype-specific.[72] The holoproteins formed a series of related polymers differing quantitatively and often qualitatively in polypeptide composition. At least seven constituent polypeptides (75, 50, 30, 24, 18, 14 and 12 kdaltons) of vicilin and two major (40 and 20 kdaltons) and four minor components (37, 27, 25 and 18 kdaltons) of legumin were recognized on sodium dodecylsulphate-polyacrylamide gels (SDS-PAG). The heterogeneity at the holoprotein level both within and between genotypes is reflected in different proportions of these polypeptides and, in one case, in the lack of a particular component (75 kdaltons). The inheritance of the electrophoretic variants examined so far has proved to be additive at both the holoprotein and polypeptide level.[71,72] The glycoprotein nature of legumin and vicilin[5] may contribute to some of the heterogeneity observed, and there is evidence that even within a particular size class of subunit not all of the molecules are glycosylated (R.A. Davey and W.F. Dudman, personal communication).

Cereals

The level of complexity seen in the legume storage proteins is more than matched by the situation in cereals. Among the major cereal crops, prolamins (alcohol-soluble) and glutelins (alkali- and acid-soluble) make up about 90% of the seed protein in approximately equal proportions. The main exceptions are rice and oats in which glutelin and globulin respectively are predominant.[42] Closer investigation has shown that these broad protein classes are in fact complex mixtures and that the distinction between some classes is by no means clear-cut. The current status of wheat storage proteins has recently been reviewed in detail by Konzak[42] and Kasarda *et al.*[39] and of maize proteins by di Fonzo *et al.*[26]

The prolamin fraction of wheat (gliadin) can be sub-divided into four main groups of holoproteins based on their electrophoretic behaviour in aluminium lactate buffer at pH3.2 on non-dissociating polyacrylamide gels.[39] As with the legume proteins, new fractionation techniques have revealed approximately 15 separate components in the total gliadin fraction of a single variety, a combination of gel electrophoresis and isoelectric focusing revealed a two-dimensional profile of the gliadin fraction with 46 different components.[78] Given the variation that exists between lines, this suggests that a vast number of component proteins may contribute to the gliadin fraction. Most, but not all gliadins consist of single polypeptide chains with a predominant component of 36.5 and other components of 44.2, 69.3 and 78.1 kdaltons.[9]

Although there is great diversity in the gliadins, there is also evidence of close similarity between the components. Evidence from peptide mapping, amino acid composition and sequencing of the N-terminal end of individual gliadins is consistent with the existence of common peptide sequences among some of the gliadins. This situation led Kasarda *et al.*[39] to speculate that "(gliadins) may represent the most heterogeneous group of closely related proteins known".

The glutelin fraction of wheat (glutenin) is also complex and difficult to define. Kasarda *et al.*[39] define it simply as "the least soluble half (or even smaller fraction) of the total protein of wheat flour". Unlike the gliadins, the holoproteins of the glutelin fraction are polymeric molecules made up from 15 component polypeptides whose molecular weights range from 11 to 133 kdaltons. These polypeptides associate in complexes which have molecular weights ranging from 50 to 2000 kdaltons.[9] Although it is generally accepted that the subunits in these large complexes are linked through intermolecular disulphide bonds, Kasarda *et al.*[39] have proposed an alternative model in which disulphide links are largely intramolecular and the subunits are held together through secondary bonding forces such as hydrogen bonding, ionic bonding and hydrophobic bonding.

The storage proteins of other cereals are also highly complex. For instance, the prolamin fraction of maize (zein) is a class of macromolecules which are heterogeneous with regard to size and charge. It was found to consist of three major subunits of molecular weight 23, 21 and 13.5 kdaltons and a minor fraction of 9.6 kdaltons on SDS–PAG following reduction with 2-mercaptoethanol. However, isoelectric-focusing of the zein fraction revealed up to 28 well-defined bands.[29]

In spite of the above evidence of heterogeneity within the different classes of storage protein it is worth remembering the remarkable degree of similarity between members of these classes as observed by earlier workers. These overall analogies of size class, solubility class and amino acid composition within the legumes and within the cereals presumably reflect the criteria that storage proteins must meet in order to fulfil their biological role as reserve substance for the germinating seed.

HOW FLEXIBLE IS THE COMPOSITION OF THE STORAGE PROTEIN FRACTION?

While individual storage proteins must meet certain specifications as discussed above, it seems reasonable to assume that the proportions of the individual protein classes that make up the total storage protein fraction of a given seed might be variable within much wider limits. Given man's need to manipulate the overall composition and level of the seed protein fraction, it is interesting to consider evidence on the limits within which these proportions can be varied without affecting seed formation. The existing evidence indicates that in fact a great deal of variation can be tolerated.

Wide variations in the composition of the storage protein fraction within a genotype can occur as a result of changes in the environment. The most striking and well-documented examples are those due to changes in nutrient supply. For example, a deficiency of sulphur in *Lupinus angustifolius*[10] and *P. sativum*[62] caused the virtual elimination of some globulin fractions, namely α-conglutin and γ-conglutin, in lupins and legumin in peas, all of which are relatively rich in sulfur-amino acids. In addition, in peas there was a selective reduction in one of the vicilin group of holoproteins. In all cases the affected protein species is relatively rich in sulphur-amino acids and there is a compensatory rise in the amount of the other storage protein species (conglutin β in lupins and the other vicilins in peas). Thus at moderate sulphur deficiency, while the levels of protein-sulphur are decreased there is relatively little effect on total seed protein per unit dry matter. Supra-optimal levels of sulphur nutrition do not result in above normal levels of α-conglutin or legumin. Other nutrient deficiencies also have a striking effect on pea seed composition. A deficiency of potassium or phosphorus resulted in a 2–2½ fold increase in legumin as a proportion of total seed protein.[62]

This kind of variation in storage protein composition with nutritional status is not confined to legumes. Different levels of nitrogen supply resulted in a change in the relative proportion of high and intermediate molecular weight species of hordein, the prolamin fraction of barley,[41] while severe sulphur deficiency caused marked reduction or elimination of about one quarter of the gliadins and alterations in some of the albumins and glutelins of wheat (C.W. Wrigley, personal communication).

Temperature is another environmental factor which can affect the relative level of individual storage protein fractions. Lower growth temperatures (15°C vs 25°C) resulted not only in an increase in total protein content of pea seeds but also, in some genotypes, in a marked increase in the proportion of legumin (from 30 to 48%) (A. Millerd and J.A. Thomson, personal communication).

Equally striking examples of flexibility in the composition of the storage protein fraction is seen between related genotypes. Perhaps the most outstanding examples are the high-lysine lines of maize and barley. The opaque-2 mutation in maize causes a reduction in the group of polypeptides that make up the lysine poor zein fraction and a compensatory increase in other fractions (especially albumins and globulins) which are richer in lysine.[26, 68] An analogous situation exists in the Hiproly[53] and Riso 1508[13] mutants of barley where once again there is a marked reduction in the contribution of its prolamin fraction (hordein) to the total seed protein. This involves a selective reduction in the lysine-poor hordein components and an increase in lysine-rich glutelin fractions as well as in free lysine.

These examples make it clear that a wide range of variation in the relative amounts of the component proteins can be tolerated within a given genotype and between closely related genotypes without affecting the plant's ability to set viable seed.

CELLULAR EVENTS ASSOCIATED WITH STORAGE PROTEIN SYNTHESIS AND DEPOSITION

Many subcellular changes have been monitored both by microscopy and by chemical methods during seed development. The general conclusions are that cell division is the major activity during the early stages of embryonic growth and that this is followed by a cell expansion phase in which reserve synthesis and deposition occurs. The growth rate of the seed and the rate of accumulation of reserves is generally sigmoidal[55] although a diauxic mode of growth has been proposed for some dicotyledonous plants.[19] It is interesting to remember that by the time cereal seeds have reached maturity the intracellular organization of their main storage tissue, the starchy endosperm, is lost and the tissue is essentially dead. In legumes the cotyledon cells retain their organization through to maturity, and in some cases, on germination the cotyledons expand and develop into photosynthetically active organs.

Seed development has been analysed in detail in many species especially with regard to fresh weight, dry weight, nitrogen content, nucleic acids, carbohydrate and lipid levels, chlorophyll and minerals, and the development of organelles such as vacuoles, chloroplasts, starch grains, mitochondria, dictyosomes, endoplasmic reticulum (ER), nuclei, ribosomes and protein bodies.[38, 3, 55, 15, 48, 25, 65, 20] The relation between many of the observed changes and protein accumulation is not clear and we will discuss only some of the more obvious interactions.

During the phase of storage protein deposition in legume cotyledons there is a parallel increase in DNA content[50, 64, 67] which is not associated with cell division. The function of this "extra" DNA and its relation, if any, to storage protein synthesis is not clear. DNA level could not be simply correlated with protein level in a number of legumes, including two lines of soybean with different DNA levels but similar protein levels.[46] Renaturation kinetics indicated that in *V. faba* there was endoreduplication of the total genome rather than amplification of selected genes.[50] However, this "extra" DNA does not appear to be fully utilized as a template because DNA and RNA, along with protein, increase in a parallel and approximately linear fashion during this phase of seed development, i.e. each increment in DNA per cell does not lead to a corresponding increase in the rate of RNA accumulation, most of which is ribosomal and transfer RNA.[48] Consistent with this, the endogenous RNA-synthesizing activity (per unit of DNA) of isolated pea nuclei decreased progressively during the period of DNA endoreduplication, although its template capacity remained constant.[48] There is as yet no strong evidence to support the suggestion of a gene dosage effect for the "extra" DNA[64] although the possibility cannot be eliminated at this stage. The DNA could have a reserve function.[48] Smith[66] showed that cotyledonary DNA levels dropped during germination of field peas, but, on the other hand, Beevers and Guernsey[7] observed a slight increase in DNA levels in garden pea cotyledons during germination. In the cereals wheat and maize, DNA level does not increase after cessation of cell division.[37, 36] Thus the function of this "extra" DNA in legumes must await further studies.

Other aspects of nuclear structure and function during seed development have received relatively little attention except for the occasional reference to the lobed nature of the organelle observed during protein accumulation. However, Fambrough *et al.*[28] noted that there were marked changes in the ratio of histone I to other histones in the nuclei of pea seeds during seed development. Whether these changes can be correlated with transcription of storage protein genes also remains to be seen.

The increase in total RNA per cell which accompanies the onset of storage protein synthesis has been interpreted, without controversy, as representing the increased requirement for ribosomes and tRNAs to translate the mRNAs presumed to arise also during the phase of protein accumulation. The biosynthesis, lifetime and degradation of the various RNAs involved in storage protein synthesis is yet to be investigated in detail. Available data[69, 58, 60, 75] suggest that the ribosomes have a relatively long lifetime because protein synthesis continues for some days after the synthesis of rRNA has stopped. As mentioned above, RNA levels in legume cotyledons increase in approximately linear fashion in parallel with protein accumulation. Protein accumulation can probably be taken as an index of rate of protein synthesis since short-term *in vivo* labelling studies showed that in pea cotyledons the storage proteins were the major proteins being synthesized. There was no evidence of large-scale synthesis of proteins with rapid turnover (our unpublished work). The linear rate of protein accumulation raises the question of the efficiency of this RNA which is largely ribosomal and tRNA. It would appear that the efficiency of the total RNA population must decline during seed development since the rate of protein accumulation does not increase. As with DNA, one could suggest that some of the RNA is not involved in protein synthesis but might instead have a reserve function. In some species with high cotyledonary levels of RNA there is a decline in RNA during germination with characteristics that would suggest translocation of soluble nucleotides to the developing axis.[56, 7]

Together with the chemical analyses, the study of subcellular structures has provided some correlations between protein deposition and the activity or appearance of ER, dictyosomes and protein bodies. The onset of protein deposition is usually associated with an extensive proliferation of the ER[51, 55, 15, 25, 77, 20] and most observers agree that ribosomes become attached to the ER (RER) during this phase. Possible exceptions to this general rule may be pea and bean. Bain and Mercer[3] found little ultrastructural evidence for RER during the rapid protein synthesis phase in peas and Sun *et al.*[70] using subcellular fractionation found that the majority of ribosomes were free in beans synthesizing storage proteins. Nevertheless, a dogma emerged which stated that storage protein synthesis occurred on RER[2] but direct evidence in support of this has only recently been obtained.[43, 14, 44]

A role for dictyosomes in the accumulation of protein reserves has not been strongly supported although a case has been made for them in wheat, some oilseeds and in cowpeas.[18, 25, 34] However, in most legumes the dictyosomes are considered to be relatively few and inactive during protein deposition[3, 55, 15] and Rest and Vaughan[63] emphasized the role of ER rather than dicytosomes in the oilseed *Sinapis alba*.

In describing the protein bodies of seeds during development and especially during protein accumulation a number of cellular organelles have been mentioned repeatedly. Large vacuoles, vesicles derived from dictyosomes and the lumen of the ER have all been suggested as sites of storage deposition, depending upon the stage of seed development and the species being studied. Within the one species large vacuoles are often the site of protein deposition early in development while at later stages deposits are frequently found in small bodies of doubtful origin, but sometimes thought to arise from the ER or dictyosomes.[15, 4, 34, 20] A quantitative analysis of the number and size of protein bodies of cowpea and pea has shown that there is a progressive decrease in their size and increase in number throughout development.[34, 20] However, these studies have not revealed how the small bodies arise, nor what happens to the large ones, and what relationship, if any, there is between the two categories. It is possible that some of the structures that have been called protein bodies in developing seeds (often only on the basis of their electron-dense material) may in fact be nothing of the sort. This may be resolved with better techniques such as immuno-cytochemistry. In any case the current picture of protein body development is complex and controversial.

LOGISTICS OF STORAGE PROTEIN SYNTHESIS AND DEPOSITION

Here we are concerned with the timing and site of storage protein synthesis and the mechanism of sequestration of these reserves. The timing of storage protein synthesis in relation to seed development has been studied using techniques with varying levels of resolving power to monitor the component proteins. These include solubility classes, ultracentrifugation, electrophoresis, serology, and ion-exchange chromatography, many of which have been supplemented with radioisotopic labelling to enhance their resolution. Light and electron microscopy has also been used to survey the deposition of protein-like material during seed development.

In general, the syntheses of the various storage protein components of a particular species are initiated at different stages of development and the proteins are accumulated at different rates. Examples from the cereals and legumes can be used to illustrate these points. In barley the albumins (non-storage proteins) are synthesized early and decline late in development. The glutelins and hordeins are synthesized simultaneously and before the globulins.[13] Analysis of the subunits of the proteins during development indicated that there was tight coordination of subunit synthesis in the hordeins but that the individual subunits of the glutelins and globulins possessed a distinctive timing and rate of synthesis. In maize the glutelin fraction had a linear time-course of synthesis while that of zein was sigmoidal. There appeared to be coordinated synthesis of the subunits of both zein and glutelins of maize.[26] The prolamins of wheat were synthesized with a sigmoidal time-course and the subunits appeared to be synthesized at a fairly constant ratio while the non-storage proteins varied markedly during development.[30] When labelled glycine and sulphate were supplied as precursors, it was found that wheat storage proteins were synthesized independently of cytoplasmic (non-storage) proteins, i.e. there was no pool of storage protein precursors in the soluble phase of the cell.[30, 32]

The timing and rate of storage protein synthesis in legumes has also been studied in detail and we will illustrate this with peas and beans. Danielsson[23] found that at early stages of development in peas a number of major holoproteins were evident in the analytical ultracentrifuge. Their ratios changed with development until, at a certain stage, the combination of legumin (12S) and vicilin (7S) was seen in a pattern typical of the mature seed. Vicilin synthesis was initiated before legumin and their rates of synthesis were different. These observations have been greatly extended and refined to include high resolution separations of the holoproteins of vicilin and legumin during pea seed development.[49] Using two-dimensional immunoelectrophoresis to fractionate the total protein extracts of immature seeds, at least four distinct, non-cross-reacting vicilin holoprotein species were detected. The electrophoretic mobility of one of these species varied with stage of development, suggesting the existence of a precursor form. The individual vicilin holoproteins and legumin appeared in a definite sequence which was similar in a number of genotypes, although the relative amounts of the holoproteins were genotype-specific. There were also quantitative variations in the subunit composition of some of the holoproteins during development. One of the remarkable features of this work is the qualitatively similar polypeptide composition of the various vicilin holoproteins which possess such strikingly different immunological specifications. There were major quantitative differences and some minor qualitative differences in the polypeptide compositions of the different vicilin species but it is not known which of these factors is responsible for conferring unique antigenic specificity on such closely related proteins. This work shows that the timing of storage protein synthesis is complex and that a multitude of controls must be needed to cover the whole spectrum of component proteins.

The polypeptide composition of the holoproteins of *P. vulgaris* has been monitored during development.[70] G-2 protein (corresponding to pea vicilin) appeared before G-1 but was quantitatively a minor component. The three major polypeptides of G-1 maintained a constant ratio to one another during development while the ratios of the three major polypeptides of G-2 varied during maturation. These results are consistent with G-1 being a homogeneous holoprotein while G-2 may consist of a series of holoproteins (as described above for pea).

Refined characterisation of the storage proteins has greatly aided the analysis of timing and rate of synthesis of storage proteins but so far most of these developments have not affected the state of our knowledge of sites of synthesis and the mechanisms by which storage proteins are assembled and deposited. It is generally agreed that storage protein occurs in protein bodies of most mature seeds[1, 73] and that some storage proteins, such as the glutelins in maize, are deposited outside protein bodies. The origin of mature protein bodies and how they are related to the large vacuoles, the many vesicles or the distended RER segments of immature seeds is something of a mystery as has been discussed earlier. Although the identification of the deposits seen in the various organelles during development leaves something to be desired, vicilin and legumin have been identified in membrane-enclosed bodies of immature seeds of *V. faba*[33] and storage proteins have been identified in organelles isolated from immature wheat.[31] If we assume that these protein deposits are all storage protein then there are many interesting questions that arise. For example, (1) does the site of protein synthesis vary with developmental stage, or with the class of storage protein, (2) what is the relationship between protein deposited in large vacuoles of immature seeds and that in smaller protein bodies of more mature seeds, (3) how are the proteins transported to the storage organelles and (4) where does protein maturation, including holoprotein assembly, occur? We cannot resolve most of these questions at present but recent data on sites of synthesis of some of the storage proteins may help to answer some of them.

Prolamin synthesis in cereals probably occurs on RER.[43, 14, 44] Although the possibility has been raised that storage protein synthesis occurs uniquely in or on protein bodies[52] it now seems likely[44] that protein bodies are continuous with the RER and may represent specialized portions of the ER given over to the deposition of reserves, as proposed by Khoo and Wolf.[40] The ER may have a role in maturation of the protein also, since it was shown that a precursor-zein molecule could be converted to authentic zein if translation occurred in the presence of the membrane.[44] Thus, it is possible that synthesis, maturation and deposition of zein may all occur on and in ER. However, it should be remembered that the prolamins are only a fraction (at most 50%) of the storage proteins of barley and maize and we have no data on synthesis or deposition of the non-prolamin fraction.

As discussed earlier, the site of storage protein synthesis in legumes has been deduced to be on the RER by correlations between the deposition of seed protein and the appearance of RER in the same cells. This was supported by Bailey *et al.*[2] who showed by *in vivo* labelling studies on cotyledon slices that legumin was probably synthesized on RER in *V. faba* and that the protein was somehow transported to protein bodies within 25 min of synthesis. They could not decide whether other organelles were involved in any aspect of synthesis or deposition because of the low specific activity of the protein and the high background. However, by subcellular fractionation Sun *et al.*[70] have found that the majority of polysomes are free in the cytoplasm of *P. vulgaris* during the rapid phase of storage protein synthesis, and little RER was found in ultrastructural studies of the corresponding phase in pea cotyledons.[3, 20] Nevertheless, our cell-free studies suggest that the majority of the polysomes are membrane-bound during the phase of rapid storage protein synthesis in peas (unpublished work) and that the storage proteins may by synthesized as precursors. This could indicate, by analogy with zein and with animal secretory proteins, that they must undergo post-transitional modification on ER or some other membrane and that they may be synthesized on membranes. Furthermore, it is known that pea legumin and vicilin are glycoproteins[5] and that the RER is probably the site of a glucosyl transferase capable of producing a lipid-linked sugar intermediate used in glycoprotein formation,[54] thus confirming a possible role for ER at least in the maturation of pea storage proteins. Thus, the bulk of the evidence favours the notion that the RER is involved at least as the site of synthesis of some storage proteins and that it may also have a role in their maturation. The site of final assembly and deposition is more controversial, mainly because of the little information currently available.

CELL-FREE STUDIES ON STORAGE PROTEIN SYNTHESIS

A number of investigators have set out to reconstitute cell-free systems from developing seeds capable of synthesizing storage proteins, and thus study the factors responsible for regulating the transcriptional and translational apparatus during development.

One of the first of such systems, and certainly the most ambitious, was the attempt by Bonner *et al.*[12] to set up a coupled transcription–translation system using chromatin from developing pea cotyledons and transcription and translation components from *E. coli.* Although with hindsight their procedures could be criticized, their data were consistent with the idea that the genetic information for storage proteins was derepressed in cotyledons but not in other pea tissues. Their approach could well serve as a model for future investigations using more homologous components.

Most cell-free studies have been concerned only with the translation apparatus. Early studies dealt mainly with total amino acid incorporating activity of systems from developing seeds with no rigorous identification of specific storage protein products being achieved.[61, 59, 8] However, concomitant with the increased use of cell-free systems greater efforts were made to identify specific translation products and thus attempt to assay for defined mRNA species. Criteria such as the ratio of incorporation of selected characteristic amino acids,[47] solubility of *in vitro* products in unique solvents such as 70% ethanol for the prolamins of cereals,[43, 16, 14] co-migration with authentic storage protein polypeptides on SDS-PAG,[6] precipitation with specific antiseum[70] and mapping of tryptic peptides[35] have all been employed.

It should be noted that co-migration of *in vitro* translation products with authentic storage protein subunits must be critically assessed. Earlier reports of apparent coincidence of *in vitro* products with zein subunits[43] in sliced SDS–PAG have since been modified following re-examination by autoradiography. The two major alcohol-soluble *in vitro* products from the maize system proved to be 1000–2000 daltons larger than authentic zein subunits.[17, 44] For at least one of the products the additional sequence is at the amino-terminal end.[17] This is presumptive evidence that, like many secreted animal proteins,[11] these storage proteins which are sequestered in protein bodies, are synthesized in a precursor form with a signal sequence which is involved in recognition and transport through a membrane. Consistent with this, *in vitro* translation in the presence of maize ER resulted in two alcohol-soluble products which were slightly smaller than those synthesized by similar polysomes when detached from the membranes.[44] A lack of precise coincidence on SDS-PAG between *in vitro* translation products and authentic pea storage proteins but a clear coincidence between many of their respective tryptic peptides suggested that precursor forms are also synthesized in a cell-free translation system from peas.[35] It is possible that on closer examination similar interpretations may apply to cell-free translation systems from other developing seeds.

In general two kinds of cell-free translation system have been used to study seed protein synthesis. The most common is the polysomal "run-off" system in which nascent polypeptide chains already associated with polysomes *in vivo* are completed *in vitro* in the presence of the necessary added soluble factors obtained either from the homologous tissue or from an entirely different source. This system provides information on the relative activity of specific mRNAs in the population but not necessarily on the absolute level of each mRNA. The second cell-free translation system involves the use of a heterologous set of ribosomes and soluble factors, commonly prepared from wheat germ, combined with the mRNA population isolated from the developing seed. In this case it is often assumed that the translational components are not limiting and that activity measured is indicative of relative levels of mRNA in question. While this is true in general, it should be noted that with mixed populations of mRNA the heterologous sytem is known to select certain mRNAs preferentially depending upon the ionic conditions, RNA concentration, and physical characteristics of the mRNA such as presence of "cap" structures and overall size. Qualitative differences may also arise between homologous and heterologous systems due to their differential processing capability (for further discussion see references 35 and 44). The spectrum of polypeptide products produced by the two types of translation system from the same seed can be either very similar, as for the soybean systems,[6] or rather distinctive, as for the pea systems,[35] on SDS-PAG.

With one exception, the messenger RNAs for storage proteins have not been extensively characterized. They appear to be typical eukaryotic messengers in that they have a polyA "tail" and an uncharacterized "cap" structure (see references 17 and 45 for zein, reference 6 for soybean glycinin; Higgins and Spencer, unpublished work for pea seed proteins). In the latter case a similar pattern of *in vitro* products was obtained with both the bound and non-bound RNA fractions following passage through oligo (dT)-cellulose. The mRNA for the two major zein subunits from maize has been purified to a high degree[17, 44] and found to be approximately 1.1–1.2 kbases long and monocistronic. The length of the mRNAs far exceeds that needed to code for the sequences of the two zein precursors (561 and 684 bases) suggesting that, in common with other eukaryotic messengers, they possess untranslated sequences which may be at both the 5′ and 3′ ends of the molecule (in addition to the "cap" and "tail" sequences). The further complexity of the zeins which was revealed by isoelectric focusing[29] points to the likelihood of a family of closely related mRNAs within these purified fractions. Acquisition of purified mRNA for specific storage proteins opens the way to a wide range of exciting lines of enquiry. Such questions as the lifetime of mRNAs, their localization in the genome, the number of copies in the genome and the regulation of their transcription can now be posed.

Much remains to be done on translational control factors particularly in relation to the synthesis of specific storage proteins. The translational efficiency of cell-free systems from seeds varies during their development and a start has been made in distinguishing changes in ribosomal efficiency from that of the supernatant factors.[61, 58, 59, 8] The recent report by Viotti *et al.*[74] that an increase in the levels of certain tRNAs can be correlated with the onset of zein synthesis in developing maize seeds also promises to give important new leads on translational control. Other features of the translational system for storage proteins, such as the aminoacyl-tRNA synthetases, initiation, elongation and termination factors, which may change with the onset of storage protein synthesis have been investigated only slightly so far.[61, 60, 76]

Cell-free studies could provide exciting new information on a number of aspects of seed protein synthesis in the near future. These include further characterization of precursor forms, the function of these precursors in the assembly and deposition of storage proteins and the details of their conversion to mature polypeptide, isolation of other mRNAs for specific storage proteins, and the analysis of the role of mRNAs and other translational components in the regulation of storage protein synthesis. Given the marked increase in research activity in this field it seems certain that the answers to these and many other questions will be forthcoming in the near future.

REFERENCES

1. ALTSCHUL, A.M., SNOWDEN, J.E., MANCHON, D.D. and DECHARY, J.M. Intracellular distribution of seed proteins. *Archs. Biochem. Biophys.* **95** (1962) 402–404.

2. BAILEY, C.J., COBB, A. and BOULTER, D. A cotyledon slice system for the electron autoradiographic studies of the synthesis and intracellular transport of seed storage protein of *Vicia faba. Planta* **95** (1970) 103–118.

3. BAIN, J.M. and MERCER, F.V. Subcellular organization of the developing cotyledons of *Pisum sativum* L. *Aust. J. biol. Sci.* **19** (1966) 49–67.

4. BARLOW, K.K., LEE, J.W. and VESK, M. Morphological development of storage protein bodies in wheat. In R.L. BIELESKI, A.R. FERGUSON and M.M. CRESWELL (Eds) *Bulletin* 12, *The Royal Society of New Zealand* (1974) pp. 793–797.

5. BASHA, S.M.M. and BEEVERS, L. Glycoprotein metabolism in the cotyledons of *Pisum sativum* during development and germination. *Pl. Physiol.* **57** (1976) 93–97.

6. BEACHY, R.N., THOMPSON, J.F. and MADISON, J.T. Isolation of polyribosomes and messenger RNA active in *in vitro* synthesis of soybean seed proteins. *Pl. Physiol.* **61** (1978) 139–144.

7. BEEVERS, L. and GUERNSEY, F.S. Changes in some nitrogenous components during the germination of pea seeds. *Pl. Physiol.* **41** (1966) 1455–1458.

8. BEEVERS, L. and POULSON, R. Protein synthesis in cotyledons of *Pisum sativum.* 1. Changes in cell-free amino acid incorporation capacity during seed development and maturation. *Pl. Physiol.* **49** (1972) 476–481.

9. BIETZ, J.A. and WALL, J.S. Wheat gluten subunits: molecular weights determined by sodium dodecyl sulfate-polyacrylamide gel electrophoresis. *Cereal Chem.* **49** (1972) 416–430.

10. BLAGROVE, R.J., GILLESPIE, J.M. and RANDALL, P.J. Effect of sulphur supply on the seed globulin composition of *Lupinus angustifolius. Aust. J. Plant Physiol.* **3** (1976) 173–184.

11. BLOBEL, G. and DOBBERSTEIN, B. Transfer of proteins across membranes. 1. Presence of proteolytically processed and unprocessed nascent immunoglobulin light chains. *J. Cell Biol.* **67** (1975) 835–851.

12. BONNER, J., HUANG, R-C. and GILDEN, R.V. Chromosomally-directed protein synthesis. *Proc. natn. Acad. Sci. U.S.A.* **50** (1963) 893–900.

13. BRANDT, A. Endosperm protein formation during kernel development of wild type and a high-lysine barley mutant. *Cereal Chem.* **53** (1976) 890–901.

14. BRANDT, A. and INGVERSEN, J. *In vitro* synthesis of barley endosperm proteins on wild type and mutant templates. *Carlsberg Res. Commun.* **41** (1976) 311–320.

15. BRIARTY, L.G., COULT, D.A. and BOULTER, D. Protein bodies of developing seeds of *Vicia faba. J. Exptl. Bot.* **20** (1969) 358–372.

16. BURR, B. and BURR, F.A. Zein synthesis in maize endosperm by polyribosomes attached to protein bodies. *Proc. natn. Acad. Sci., U.S.A.* **73** (1976) 515–519.

17. BURR, B., BURR, F.A., RUBENSTEIN, I. and SIMON, M.N. Purification and translation of zein messenger RNA from maize endosperm protein bodies. *Proc. natn. Acad. Sci. U.S.A.* **75** (1978) 696–700.

18. BUTTROSE, M.S. Ultrastructure of the developing wheat endosperm. *Aust. J. biol. Sci.* **16** (1963) 305–317.

19. CARR, D.J. and SKENE, K.G.M. Diauxic growth curves of seeds, with special reference to french beans (*Phaseolus vulgaris* L.) *Aust. J. biol. Sci.* **14** (1961) 1–12.

20. CRAIG, S., GOODCHILD, D.J. and HARDHAM, A.R. Structural aspects of protein accumulation in developing pea cotyledons with emphasis on qualitative and quantitative changes in parenchyma cell vacuoles. *Aust. J. Plant Physiol.* **6** (1979) 81–98.

21. DANIELSON, C.E. Plant proteins. *Ann. Rev. Pl. Physiol.* **7** (1956) 215–236.

22. DANIELSSON, C.E. Seed globulins of the Gramineae and Leguminosae. *Biochem. J.* **44** (1949) 387–400.

23. DANIELSSON, C.E. A contribution to the study of the synthesis of the reserve proteins in ripening pea seeds. *Acta Chem. Scand.* **6** (1952) 149–159.

24. DERBYSHIRE, E., WRIGHT, D.J. and BOULTER, D. Legumin and vicilin, storage proteins of legume seeds. *Phytochemistry* **15** (1976) 3–24.

25. DIECKERT, J.W. and DIECKERT, M.C. The chemistry and cell biology of the vacuolar proteins of seeds. *J. Food Sci.* **41** (1976) 475–482.

26. DI FONZO, N., FORNAZARI, E., SALAMINI, F. SDS-protein subunits in normal, *opaque-2* and *floury-2* maize endosperms. *Maydica* **22** (1977) 77–88.

27. DUDMAN, W.F. and MILLERD, A. Immunochemical behaviour of legumin and vicilin from *Vicia faba*: survey of related proteins in the Leguminosae Subfamily Faboideae. *Biochem. Systematics and Ecol.* **3** (1975) 25–33.

28. FAMBROUGH, D.M., FUJIMURA, F. and BONNER, J. Quantitative distribution of histone components in the pea plant. *Biochemistry* **7** (1968) 575–584.

29. GIANNAZZA, E., RIGHETTI, P.G., PIOLI, F., GALANTE, E. and SOAVE, C. Size and charge heterogeneity of zein in normal and *opaque-2* maize endosperms. *Maydica* **21** (1976) 1–17.

30. GRAHAM, J.S.D. and MORTON, R.K. Studies of proteins of developing wheat endosperm: separation by starch gel electrophoresis and incorporation of [^{35}S]-sulphate. *Aust. J. biol. Sci.* **16** (1963) 357–365.

31. GRAHAM, J.S.D., MORTON, R.K. and RAISON, J.K. Isolation and characterization of protein bodies from developing wheat endosperm. *Aust. J. biol. Sci.* **16** (1963) 375–383.

32. GRAHAM, J.S.D., MORTON, R.K. and RAISON, J.K. The *in vivo* uptake and incorporation of radioisotopes into proteins of wheat endosperm. *Aust. J. biol. Sci.* **17** (1964) 102–114.

33. GRAHAM, T.A. and GUNNING, B.E.S. Localization of legumin and vicilin in bean cotyledons using fluorescent antibodies. *Nature* **228** (1970) 81–82.

34. HARRIS, N. and BOULTER, D. Protein body formation in cotyledons of developing cowpea (*Vigna unguiculata*) seeds. *Ann. Bot.* **40** (1976) 739–744.

35. HIGGINS, T.J.V. and SPENCER, D. Cell-free synthesis of pea seed proteins. *Pl. Physiol.* **60** (1977) 655–661.

36. INGLE, J., BEITZ, D. and HAGEMANN, R.H. Changes in composition during development and maturation of maize seeds. *Pl. Physiol.* **40** (1965) 835–839.

37. JENNINGS, A.C. and MORTON, R.K. Changes in nucleic acids and other phosphorus-containing compounds of developing wheat germ. *Aust. J. biol. Sci.* **16** (1963) 332–341.

38. JENNINGS, A.C., MORTON, R.K. and PALK, B.A. Cytological studies of protein bodies of developing wheat endosperm. *Aust. J. biol. Sci.* **16** (1063) 366–374.

39. KASARDA, D.D., BERNARDIN, J.E. and NIMMO, C.C. Wheat proteins. In Y. POMERANZ (Ed) *Advances in Cereal Science and Technology.* American Association of Cereal Chemists, St. Paul (1976) pp. 158–236.

40. KHOO, U. and WOLF, M.J. Origin and development of protein granules in maize endosperm. *Am. J. Bot.* **57** (1970) 1042–1050.

41. KOIE, B., INGVERSON, J., ANDERSON, A.J., DOLL, H. and EGGUM, B.O. Composition and nutritional quality of barley proteins. In *Evaluation of Seed Protein Alteration by Mutation Breeding.* International Atomic Energy Agency, Vienna (1976) pp. 55–61.

42. KONZAK, C.F. Genetic control of the content, amino acid composition and processing properties in wheat. *Advan. in Genetics* **19** 407–582.

43. LARKINS, B.A. and DALBY, A. *In vitro* synthesis of zein-like protein by maize polyribosomes. *Biochem. biophys. Res. Commun.* **66** (1975) 1048–1054.

44. LARKINS, B.A. and HURKMAN, W.J. Synthesis and deposition of zein in protein bodies of maize endosperm. *Pl. Physiol.* **62** (1978) 256–263.

45. LARKINS, B.A., JONES, R.A. and TSAI, C.Y. Isolation and *in vitro* translation of zein messenger ribonucleic acid. *Biochemistry* **15** (1976) 5506–5511.

46. MADISON, J.T., THOMPSON, J.F. and MUENSTER, A-M.E. Deoxyribonucleic acid, ribonucleic acid, protein and uncombined amino acid content of legume seeds during embryogeny. *Ann. Bot.* **40** (1976) 745–756.

47. MEHTA, S.L., LODHA, M.L., MALI, P.C., SINGH, J. and NAIK, M.S. Characterization of polysomes and incorporation *in vitro* of leucine and lysine in normal and *opaque-2 Zea mays* endosperm during development. *Phytochemistry* **12** (1973) 2815–2820.

48. MILLERD, A. and SPENCER, D. Changes in RNA-synthesizing activity and template activity in nuclei from cotyledons of developing pea seeds. *Aust. J. Plant Physiol.* **1** (1974) 331–341.

49. MILLERD, A., THOMSON, J.A. and SCHROEDER, H.E. Cotyledonary storage proteins in *Pisum sativum.* III. Patterns of accumulation during development. *Aust. J. Plant Physiol.* **5** (1978) 519–534.

50. MILLERD, A. and WHITFELD, P.R. Deoxyribonucleic acid and ribonucleic acid synthesis during the cell expansion phase of cotyledon development in *Vicia faba* L. *Pl. Physiol.* **51** (1973) 1005–1010.

51. MORTON, R.K., PALK, B.A. and RAISON, J.K. Intracellular components associated with protein synthesis in developing wheat endosperm. *Biochem. J.* **91** (1964) 522–528.

52. MORTON, R.K. and RAISON, J.K. A complete intracellular unit for incorporation of amino acid into storage protein utilizing adenosine triphosphate generated from phytate. *Nature* **200** (1963) 429–433.

53. MUNCK, L. Barley seed proteins. In G.E. INGLETT (Ed) Symposium: *Seed Proteins.* Avi Publishing Co., Westport, CT (1972) 144–164.

54. NAGAHASHI, J. and BEEVERS, L. Subcellular localization of glycosyl transferases involved in glycoprotein biosynthesis in the cotyledons of *Pisum sativum* L. *Pl. Physiol.* **61** (1968) 451–459.

55. OPIK, H. Development of cotyledon cell structure in ripening *Phaseolus vulgaris* seeds. *J. Exptl. Bot.* **19** (1968) 64–76.

56. OOTA, Y. and OSAWA, S. Migration of storage RNA from cotyledons into growing organs of bean seed embryos. *Experientia* **10** (1954) 254–256.

57. OSBORNE, T.B. The vegetable proteins. In R.H. PLUMMER and F.G. HOPKINS (Eds) *Monographs on Biochemistry.* Longmans, Green, London (1924).

58. PAYNE, P.I. and BOULTER, D. Free and membrane-bound ribosomes of the cotyledons of *Vicia faba.* 1. Seed development. *Planta* **84** (1969) 263–271.

59. PAYNE, E.S., BROWNRIGG, A., YARWOOD, A. and BOULTER, D. Changing protein synthetic machinery during development of seeds of *Vicia faba. Phytochemistry* **10** (1971) 2299–2303.

60. POULSON, R. and BEEVERS, L. RNA metabolism during development of cotyledons of *Pisum sativum. Biochim. biophys. Acta* **308** (1973) 381–389.

61. RABSON, R., MANS, R.J. and NOVELLI, G.D. Changes in cell-free amino acid incorporating activity during maturation of maize kernels. *Archs. Biochem. Biophys.* **93** (1961) 555–562.

62. RANDALL, P.J., THOMSON, J.A. and SCHROEDER, H. Cotyledonary storage proteins in *Pisum sativum.* IV Effects of sulphur, phosphorus, potassium and magnesium deficiencies. *Aust. J. Plant Physiol.* **6** (1979) 11–24.

63. REST, J.A. and VAUGHAN, J.G. The development of protein and oil bodies in seeds of *Sinapis alba* L. *Planta* **105** (1972) 245–262.

64. SCHARPÉ, A. and VAN PARIJS, R. The formation of polyploid cells in ripening cotyledons of *Pisum sativum* L. in relation to ribosome and protein synthesis. *J. Exptl. Bot.* **24** (1973) 216–222.

65. SIMMONDS, D.H. and CAMPBELL, W.P. Morphology and chemistry of the rye grain. In W. Bushuk (Ed) *Rye: Production, Chemistry and Technology*. Amer. Assoc. Cereal Chemists. St. Paul, MN (1974) pp. 63–110.

66. SMITH, D.L. Nuclear changes in cotyledons of *Pisum arvense* L. during germination. *Ann. Bot.* **35** (1971) 511–521.

67. SMITH, D.L. Nucleic acid, protein and starch synthesis in developing cotyledons of *Pisum arvense* L. *Ann. Bot.* **37** (1973) 795–804.

68. SOAVE, C., RIGHETTI, P.G., LORENZONI, C., GENTINETTA, E. and SALAMINI, F. Expressivity of the *opaque-2* gene at the level of zein molecular components. *Maydica* **21** (1976) 61–75.

69. STURANI, E. and COCUCCI, S. Changes in the RNA system in the endosperm of ripening castor bean seeds. *Life Sci.* **4** (1965) 1937–1944.

70. SUN, S.M., MUTSCHLER, M.A., BLISS, F.A. and HALL, T.C. Protein synthesis and accumulation in bean cotyledons during growth. *Pl. Physiol.* **61** (1978) 918–923.

71. THOMSON, J.A. and SCHROEDER, H.E. Cotyledonary storage proteins in *Pisum sativum*. II. Hereditary variation in components of the legumin and vicilin fractions. *Aust. J. Plant Physiol.* **5** (1978) 281–294.

72. THOMSON, J.A., SCHROEDER, H.E. and DUDMAN, W.F. Cotyledonary storage proteins in *Pisum sativum*. I. Molecular heterogeneity. *Aust. J. Plant Physiol.* **5** (1978) 263–280.

73. VARNER, J.E. and SCHIDLOVSKY, G. Intracellular distribution of proteins in pea cotyledons. *Pl. Physiol.* **38** (1963) 139–144.

74. VIOTTI, A., BALDUCCI, C. and WEIL, J.H. Adaptation of the tRNA population of maize endosperm for zein synthesis. *Biochim. biophys. Acta* **517** (1978) 125–132.

75. VIOTTI, A., SALA, E., ALBERI, P. and SOAVE, C. RNA metabolism and polysome profiles during seed development in normal and *opaque-2* maize endosperms. *Maydica* **20** (1975) 111–124.

76. WELLS, G.N. and BEEVERS, L. Protein synthesis in cotyledons of *Pisum sativum* L. Messenger RNA-independent formation of a methionyl-tRNA initiation complex. *Archs. Biochem. Biophys.* **170** (1975) 384–391.

77. WOLF, W.J. Soybean ultrastructure and its relationship to processing. In G.E. INGLETT (Ed) Symposium: *Seed Proteins*. Avi Publishing, Westport, CT (1972) pp. 231–241.

78. WRIGLEY, C.W. and SHEPHERD, K.W. Electrofocusing of grain proteins from wheat genotypes. *Ann. N. Y. Acad. Sci.* **209** (1973) 154–162.

POLLEN ANTIGENS, ALLERGENS AND ENZYMES

BARBARA J. HOWLETT, H.I.M. VASANTHE VITHANAGE

and

R. BRUCE KNOX

School of Botany, University of Melbourne, Parkville, Vic. 3052, Australia

INTRODUCTION

The pollen grain is the structure that houses the male gametes of flowering plants during their journey from the anther to a receptive stigma for pollination, and produces a pollen tube that transports the sperm cells to the embryo sac for fertilization. It contains two or three cells at maturity within a common wall. This protects the grain during transport by animal vectors or air currents, and its surface carries a range of macromolecules presumably enabling the pollen to be recognised and accepted as compatible or rejected as incompatible at the stigma surface. After germination, the grains need adequate reserves to provide for the initial growth of the pollen tube, and the necessary enzymes to penetrate the stigma surface and grow through the style to the ovule.

It is the purpose of this review to consider the present state of knowledge of pollen macromolecules, proteins and glycoproteins, some of which are antigenic in rabbits, allergenic in susceptible humans, or possess demonstrable enzymic activities. These categories are not mutually exclusive. It is well known that most allergens and enzymes are antigens; some may possess all three characteristics although this has not been demonstrated experimentally. Their importance in biology and agriculture lies in their vital role in plant reproduction, and in medicine because of their role in allergic diseases. We will discuss their physicochemical characteristics, origin and location within the pollen grain.

POLLEN STRUCTURE AND FUNCTION

In two-thirds of flowering plant families, the pollen grains comprise two cells at maturity: a vegetative cell to regulate tube growth and a generative cell that will divide on germination to produce the two sperm cells. In the remainder of families, all three cells are present at maturity, the sperm cells being contained wholly within the vegetative cell, separated not by walls but by double plasma membranes. The remarkable feature of pollen grains is therefore the wall which encloses these separate cells. It is the wall that makes it possible for pollen grains to function as free cells, together with the reserves stored in the cytoplasm and their state of temporary dormancy, which is akin to seeds in terms of the low respiratory and metabolic rate, and water content. General reviews of pollen structure have recently appeared.[53, 54, 101]

Elucidation of the structure of the pollen grain wall has been made possible during the past 20 years with the advent of the transmission electron microscope, which has clearly delineated the two major layers: the outer exine and inner intine. The exine carries all the intricate patterning attributes specific to the different types of pollen, and is made of sporopollenin, a wall polymer remarkably resistant to degradation. It contains a variety of crypt-like cavities or voids which have direct access to the surface of the grain via micropores.[29, 33] In most pollen types that have been examined, these exine cavities are filled with proteins, glycoproteins, lipids and pigments, and the surface of the grain is coated with a sticky material, the pollenkitt or tryphine, which aids some pollen grains to adhere in clumps for pollination. Control of exine synthesis resides with the diploid genome of the parent plant, and not with the pollen grain, both in terms of the determination of pattern, and production of the wall polymer.[31, 32, 33]

The intine consists of a matrix of polysaccharides, mainly pectocellulosic in nature.[98] It is produced by the pollen protoplast and deposited during pollen development on the inner surface of the exine, and is especially thickened at the germinal apertures where it forms the tip for a potential pollen tube. In most grains, the intine is exposed to the surface at these apertures as it is not covered by exine.

POLLEN ANTIGENS

An antigen is defined as a substance, generally a protein, glycoprotein or polysaccharide, that will induce antibody formation in animals such as the rabbit, and that will react specifically with the antibodies raised against it.[96] Antigenic activity in a rabbit provides a convenient means of detecting a range of such macromolecules that may not possess any known catalytic activity and so be detected as enzymes. In this section, we will review what is known of the pollen antigens and will start by considering briefly the techniques currently available to detect and characterize antigens.

Nature and detection

The remarkable specificity with which antigens can be recognized by their specific antibodies provides a very powerful tool for their isolation and characterization, as well as for their localization using labelled antibodies. Most proteins and glycoproteins with molecular weights greater than 10,000 and some polysaccharides can act as antigens. They contain many different antigenic determinants which are the sites in the molecule that induce antibody formation. Proteins and glycoproteins are usually better antigens than polysaccharides.[96] Antigens must come from a foreign source, either a different species or different animal of the same species that is being immunized; hence, plant proteins are usually excellent antigens. Antigenic activity may often provide the sole specific means of identifying particular proteins or glycoproteins in a complex mixture.

Antisera to pollen antigens have usually been raised to extracts of water- or buffer-soluble components, either by using extracts containing cytoplasmic and wall components obtained by a combination of freeze-thawing, -pressing and macerating (e.g.[4, 69]), or by preparing short or long term diffusates from whole grains. Short term diffusates prepared from living pollen grains in isotonic media are considered to provide antigen extracts containing surface wall components free of any cytoplasmic contaminants.[39, 62] Such short term diffusates of *Gladiolus gandavensis* pollen, after dialysis to remove low molecular weight components, contain about equal quantities of protein and carbohydrate, and trace amounts of lipid.[11] The techniques for raising antisera to pollen and other plant antigens have recently been reviewed, and standard schedules given.[55]

Antigens are detected by their ability to bind and precipitate their specific antibodies, either in agar or agarose gels or in fluid. The most common technique immunodiffusion, was developed by Ouchterlony[88] and involves double diffusion of antigen and antibody from wells cut in agar gel. At various points between the two wells, precipitation of antigen/antibody complexes occurs. The strength or titre of antibodies can be estimated by a dilution series of antiserum against a known concentration of antigen by double diffusion or by the micro-precipitin ring test (see[9, 69])

Double diffusion alone is not sensitive enough to enable a determination of the total number of antigens in a mixture. This can be estimated after separation of the antigens by electrophoresis in agarose gel[95] and their diffusion against antisera forming arcs of antigen–antibody precipitates. The relative quantities of antigens in a mixture can be estimated using a two-dimensional crossed-immunoelectrophoretic technique, the Laurell method, as adapted for grass pollen antigens.[73] When antigens have been purified, quantitation of antigen concentration in different extracts can be obtained by radial immunodiffusion into an agar gel containing specific antiserum, or by rocket electrophoresis where the antigen is electrophoresed into agar gel containing antiserum.[26] This latter method is extremely rapid, producing visible precipitin arcs within 30 min, whereas immunodiffusion may take 24 h or longer. Most plant extracts are complex mixtures of antigens, and pollen is no exception. Up to sixteen antigens have been detected in extracts of grass pollen[2, 3, 43] and about nine antigens have been detected in short term surface diffusates of poplar, *Cosmos* and *Gladiolus* pollen,[12, 38, 62] but almost twice as many have been detected in grass and ragweed pollen using crossed-immunoelectrophoresis.[68a, 73a]

The number of antigens in an extract depends upon the antigenicity of the molecules injected into the rabbits and the sensitivity of the method of detection. Increased sensitivity on a micro-scale can be achieved by using antigens or antisera labelled with ^{125}I followed by autoradiography or use of a gammacounter. An indication of both the number of antigens present and their molecular weights can be obtained by immunoprecipitation of ^{125}I labelled antigens followed by polyacrylamide gel electrophoresis in sodium dodecyl sulphate.[12] Enzymes such as peroxidase conjugated to the antiserum also increase sensitivity since the cytochemical reaction product may be visualized in gels or quantitated using spectrophotometry.[82] Very few antigens other than glycoproteins have been investigated in pollen. However one polysaccharide antigen, an arabinogalactan from ragweed pollen, was found to be antigenic in mice, and the antibody produced had a specificity for galactose.[48]

Specificity of pollen antigens

The immunodiffusion methods developed by Ouchterlony[88] provide a convenient means of assessing the relationships between antigens in mixtures such as pollen diffusates. Several diffusates can be tested against a single antiserum, and reactions of identity, partial identity or non-identity detected by the appearance of the precipitin arcs. This has been employed by Wodehouse[110, 111] in studies of pollen antigens of grasses, ragweeds and various Chenopods. He detected a surprising degree of cross-reactivity between pollens of different species. In grasses, Watson and Knox[108] found taxonomically-ordered variation in pollen diffusates by testing species belonging to all the major groups against antiserum raised to ryegrass, *Lolium perenne.* Several antigens were common to pollen of Festucoid grasses, but antigens from other grass groups were generally quite different. The question must also be asked as to whether these antigens are unique to pollen or occur in other tissues of the plant. This has been approached by at least two groups. In 1960, Linskens[71] showed that certain pollen antigens of *Petunia hybrida* were shared by the stigma. Recently, Clarke and coworkers have demonstrated the presence of antigens common to pollen and other somatic tissues of *Gladiolus gandavensis*[12] and *Prunus avium.*[91] An interesting feature of these results is that the pollen expressed a wider range of antigens than any somatic tissue tested. In *Prunus*, there were 7 pollen antigens, 5 stigma antigens, 3 leaf antigens and 1 stem antigen detected. However callus cells derived by tissue culture from the somatic tissues expressed a greater number of antigens than the parental organ, and some of these antigens were secreted into the culture medium in liquid suspension cultures. This observation, taken together with the presence of pollen antigens in somatic tissues, suggests that these antigens may be involved in cell–cell interactions of both somatic and sexual recognition systems.[11] It is also possible that in the case of the somatic tissues, the reduced number of antigens is due to the presence of interfering compounds such as phenolics in the tissue homogenates used as antigen preparations. These could bind the antigens, and prevent them being accessible to the animal's immune system.

Localization of antigens

Antigens have been localized within pollen grains using immunofluorescence methods, which rely on the specificity of the antibody for its antigen at its original site within the grain.[21, 81] In a series of studies, antigens to diffusates from several types of pollen have been located within pollen-wall sites, either within the exine cavities or in the intine.[51, 61, 63] Antigen E of ragweed has been shown to be present in both sites.[38, 59] These tests have also revealed the extreme mobility of the antigens: when pollen grains are hydrated, no aqueous fixative has yet been found that will retain the exine proteins within their wall sites. These are released immediately on moistening, and several minutes later, the intine proteins are emitted more slowly from around the germinal apertures from which the pollen tube is produced. Methods for antibody labelling and detection by immunoelectron microscopy have been reviewed and their applications for plant cells established.[55] However, they have not yet been applied to the localization of pollen antigens since the rapid mobility of the antigens makes standard methods difficult to use.

The S-antigen of pollen

Pollen antigens first became the object of intense biological interest in 1929 when East[18] proposed their role in the puzzling genetic phenomenon of self incompatibility. East had observed that pollen from one plant is unable to effect self-fertilization, even though it is fertile on another individual of the same population. He carried out genetic experiments which revealed that the process was controlled by a series of alleles of the S gene, the presence of identical alleles in pollen and style leading to mutual inhibition. East argued that since each S-allele produced a different and highly-specific reaction, this indicated the presence of an antigen-like pollen protein as the gene product which reacted with a complementary stylar antibody-like protein of similar specificity.

This suggestion was much ahead of its time, and it was not taken up until 20 years later when Lewis[69] in London began a long series of experiments aimed at detecting S-allele specific antigens in pollen of the evening primrose, *Oenothera organensis.* He raised antisera in rabbits to long term extracts of pollen from nine different clones, each being heterozygous for pairs of combinations of four S-alleles. The antisera produced had very high titres against specific antigens, producing detectable precipitins in dilutions as low as 1 in 1280. Specificity is the key to the hypotheses concerning the mechanism of self incompatibility, and immuno-absorption provided a simple test of specificity. Lewis absorbed his antisera with extracts of pollen containing other combinations of S-alleles so that any common components would be precipitated, leaving only antibodies specific to the pair of S-alleles under test. The absorbed antisera were tested against their homologous antigens and against antigens containing a common S-allele, and others with no common S-alleles. High titres were obtained against homologous antigens, lower titres against antigens with only one common S-allele, and very low titres against antigens with different S-alleles. These pioneering tests revealed the correlation between antigen specificity and S-alleles in pollen of *Oenothera.*

Later, Linskens[71] at Nijmegen in Holland used similar tests to demonstrate S-allele specificity in the pollen of *Petunia hybrida* which also possesses a well-developed self incompatibility system. Linskens revealed that in *Petunia* S-specific antigens occur in both pollen and style, but he did not attempt absorption experiments. The techniques of immunodiffusion and immunoelectrophoresis became widely available in the 1960's, so that the number of pollen antigens and their precise immunological relationships could be established. Makinen and Lewis [75] demonstrated that the S-antigen of *Oenothera* comprised nearly 20% of extractable pollen protein, and diffused from the pollen into isotonic medium within 30 min, an observation that suggests it may be located within the pollen walls. In further experiments, single pollen grains were shown to produce radial immunoprecipitates when sprinkled on antibody-containing agar gels.[70] Using gels containing antisera to S_6 pollen, pollen from S_2S_3 anthers failed to produce precipitates, while half the pollen grains from S_3S_6, and nearly all the pollen from S_6S_6 anthers produced precipitates, again confirming the specificity of the S-antigen.

Following these early demonstrations of S antigen specificity in pollen, there has been little progress made in its isolation or characterization. Immunoelectrophoretic studies of the spectrum of pollen antigens in specific S-genotypes of cabbage and marrow-stem kale, *Brassica oleracea*, have failed to reveal the presence of a specific antigen with differing electrophoretic mobilities in pollen from different S-alleles[63, 83, 84] despite the demonstration of striking S-allele specificity in a stigma antigen. The stigma S-antigen has been shown to possess a high iso-electric point, periodate-sensitive vicinal hydroxyl groups, and to be a glycoprotein with an affinity for the lectin concanavalin A.[85] The existence of antigens common to both pollen and stigma has recently been confirmed in *Gladiolus gandavensis*, which has a breeding system in which there are no barriers to self or cross matings. The common antigen has a molecular weight of 40,000, and was detected in immunoprecipitates of ^{125}I-labelled antigens after fractionation by SDS-polyacrylamide gel electrophoresis.[12] The application of such techniques to S-allele specific pollen diffusates should enable the identification and characterization of the putative S-antigen.

POLLEN ALLERGENS

Interest in pollen has been greatly stimulated by the discovery that it is a causative agent of hayfever and seasonal asthma in man. In 1831, Elliotson noted that the worst of catarrh symptoms accompanied the maturing of the grasses during the haymaking season. These observations were confirmed and extended by Blackley[8] who not only demonstrated the role of pollen in initiating allergic disease, but also the presence of the allergenic pollens in the atmosphere. Kamann[47] and later Noon[86] found that protein fractions extracted from grass pollen produced skin reactions in sensitive patients. Since then considerable efforts have been made to understand more about the nature of these proteins and glycoproteins, which are termed *allergens* because they are capable of eliciting the formation of specific skin-sensitizing or reaginic antibodies, Immunoglobulin E (IgE), in susceptible humans. Ishizaka and co-workers[41, 42] demonstrated that when certain genetically predisposed individuals become exposed to allergens, specific IgE is formed and may bind to basophil and mast cells. Upon subsequent exposure, the allergen may bind to pairs of adjacent IgE molecules on the surface of the mast cells, resulting in degranulation and the release of histamine and other substances causing contraction of smooth muscle of the bronchioles, vasoconstriction and other symptoms of allergic disease.

Allergens are now known to be present in many airborne pollen types, in fungal spores, house-dustmites, animal danders, fish, eggs, milk and other foods.

The most notable pollens that cause allergic disease are dispersed in air currents for pollination and not transferred by animal vectors. Airborne pollens exhibit a marked seasonal periodicity, with tree pollens appearing first in winter and spring, followed by grasses and herb pollens in spring and early summer, and in North America by ragweed in autumn (for review[54]). All these pollens can reach high concentrations in the atmosphere, and a proportion can travel long distances before impacting on a stigma or the human upper respiratory tract.

Assays for allergens

There are several biological assays for pollen allergens: (1) direct skin testing of patients, using a small sample of the suspected allergen usually in a buffered saline extract, which is pricked intradermally on the forearm, giving a swollen red wheal in allergic patients. This can be compared with a histamine control to give an indication of the allergenicity of the sample; (2) histamine release assay where leukocytes of allergic individuals are incubated in the presence of the suspected allergen and the amount of histamine released is measured;[87] (3) the radio-allergo-sorbent test (RAST) in which suspected allergens are chemically conjugated to paper discs, and incubated in allergic patient's serum. After washing, the discs are exposed to ^{125}I-labelled antiserum raised against human IgE, and the counts bound to the disc indicate the amount of allergen-specific IgE, which is related to the allergenicity of the pollen extract.[44]

RAST-inhibition is also a useful test. In this case, a soluble allergen is present, which competes with that on the disc for the IgE in the allergic serum. By using a series of increasing concentrations of soluble allergen, the reduction in binding to the disc can be measured, and gives an indication of the allergenicity of the soluble allergen with reference to the known allergen insolubilized in the disc.[6, 20]

Ragweed pollen allergens

King and co-workers[50] have purified from saline extracts of low ragweed, *Ambrosia elatior*, pollen two major acidic protein allergens, Antigens E and K, with molecular weights of 38,000, which contain less than 1% carbohydrate, account for 6% and 3% of the soluble pollen protein, and show partial immunological identity. Three other basic protein allergens have been isolated from ragweed pollen: Antigen Ra3,[103] Ra5,[66] and Ra4 or BPA-R.[24, 92] These have molecular weights of 11,000, 5,000 and 23,000 respectively, and account together for less than 1% of the soluble pollen protein. Of these allergens, Antigen E is by far the most active. It consists of two polypeptide chains (MW 26,000 and 13,000) which are non-covalently associated. The two chains, when separated, have very little allergenic or antigenic activity. Several other chemical modifications have been carried out, and have recently been reviewed.[49] The amino acid sequence of Antigen Ra5 has been determined[80] and consists of 45 amino acids, 8 of which are cysteine. Another interesting feature is that of the 5 amino acids at the C terminal end of the polypeptide 3 are lysine, which makes this region very strongly alkaline.

Grass pollen allergens

The extracts of grass pollen used by Noon[86] were sterilized by autoclaving before being used in tests on patients. This revealed their remarkable heat stability, which has been confirmed by Augustin and co-workers[1–3] who isolated a series of allergens from cocksfoot (*Dactylis glomerata*) and timothy (*Phleum pratense*) pollen, called the I antigens. Each of the allergens proved to be antigenic in rabbits. The major antigen, Antigen A, however, proved to have much less allergenic activity than the I antigens.[3]

Ryegrass (*Lolium perenne*) pollen contains four allergens which have been characterized by Johnson and Marsh.[45, 46, 78] Two of them are acidic glycoproteins, called the Groups I and II allergens, with MW of 32,000 and 11,000 respectively. Each one has been isolated in multiple electrophoretic forms, which in the case of Group I allergens appear to differ in amino acid composition only in the ratio of glutamate to glutamine. Groups III and IV allergens proved to be basic, with MW of 11,000 and 50,000 respectively. Antigens with allergenic activity have been detected in extracts of timothy and ryegrass pollen by their binding to human IgE using crossed radioimmunoelectrophoresis.[73a]

Malley and Harris[77] isolated two acidic allergens from timothy pollen, with MW of 30,000 and 16,000, which were active in skin tests of timothy sensitive patients at concentrations as low as 10^{-9}g.ml^{-1}.

Tree pollen allergens

In general, pollen of tree species is not as important a source of pollen allergens as herbaceous plants, and indeed, pollen of *Pinus* spp., which often falls as yellow rain, has little demonstrable allergenic activity.[109] However, in northern Europe, the pollen of birch, *Betula verrucosa*, and alder, *Alnus glutinosa*, are equally important as the grasses. Birch pollen contains an acidic protein allergen, MW 20,000, which has been partially purified.[4] This antigen cross-reacts immunologically with the alder allergen.[27]

Pollen of *Pinus* and other gymnosperms, and even the spores of lower plants, are often common constituents of the atmosphere in temperate climates, and yet are only rarely implicated in allergic disease.[109] In the case of pine pollen, the surface of the grains appears particularly hydrophobic, making the release of potentially allergenic proteins in either aqueous solutions or in the nasal mucosa unlikely. Also, gymnosperms are less advanced than flowering plants, and it is likely that there is a corresponding decrease in antigenic complexity of the pollen macromolecules, so that they may not possess the number of allergenic determinants necessary to elicit the allergic response.

Nature and characteristic features

These allergens represent a diverse range of proteins and glycoproteins with molecular weights ranging from 5000 to 50,000, and consisting in some cases of several polypeptide chains. There have been very few reports of molecules other than proteins or glycoproteins possessing allergenic activity in pollen grains. One such case is an intine polysaccharide, a pectin, from *Cryptomeria japonica* which apparently causes allergic disease in Japan.[25, 40]

The most potent pollen allergens are acidic, including Antigen E of ragweed, and Group I allergen of ryegrass, but others are basic. Berrens[7] suggested that allergenic activity might be associated with the occurrence of *N*-glycosidic linkages between lysine in the polypeptide chains and sugars. However, some allergens such as Antigen E in ragweed, contain very little carbohydrate, 0.8%,[50] and Johnson and Marsh[46] showed that when the 5% carbohydrate is removed from Group I allergen of ryegrass, the allergenic activity is unimpaired. It seems unlikely that the presence of *N*-glycosidic linkages is a prerequisite for allergenic activity. Amino acid compositions of major allergens, Group I of ryegrass and Antigen E of ragweed, have been determined.

Similarly allergens from other sources, milk, codfish, animal danders, have a wide range of properties and molecular weights from 5000 to 65,000. King[49] has suggested that there is such an upper size limit for allergens since they must be absorbed across the plasma membrane either in the nasal mucosa, the respiratory tract or the gastro-intestinal tract. The lower limit may reflect a minimum level of complexity of allergenic determinants essential to invoke a response, and the necessity for an allergen to be bivalent and bridge adjacent IgE molecules on the mast cell surface.

Specificity and function of pollen allergens

In a family such as the Gramineae, all of which shed their pollen into the atmosphere, the question must be raised as to whether each species has its own unique allergens, or whether allergenic determinants are shared between closely related species and genera. This question has been approached using the technique of RAST inhibition.[6, 20] The test is based on competition for binding sites between a solid phase allergen insolubilized on a disc, and soluble allergens from a different sample or species (see Assays for allergens, above). When extracts of pollen of other species of ragweed were allowed to compete with insolubilized *Ambrosia elatior* allergens, or various grass species allowed to compete with insolubilized allergens of june grass, *Poa pratensis*, the degree of inhibition indicates the degree to which allergenic determinants are shared. The various species of ragweed showed a similar affinity to *A. elatior*, while false ragweed, *Franseria*, had little inhibitory effect indicating that it had quite different determinants. Among the grasses, differing degrees of inhibition were obtained dependent on the taxonomic relationships. Meadow fescue, cocksfoot and rye-grass were as effective as june grass in the tests, while timothy and sweet vernal had little inhibitory effect, and belong to different groups within the Festucoid grasses. The function of allergens in nature is not known, although they may be concerned with the early events of pollen germination on the stigma. They may possess recognition functions, provide a source of reserve protein for the germinating pollen tube or have enzyme activity. No conclusive evidence supporting any of these functions is currently available, although they have been implicated in pollen–stigma recognition.[38]

Localization and release of allergens

The kinetics of release of antigens from birch pollen, including the birch allergen, have been established.[4, 5] Using immunofluorescence methods, the major birch allergen was found to be emitted within seconds of moistening. Similar results were obtained for ragweed allergens, using antisera specific to Antigen E.[38, 59] On moistening, Antigen E was emitted immediately from the exine surface and from the cavities via micropores; while later emission occurred from the intine at the germinal apertures. Marsh[78] has used extracts measured by skin testing and leukocyte histamine release assays to demonstrate that only 10% of the total Antigen E of ragweed pollen is released within the first 90 min of extraction. The immunofluorescence tests[38, 59] demonstrate that most Antigen E is located in the polysaccharide matrix of the intine layer, and release *in vitro* may be retarded by the alkaline extraction buffer used.

The antigens of grass pollen are also located in pollen wall sites[60] although immunofluorescence tests have not as yet been made with antisera specific to particular allergens. No other allergens, with the exception of the intine polysaccharide of *Cryptomeria*,[25] have been located within the pollen grains. The presence of trace quantities of Antigen E in leaves of ragweed has been demonstrated,[67] although the results could be attributable to pollen rain on the leaf surfaces.

POLLEN ENZYMES

Mature pollen is a source for a large number of enzymes. Many appear in solution when the pollen is moistened, and Green[23] first demonstrated the property of moistened pollen to hydrolyse starch. Since then a variety of hydrolases, transferases, dehydrogenases, oxidases, ligases and lyases have been detected.[10, 74, 90] Several of these are known to be stored in wall sites,[57, 58, 102] while others occur in the cytoplasm.[22]

It is likely that some of these enzymes are involved in pollen germination, especially in nutrition of the growing pollen tube and its penetration of the stigma.

Enzymes of polysaccharide metabolism

Most of the enzymes that have been studied are from ungerminated pollen and there has been no good evidence for increased synthesis during germination, except in the case of some enzymes of sugar metabolism.[68, 99] In one case, Southworth and Dickinson[100] found a two-fold increase in the synthesis of callose after germination of *Lilium longiflorum* pollen. Callose is an important constituent of pollen tube walls and it is not surprising that its synthesis is increased during germination.

Dickinson and co-workers[14, 15] have also studied the activities of several other enzymes involved in sugar and polysaccharide metabolism in both germinating and non-germinating lily pollen. The site of these enzymes in the grain is not known although they were usually extracted from ground homogenates of pollen, and so cytoplasmic and wall components would be present. Pectinase, has been detected in ungerminated *Nicotiana tabacum* pollen.[64] This enzyme will degrade commercial preparations of *Citrus* pectin to galacturonic acid.

Cutinases in pollen

An enzyme which has been implicated in penetration of the stigma surface is cutinase, which hydrolyses the long chained hydroxy-monocarboxylic acids, the major constituent of the stigma cuticle in many plants. Linskens and Heinen[72] have shown that only pollen of plants with waxy stigmatic cuticles contained the enzyme, which was absent in pollen of species which lacked such structures, for example, lily pollen. Recently, Shayk and Kolattukudy[97] demonstrated cutinase activity in pollen of *Nasturtium* which appeared to be synthesized before pollen germination, and which hydrolysed the stigmatic cuticle. The authors considered it may be located in the pollen wall since it apparently acted on the cuticle very rapidly.

Another enzyme, concentrated mainly at the tips of the pollen tube is an alkaline phosphatase from the *Calotropis procera.*[76]

Pollen enzyme polymorphism

Many of the pollen enzymes when examined after polyacrylamide gel electrophoresis appear as multiple forms of differing charge and/or molecular weight.[74, 101] Esterase isoenzymes have been subjected to extensive genetic analysis in maize, *Zea mays.*[79] and nine loci have been demonstrated to control them. The forms show a certain amount of tissue specificity, variation in substrate utilization, and in response to several inhibitors and activators. In *Hemerocallis*, multiple forms of esterase are present in different tissues tested and two are specific to pollen.[10] Recently, 9 multiple forms of esterase and 12 of peroxidase have been detected in diffusates of birch pollen.[65] In pollen of sunflower, *Helianthus annuus*, 6 multiple forms of esterase have been demonstrated,[104] 4 of which are specific to pollen and 2 shared with either leaf or stigma; in addition, 2 multiple forms of acid phosphatase occurred, one of which is common to leaf and stigma. These multiple forms probably have a broad range of affinity for the substrate in the reaction mixture in which they were detected, since the substrate used is usually a simple molecule. However, *in vivo*, they may show greater specificity for their more complex substrates.

Localization and origin of pollen-wall enzymes

Following freeze-sectioning of fresh pollen, a range of enzymic activities has been demonstrated in the pollen-wall in species from many different families representing a diversity of pollen types.[57, 58, 102] These had not previously been detected in wall sites since conventional aldehyde fixation techniques do not prevent loss of the enzymes from the wall sites when moistened in the aqueous fixatives.[22] Heslop-Harrison and co-workers[29–36] have located the pollen-wall protein in two principal sites: (1) in the cavities of the outer sporopollenin wall layer, the *exine*, and (2) in the polysaccharide matrix of the inner layer, the intine, which is especially thickened at the germinal apertures.

The transfer of proteins from parental tapetal cells to the exine cavities has been demonstrated by transmission electron microscopy in anthers of the Malvaceae[34] and Cruciferae[16, 17, 35] The protein is seen as fibrillar material sealed within the cavities by electron-dense lipids and other components of the sticky surface pollenkitt or tryphine which is also of tapetal origin. The transfer has been demonstrated cytochemically in anther sections stained with the fluorescent protein probe 1, anilino-8-naphthyl sulphonic acid in pollen of *Iberis.*[35, 36]

Such evidence for transfer of tapetal materials to the exine is entirely qualitative. Recently, quantitative data for such transfer in two systems, *Brassica oleracea* and *Helianthus annuus*, has been obtained for the enzymes acid phosphatase and esterase.[106, 107] Freeze-sectioned anthers at different stages of development were incubated in the enzyme reaction mixtures containing the substrates α-naphthyl phosphate and α-naphthyl acetate, and the coloured reaction products examined using a scanning microscope photometer. The deposits of reaction product were found to be quantitative, so that changes in enzymic activity during development of both pollen grains and tapetal cells could be followed.

In *Brassica* the parietal tapetal cells exhibited a striking increase in esterase activity reaching a maximum just prior to tapetal dissolution.[106] However, the pollen esterase activity was low until the period of tapetal dissolution when activity showed a five-fold increase until maturity. The esterase activity was associated with the exine in mature pollen. In contrast, acid phosphatase showed two periods of accumulation during pollen development, associated first with intine incorporation early in pollen development and later in the cytoplasm of maturing pollen. Acid phosphatase activity in the tapetum was low throughout anther development. These quantitative data support the conclusion that this exine enzyme is a product of the diploid parental cells of the tapetum, and that the intine proteins are products of the haploid pollen protoplast.

In sunflower, *Helianthus annuus*, these enzymes showed different patterns of activity and sites of accumulation.[107] Acid phosphatase occurred in both intine and exine sites, while esterase was present largely in the intine. In the tapetum, acid phosphatase showed high activity in the parietal tapetal cells and in the plasmodium until its dissolution. At this time, accumulation of acid phosphatase occurred on the pollen surface filling the exine cavities, accompanied by a marked increase in activity. Thus, acid phosphatase had a characteristic sporophytic pattern of accumulation in sunflower,[107] while esterase exhibited a pattern typical of gametophytic synthesis. These differences probably reflect the broad range of enzymes which can act on the simple synthetic substrates in the reaction mixtures. In spite of these limitations, the data suggests sporophytic synthesis of exine proteins, although no direct biochemical or immunological evidence for such transfers has been obtained. This is probably because of the great difficulties in separating fractions from tapetum and pollen in the developing anther. This has recently been achieved by Herdt *et al.*,[28] who have detected a range of enzymes involved in phenylpropanoid metabolism in the tapetal fraction of developing anthers. These enzymes have been implicated in the synthesis of the exine.

POLLEN–STIGMA INTERACTIONS

In the 1970's, interest in S-gene interactions in pollen and style has changed emphasis, and attempts have been made to implicate pollen antigens in the recognition reactions that mediate acceptance or rejection of the pollen at the stigma surface, within the stigma and style, or at the sperm/egg interface. In 1970, Knox, Heslop-Harrison and Reed[60] demonstrated the extracellular location of the diffusible pollen antigens of *Gladiolus* and ragweed within the inner intine and outer exine layers of the wall. This was followed by the demonstration of the binding of the pollen antigens to the stigma surface during pollination in grasses[60] and in *Cosmos* and ragweed, using immunofluorescence methods. The antigens appear to be strategically placed for communication with the stigma, perhaps as mediators of cell recognition.

The question of which of the many pollen antigens bind to the stigma surface could not be resolved with these methods (which utilized antisera raised to the complete pollen diffusate);

although in ragweed and *Cosmos* tests with specific anti-Antigen Esera demonstrated that this pollen antigen did bind to the stigma surface. In *Cosmos bipinnatus*, which has a well-defined self incompatibility system, the specificity of binding was examined again using immunofluorescence methods: diffusates from both self and compatible pollen bound to the stigma surface.[37] Supporting evidence has been obtained in a recent study of self incompatibility in grasses, where FITC-labelled self and compatible pollen protein diffusates were equally effective in binding to the stigma surface.[105] In a quantitative study of pollen protein interactions at the stigma surface of *Gladiolus*, considerable non-specific binding was found, since ^{125}I-labelled self and foreign pollen proteins, and even animal proteins, bound equally well.[156] The stigma surface appears to possess a rather general adhesive capacity for macromolecules.[11] Specificity has, however, been detected in the binding of the lectin concanavalin A to the stigma surface.

Pollen antigens have been more conclusively implicated in stigma recognition reactions in the course of a series of experiments in which the pollen or stigma surfaces have been modified to examine their role in the process of both interspecific and self pollination. Hybridization is normally unsuccessful between the American black cotton-wood, *Populus deltoides*, and the white poplar, *P. alba*. However, when killed self compatible pollen or self pollen-wall protein diffusates were applied to *P. deltoides* stigmas, followed by incompatible *P. alba* pollen, hybrid seeds were produced.[62] In the absence of the killed self pollen or diffusate, no seed set occurred, suggesting that a recognition factor in the self pollen diffusate results in the acceptance of the normally incompatible pollen. The diffusates contained proteins and glycoproteins since they had been fractionated by ammonium sulphate precipitation and dialysed before use.

Pollen manipulation experiments have since been extended to a number of self incompatibility systems, especially *Cosmos bipinnatus*,[39] various cultivars of apple[4, 13] and *Petunia hybrida*,[94] where the presence of killed compatible pollen has enabled the stimulation of self seed set. Furthermore, *Cosmos* stigmas treated with diffusate of compatible pollen also accepted self pollen to the same extent as in the pollen mix experiments, resulting in self seed production.[39] When the diffusate was heated to 60°C for 10 min, the treatment was still successful. A protein, antigenically similar to Antigen E of ragweed, was present in this heated diffusate, perhaps implicating the allergen in pollen–stigma interactions. Such treatments of pollen were ineffective in overcoming self incompatibility in *Lilium longiflorum*[19] and *Nicotiana alata.*[89] γ-Irradiated pollen of *Nicotiana* showed partial transfer of genetic characteristics suggesting that some kind of transformation instead of fertilization may have occurred.

In experiments with self incompatible *Brassica oleracea*, Roggen[93] demonstrated that washing self pollen in certain organic solvents, or coating self pollen with the lipidic pollenkitt or tryphine of compatible types would all stimulate self seed production. While the interpretation of these experiments is open to question,[11] nevertheless they provide an experimental approach which could lead to the demonstration of the active pollen fraction involved in the stigmatic acceptance or rejection process.

In the crucifers *Iberis* and *Brassica* spp. the exine proteins released from the pollen in short term surface washes have been shown to trigger the production of the 1,3-β-glucan callose in the walls of self stigma papillae but not compatible stigma papillae, mimicking the rejection callose produced in the stigmas adjacent to rejected self pollen.[32, 35, 36] The pollen diffusates used contained at least eight antigens[63] among the many components likely to be present. Again, the use of this experimental approach could provide the evidence for a particular pollen fraction to initiate the rejection response.

CONCLUSIONS

1. *Pollen grains express a wider range of antigens than other plant cell types*

Pollen grains contain many antigens, most of which are proteins or glycoproteins. More antigens have been detected in pollen extracts than in any other plant tissue. At least nine diffusible antigens are located in the pollen wall, either in cavities and at the surface of the outer exine layer or in the polysaccharide matrix of the inner intine layer. Non-diffusible antigens occur in cytoplasmic sites. In grass pollen, 16 antigens have been detected in extracts likely to originate from wall sites. The wall antigens bind to the stigma surface at pollination, and may exhibit S-allele specificity in self incompatibility.

2. *Pollen allergens elicit specific Immunoglobulin E antibody formation and binding in man*

Many common airborne pollen types contain allergens. These comprise multiple molecular forms of proteins and glycoproteins. They are antigenic in rabbits, and different species of pollen may share allergenic determinants. Allergens account for up to 10% of the diffusible protein of pollen. They trigger formation of specific IgE in susceptible humans, and interact with IgE on the surface of mast and basophil cells of the upper respiratory tract. The role of allergens in nature is not understood. They may serve as recognition factors, as storage reserves during pollen germination or possess enzymic activity during pollen–stigma interactions.

3. *Pollen enzymes are involved in the sequential events of pollination*

Many enzyme activities have been detected in ungerminated pollen, and the enzymes are located both in the cytoplasm, and in the wall sites. Enzymes associated with the exine cavities have been shown to be sporophytic in inheritance, derived by transfer from the parental tapetal cells. Intine-located enzymes are gametophytic in inheritance, and are laid down together with the intine polysaccharide early in pollen development. The enzymes diffuse from germinating grains, and are associated with the pollen tube during stigma penetration and growth.

REFERENCES

1. AUGUSTIN, R. Grass pollen allergens. III. Their differentiation from the other pollen antigens by immuno-electrophoretic studies in relation to skin reactivity, enzymic digestion, heat and pH stabilities. *Immunology* **2** (1959) 230–251.

2. AUGUSTIN, R. Antigens and other allergens of grass pollens. *Proc. 5th Eur. Congr. Allergy*, Basle (1963) 137–46 (Sellostverlag der Schweizerischen Allergie-Gesellschaft).

3. AUGUSTIN, R., O'SULLIVAN, S. and DAVIES, I. Isolation of grass pollen antigens failing to induce IgE reagin formation though capable of inducing IgG antibody formation. *Int. Archs Allergy appl. Immun.* **41** (1971) 144–147.

4. BELIN, L. Separation and characterisation of birch pollen antigens with special reference to the allergenic components. *Int. Archs. Allergy. appl. Immun.* **42** (1972) 329–342.

5. BELIN, L. and ROWLEY, J.R. Demonstration of birch pollen allergen from isolated pollen grains using immunofluorescence and a single radial immunodiffusion technique. *Int. Archs. Allergy appl. Immun.* **40** (1971) 754–769.

6. BERNSTEIN, I.L., PERERA, M., GALLAGHER, J., MICHAEL, J.G. and JOHANSSON, S.G.O. *In vitro* cross-allergenicity of major aeroallergenic pollens by the radio-allergo-sorbent technique. *J. Allergy clin. Immun.* **57** (1976) 141–152.

7. BERRENS, L. The chemistry of atopic allergens. *Monogr. Allergy* **7** (1971).

8. BLACKLEY, C.H. *Experimental Researches on the Cause and Nature of Catarrhus Aestivus.*

Bailliere, Tindall & Cox, London (1873).

9. BOREHAM, P. and GILL, G.S. Serological identification of reptile feeds of *Glossina. Acta trop.* **30** (1973) 356–365.

10. BREWBAKER, J.L. Pollen enzymes and isoenzymes. In J. HESLOP-HARRISON (Ed) *Pollen: Development and Physiology*, Butterworths, London (1971) pp. 156–170.

11. CLARKE, A.E. and KNOX, R.B. Cell recognition in flowering plants. *Q. Rev. Biol.* **53** (1978) 1–28.

12. CLARKE, A.E., KNOX, R.B., HARRISON, S., RAFF, J. and MARCHALONIS, J. Common antigens and male–female recognition. *Nature Lond.* **265** (1977) 161–163.

13. DAYTON, D.F. Overcoming self-incompatibility in apple with killed compatible pollen. *J. Am. Soc. Hort. Sci.* **99** (1974) 190–192.

14. DICKINSON, D.B. and DAVIES, M.D. Nucleotide diphosphate kinase from lily pollen. *Plant Cell Physiol.* **12** (1971) 157–160.

15. DICKINSON, D.B. and DAVIES, M.D. Metabolism of germinating lily pollen: pollen enzymes. In J. HESLOP-HARRISON (Ed) *Pollen: Development and Physiology.* Butterworths, London (1971) pp. 190–193.

16. DICKINSON, H.G. and LEWIS, D. Cytochemical and ultrastructural differences between intraspecific compatible and incompatible pollinations in *Raphanus. Proc. R. Soc. B.* **183** (1973) 21–38.

17. DICKINSON, H.G. and LEWIS, D. The formation of tryphine coating the pollen grains of *Raphanus* and its properties relating to the self-incompatibility system. *Proc. R. Soc. B.* **184** (1973) 149–165.

18. EAST, E.M. Self sterility. *Biblthca genet.* **5** (1929) 331–370.

19. FETT, W.F., PAXTON, J.D. and DICKINSON, D.B. Studies on the incompatibility response on *Lilium longiflorum. Am. J. Bot.* **63** (1976) 1104–1108.

20. GLEICH, G.J., LEIFERMAN, K.M. and JONES, R.T. Use of radioallergosorbent test for the determination of the potency and cross-reactivity of allergy extracts. In RAST R. EVANS (Ed) *Advances in Diagnosis of Allergy.* Symposia Specialists, Miami (1975) 57–83.

21. GOLDMAN, M. *Fluorescent Antibody Methods.* Academic Press. New York. (1968).

22. GORSKA-BRYLASS, A. Hydrolases in pollen grains and pollen tubes. *Acta. Soc. Bot. Pol.* **34** (1965) 589–604.

23. GREEN, J.R. On the germination of pollen grain and nutrition of pollen tube. *Stain Technol.* **45** (1894) 75–79.

24. GRIFFITHS, B.W. and BRUNET, R. Isolation of a basic protein antigen of low ragweed pollen. *Can. J. Biochem.* **49** (1971) 396–400.

25. HARA, A., YAMASHITA, H. and KOBAYASHI, A. Isolation of a polysaccharide from the inner cell wall, intine, of pollen of *Cryptomeria japonica Pl. Cell Physiol.* Tokyo **18** (1977) 381–386.

26. HEJGAARD, J. and BÖG-HANSEN, T.C. Quantitative immuno-electrophoresis of barley and malt proteins. *J. Inst. Brew.* **80** (1974) 436–442.

27. HERBESTSON, S., PORATH, J. and COLLDAHL, H. Allergens from alder pollen. *Acta chem. scand.* **12** (1958) 737.

28. HERDT, E., SUTFELD, R. and WIERMANN, R. The occurrence of enzymes involved in phenyl propanoid metabolism in the tapetum fraction of anthers. *Cytobiologie* **17** (1978) 433–441.

29. HESLOP-HARRISON, J. Pollen wall development. *Science* **161** (1968) 230–237.

30. HESLOP-HARRISON, J. Scanning electron microscopic observations on the wall of the pollen grain in *Cosmos bipinnatus*, Compositae. *Proc. Engis Stereoscan Colloquium* (1969) pp. 89–96.

31. HESLOP-HARRISON, J. Sporopollenin in the biological context. In J. BROOKS *et al.* (Ed) *Sporopollenin.* Academic Press, London (1972) pp. 1–30.

32. HESLOP-HARRISON, J. Incompatibility and the pollen–stigma interaction. *A Rev. Pl. Physiol.* **26** (1975) 403–425.

33. HESLOP-HARRISON, J. Physiology of the pollen grain surface, *Proc. R. Soc. B.* **190** (1975) 275.

34. HESLOP-HARRISON, J., HESLOP-HARRISON, Y., KNOX, R.B. and HOWLETT, B. Pollen-wall proteins: 'gametophytic' and 'saprophytic' fractions in the pollen walls of Malvaceae. *Ann. Bot.* **37** (1973) 402–412.

35. HESLOP-HARRISON, J., KNOX, R.B. and HESLOP-HARRISON, Y. Pollen-wall proteins:

Exine held fractions associated with the incompatibility response in Cruciferae. *Theor. appl. Genet.* **44** (1974) 133–137.

36. HESLOP-HARRISON, J., KNOX, R.B., HESLOP-HARRISON, Y. and MATTSSON, O. Pollen-wall proteins: emission and role in incompatible responses. In I.G. DUCKETT and P.A. RACEY (Eds) *Biology of the male gamete. Biol. J. Linn. Soc.* Suppl. 1 (1975) pp. 189–202.

37. HOWLETT, B.J. Pollen-wall proteins of *Cosmos bipinnatus* M.Sc. Thesis (1973) Australian National University, Canberra.

38. HOWLETT, B.J., KNOX, R.B. and HESLOP-HARRISON, J. Pollen-wall proteins: release of the allergen Antigen E from intine and exine sites in the pollen grains of ragweed and *Cosmos. J. Cell. Sci.* **13** (1973) 603–619.

39. HOWLETT, B.J., KNOX, R.B., PAXTON, J.D. and HESLOP-HARRISON, J. Pollen-wall proteins: physicochemical characterization and role in self-incompatibility in *Cosmos bipinnatus. Proc. R. Soc. B.* **188** (1975) 167–182.

40. IDE, T., ASHIDA, T. and TORII, K. Studies on the immunologically active substances in the pollen of a Japanese cedar, *Cryptomeria japonica* D. Don II. *Proc. 17th Ann. Meeting Palynological Soc. Japan* (1976) pp. 36–37.

41. ISHIZAKA, K. and ISHIZAKA, T. Human reaginic antibodies and immunoglobulin E. *J. Allergy* **42** (1968) 330–363.

42. ISHIZAKA, K., ISHIZAKA, T. and HORNBROOK, M.M. Physicochemical properties of reaginic antibody. N. Prescence of a unique immunoglobulin as a carrier of reaginic activity. *J. Immunol.* **97** (1966) 75–85.

43. IWANAMI, Y. Physiological researches of pollen, 19. Studies on the antigen substances in the pollen grains by immunoelectrophoresis, *J. Yokohama munic. Univ.* **184** CC-59, *Biol.* **26** (1968) 1–19.

44. JOHANSSON, S.G.O. Determination of IgE and Ige antibody by RAST. In R. EVANS (Ed) *Advances in Diagnosis of Allergy: RAST* Symposia Specialists, Miami (1975) pp. 7–16.

45. JOHNSON, P. and MARSH, D.G. Allergens from common rye grass pollen. I. Chemical composition and structure. *Immunochemistry* **3** (1966) 91–100.

46. JOHNSON, P. and MARSH, D.G. Allergens from common rye grass pollen. II. The allergenic determinants and carbohydrate moiety. *Immunochemistry* 3 (1966) 101–110.

47. KAMANN, O. Zur kenntnis des Roggen-Pollens und des darin enthaltenen Heufiebergiftes. *Beitr. Chem. Physiol. Path.* **5** (1904) 346–354.

48. KIND, L.F. and NILSSON, B. Biological activity of a ragweed arabinogalactan. 1. Galactose specificity of *in vitro* and *in vivo* effects. *Immunology* **13** (1967) 477–482.

49. KING, T.P. Chemical and biological properties of some atopic allergens. *Adv. Immun.* **23** (1976) 77–105.

50. KING, T.P., NORMAN, P.S. and CONNEL, J.T. Isolation and characterization of allergens from ragweed pollen. *Biochemistry NY* **3** (1964) 458–468.

51. KNOX, R.B. Pollen-wall proteins: localization, enzymic and antigenic activity during development in *Gladiolus. J. Cell Sci.* **9** (1971) 209–237.

52. KNOX, R.B. Pollen-wall proteins: pollen stigma interactions in ragweed and *Cosmos* (Compositae) *J. Cell Sci.* **12** (1973) 421–443.

53. KNOX, R.B. Cell recognition and pattern formation. In C. GRAHAM and P.F. WAREING (Eds) *Developmental Biology of Plants and Animals*, Ch. 3.5, Blackwell, Oxford (1976) pp. 141–168.

54. KNOX, R.B. *Pollen and Allergy.* Studies in Biology, Edward Arnold, London (1979) No. 107.

55. KNOX, R.B. and CLARKE, A.E. Localization of proteins and glycoproteins by binding to labelled antibodies and lectins. In J.L. HALL (Ed) *Electron microscopy and cytochemistry of plant cells.* Elsevier, North Holland, Amsterdam (1978) Ch. 3.

56. KNOX, R.B., CLARKE, A.E., HARRISON, S., SMITH, P. and MARCHALONIS, J.J. Cell recognition in plants: Determinants of the stigma surface and their pollen interactions, *Proc. natn. Acad. Sci. U.S.A.* **73** (1976) 2788–2792.

57. KNOX, R.B. and HESLOP-HARRISON, J. Cytochemical localization of enzymes in the wall of the pollen grain. *Nature Lond.* **223** (1969) 92–94.

58. KNOX, R.B. and HESLOP-HARRISON, J. Pollen-wall proteins: localization and enzymic activity. *J. Cell Sci.* **6** (1970) 1–27.

59. KNOX, R.B. and HESLOP-HARRISON, J. Pollen-wall proteins: localization of antigenic and allergenic proteins in the pollen grain walls of *Ambrosia* spp. (ragweeds). *Cytobios* **4** (1971) 49–54.

60. KNOX, R.B. and HESLOP-HARRISON, J. Pollen-wall proteins: fate of intine-held antigens in compatible and incompatible pollinations of *Phalaris tuberosa L. J. Cell Sci.* **9** (1971) 239–251

61. KNOX, R.B., HESLOP-HARRISON, J. and REED, C. Localization of antigens associated with the pollen grain wall by immunofluorescence. *Nature Lond.* **225** (1970) 1066–1068.

62. KNOX, R.B., WILLING, R.R. and ASHFORD, A.E. Role of pollen wall proteins as recognition substances in inter-specific hybridization in poplars. *Nature Lond.* **237** (1972) 381–383.

63. KNOX, R.B., HESLOP-HARRISON, J. and HESLOP-HARRISON, Y. Pollen-wall proteins: localization and characterization of gametophytic and sporophytic fractions. In J.G. DUCKETT and P.A. RACEY (Eds) *The Biology of the male gamete. Biol. J. Linn. Soc.* **7** (1975) 177–187.

64. KROH, M. and LOEWUS, M.W. Pectolytic enzyme activity from *Nicotiana tabacum* pollen. *Phytochemistry* **17** (1978) 797–799.

65. LAINE, S. Diffusion of proteins by intact birch pollen grains: enzymic and antigenic activity. *Grana* **16** (1977) 187–189.

66. LAPKOFF, C.B. and GOODFRIEND, L. Isolation of a low molecular weight ragweed pollen allergen: Ra 5. *Int. Archs Allergy appl. Immun.* **46** (1974) 215–229.

67. LARSON, J.B. and GLEICH, G.J. Changes in the antigenic composition of the short ragweed plant during maturation. *J. Allergy clin. Immun.* **56** (1975) 112–116.

68. LENDZIAN, K. and SCHAFER, E. Invertases of pollen of *Haemanthus albiflos. Phytochemistry* **12** (1973) 1227–1228.

68a. LEE, Y.S. and DICKINSON, D.B. Characterization of pollen antigens from *Ambrosia* L. (compositae) and related taxa by immunoelectrophoresis and radial immunodiffusion. *Am. J. Bot.* **66** (1979) 245–252.

69. LEWIS, D. Serological reactions of pollen incompatibility substances. *Proc. R. Soc. B.* **140** (1952) 127–135.

70. LEWIS, D., BURRAGE, S. and WALLS, D. Immunological reactions of single pollen grains, electrophoresis and enzymology of pollen protein exudates. *J. exp. Bot.* **18** (1967) 371–378.

71. LINSKENS, H.F. Zur frage der abwehr – körper der inkompatibilitäts reaktion von *Petunia* III, *Z. Bot.* **48** (1960) 126–135.

72. LINSKENS, H.F. and HEINEN, W. Cutinase-Nachweis in Pollen. *Z. Bot.* **50** (1962) 338–347.

73. LOWENSTEIN, H., NIELSEN, L. and WEEKE, B. Purification of timothy pollen allergens followed by quantitative immunoelectrophoresis. *Int. Archs Allergy appl. Immun.* **49** (1975) 95–98.

73a. LOWENSTEIN, H. Quantitative immunoelectrophoretic methods as a tool for the analysis and isolation of allergens. *Prog. Allergy* **25** (1978) 1–62.

74. MAKINEN, Y. and McDONALD, T. Isoenzyme polymorphism in flowering plants. III. pollen enzymes and isoenzymes. *Physiologia Pl.* **21** (1968) 477–486.

75. MAKINEN, Y.L.A. and LEWIS, D. Immunological analysis of incompatibility proteins and of cross-reacting material in a self-compatible mutant of *Oenothera organensis, Genet. Res.* **3** (1962) 352–363.

76. MALIK, C.P., MEHAN, M. and VERMANI, S. Distribution of alkaline phosphatase in pollen grains and pollen tubes of *Calotrophis procera Biochem. Physiol. Pflanzen BPD* **167** (1975) 601–603.

77. MALLEY, A. and HARRIS, R.L. Biologic properties of a non-precipitating antigen from timothy pollen extracts. *J. Immun.* **88** (1967) 825–830.

78. MARSH, D.G. Allergens and the genetics of allergy. In M. SELA (Ed) The Antigens. Academic Press, London (1975) pp. 271–359.

79. MCDONALD, T. Isoenzymic studies of esterase, L-aspartate: 2-oxoglutarate aminotransferase and carbohydrase in *Zea mays* Ph.D. Thesis (1969) University of Hawaii.

80. MOLE, L.E., GOODFRIEND, L., LAPKOFF, C.B., KEHOE, J.M. and CAPRA, J.D. The amino acid sequence of ragweed pollen allergen Ra5. *Biochemistry, NY* **14** (1975) 1216–1220.

81. NAIRN, R. *Fluorescent Protein Tracing*, 2nd ed. Churchill, London (1977).

82. NAKANE, P.K. Simultaneous localization of multiple tissue antigens using the peroxidase-labelled antibody method: a study on pituitary glands of the rat. *J. Histochem. Cytochem.* **16** (1968) 557–560.

83. NASRALLAH, M.E. and WALLACE, D.H. Immunochemical detection of antigens in self-

incompatibility genotypes of cabbage. *Nature Lond.* **213** (1967) 700–701.

84. NASRALLAH, M.E., BARBER, J.T. and WALLACE, D.H. Self-incompatibility proteins in plants: detection genetics and possible mode of action. *Heredity* **25** (1970) 23–27.

85. NISHIO, T. and HINATA, K. Stigma proteins in self-incompatible *Brassica campestris,* L. and self-compatible relatives, with special reference to S-allele specificity. *Japan. J. Genet.* **53** (1978) 27–33.

86. NOON, L. Prophylactic inoculation against hayfever. *Lancet* **1** (1911) 1972.

87. OSLER, A.G., LICHTENSTEIN, L.M. and LEVY, D.A. *In vitro* studies of human reaginic allergy. *Adv. Immun.* **8** (1968) 183–231.

88. OUCHTERLONY, O. Diffusion in gel methods for immunological analysis. *Progr. Allergy* **5** (1958) 1–78.

89. PANDEY, K.K. Evolution of incompatibility systems in plants: complementary and mating locus in flowering plants and fungi. *Theor appl. Genet.* **50** (1977) 89–101.

90. PATON, J.B. Pollen and pollen enzymes. *Am. J. Bot.* **8** (1921) 471–501.

91. RAFF, J.W., HUTCHINSON, J.F., KNOX, R.B. and CLARKE, A.E. Cell recognition: antigenic determinants of plant organs and their cultured callus cells. *Differentiation* **12** (1979) 179–186.

92. ROEBBER, M., MARSH, D.G. and GOODFRIEND, L. An improved procedure for isolation of ragweed pollen allergen Ra5. *J. Immun.* **115** (1975) 303–304.

93. ROGGEN, H. Stigma application of an extract from rape pollen. (*Brassica napus L.*) affects self-incompatibility in Brussel sprouts (*Brassica oleracea L. var. gemmifera* DC) *Incompatibility Newsletter* **6** (1975) 80–86.

94. SASTRI, D.C. and SHIVANNA, K.R. Recognition pollen alters intraspecific incompatibility in *Petunia Incompatibility Newsletter* **7** (1976) 22–24.

95. SCHEIDIGGER, J.J. Une micro-méthode de l'immunoelectrophorèse. *Int. Archs Allergy* **7** (1955) 103–110.

96. SELA, M. Immunological studies with synthetic polypeptides. In F.J. DIXON and J.H. HUMPHREY (Eds) *Advances in Immunology.* **5** (1966) 29–130.

97. SHAYK, M. and KOLATTUKUDY, P.E. Production of a novel extracellular cutinase by the pollen and the chemical composition and ultrastructure of the stigma cuticle of Nasturtium. (*Tropaeolum majus*). *Plant Physiol.* **60** (1977) 907–915.

98. SITTE, P. Untersuchungen zur submicroskopischen morphologie der pollen-und sporenmembranen. *Microscopie* **8** (1953) 290–299.

99. SOOD, P.P., MALIK, C.P. and TIWARI, H.B. A histochemical study of the localization of succinic dehydrogenase in the germinating pollen grains of *Portulaca grandiflora Z. Biol.* **116** (1969) 215–220.

100. SOUTHWORTH, D. and DICKINSON, D.B. β-1,3 glucan synthetase from *Lilium longiflorum* pollen. *Plant Physiol.* **56** (1975) 83–87.

101. STANLEY, R.G. and LINSKENS, H.F. *Pollen: Biology, Biochemistry Management* (1974) Springer-Verlag, Berlin.

102. TSINGER, N.V. and PETROVSKAYA-BARANOVA, T.P. The pollen grain wall — a living, physiologically — active structure. *Dokl. Akad. Nauk SSSR* **138** (1961) 466–496.

103. UNDERDOWN, B.J. and GOODFRIEND, L. Isolation and characterization of a protein allergen from ragweed pollen. *Biochemistry, NY* **8** (1969) 980–989.

104. VITHANAGE, H.I.M.V. Pollen wall proteins and breeding systems of plants. Ph.D. Thesis (1978) University of Melbourne.

105. VITHANAGE, H.I.M.V. and HESLOP-HARRISON, J. The pollen–stigma interaction. Fate of fluorescent-labelled pollen-wall proteins on the stigma surface of rye (*Secale cereale*) *Ann. Bot.* **43** (1979) 113–114.

106. VITHANAGE, H.I.M.V. and KNOX, R.B. Pollen-wall proteins: quantitative cytochemistry of the origins of intine and exine enzymes in *Brassica oleracea. J. Cell Sci.* **21** (1976) 423–435.

107. VITHANAGE, H.I.M.V. and KNOX, R.B. Pollen development and quantitative cytochemistry of exine and intine enzymes in sunflower, *Helianthus annuus Ann. Bot.* (in press).

108. WATSON, L. and KNOX R.B. Pollen wall antigens and allergens. Taxonomically ordered variation among grasses. *Ann. Bot.* **40** (1976) 399–408.

109. WODEHOUSE, R.P. *Pollen grains* McGraw-Hill, New York (1935).

110. WODEHOUSE, R.P. Antigenic analysis by gel diffusion I. Ragweed pollen. *Int. Archs Allergy appl. Immun.* **5** (1954) 425–433.

111. WODEHOUSE, R.P. Antigenic analysis by gel diffusion. III. Pollens of the Amaranth-Chenopod group. *Ann. Allergy* **15** (1957) 527–536.

PLANT RESPONSES TO ANAEROBIC CONDITIONS IN SOIL AND SOLUTION CULTURE

M.C. DREW

Agricultural Research Council, Letcombe Laboratory, Wantage, OX12 9JT, U.K.

INTRODUCTION

When soil becomes saturated with water (waterlogged), air is displaced from the pore space fraction, and oxygen dissolved in the soil water is depleted within a few hours or days by the respiration of plant roots and soil micro-organisms.[27, 48, 83] The net flux of oxygen into the soil which is necessary to sustain their aerobic respiration is of the order of 10 litres m^{-2} day^{-1} with abundant roots and at summer temperatures.[12, 90] Diffusion, the predominant mechanism for soil aeration, can contribute little to the replenishment of oxygen in the bulk soil once the soil pores become water filled; the steady state rate of diffusion of O_2 through water to a potent O_2 sink is some 3×10^6 times slower than through an equivalent distance in air, principally due to the smaller diffusion coefficient but also because of the low solubility in water.

The anaerobic conditions resulting from interruption in the free exchange of gas between the soil and the atmosphere leads to changes in the chemical and biological properties of the soil that adversely affect the growth of plants not especially adapted to wet land conditions. Characteristic symptoms of damage to vegetative growth in herbaceous and woody plants include a slowing of leaf and shoot extension, and of the rate of dry matter accumulation by the aerial tissues and conspicuous yellowing of the older (lower) leaves suggesting premature senescence.[60, 61] Wilting,[62] leaf epinasty[51] and leaf shedding[59] are typical responses of some species. In Gramineae, tillering is suppressed. The early development of specialized, adventitious roots at the base of the stem is a widespread response in monocotyledons[109] as well as in herbaceous[51, 60, 109] and woody[19, 33, 81] dicotyledons. Some of these roots become negatively geotropic and emerge from the soil.[33, 81] The original root system usually stops extending or accumulating dry matter, and may undergo degradation, particularly at the root tip.[14, 28, 45, 84, 106] Prolonged waterlogging can lead to death of the entire plant in species not especially adapted to such conditions.[100] Temperature, however, is an important modifying factor, plants showing greater sensitivity to anaerobic conditions at warm temperatures.[101] Herbaceous plants and trees can often tolerate prolonged oxygen deficiency during the cooler temperatures of winter and spring when both respiration and growth rates are low.[33, 100]

Despite an extensive literature treating the effects of poor aeration on root function and shoot growth, the causes of injury remain obscure. This paper discusses the various possible mechanisms of injury to plants in anaerobic soil and de-oxygenated nutrient solutions, and identifies properties that confer some degree of tolerance to such conditions. The review is concerned principally with damage to vegetative growth, but it is important to recognize that crop yields can be appreciably depressed by waterlogging for only 24 h at critical stages in development, e.g. shortly before flowering (peas,[15] corn[17]) or during microsporogenesis, leading to pollen sterility (cereals[1]). It can also influence plant growth indirectly, e.g. by encouraging pathogen attack[13, 34] by inhibiting mycorrhizal association,[58] by interfering in nodulation of legumes[99] or by promoting denitrification.[83, 90]

MECHANISMS OF DAMAGE TO PLANTS

The growth and function of roots not adapted to oxygen stress depend upon aerobic respiration using oxygen supplied from the rooting medium. The minimum concentration of oxygen required at the root surface is not easy to predict with confidence. The apparent K_m of cytochrome oxidase may be as small as 2×10^{-8} M[69] and there is evidence that a maximum is attained at oxygen concentrations some 4–5 times greater than the K_m (i.e. at lower concentrations than expected from a Michaelis–Menten relation). Thus 10^{-7} M oxygen at the active site would not be expected to limit oxidative phosphorylation. This concentration is only 0.05% of that in equilibrium with air and corresponds to a partial pressure of 10^{-4} atm (10^{-5} MPa). However, because of the distances over which dissolved oxygen from the outer medium has to diffuse to reach mitochondria within cells, concentrations in the outer solution need to be appreciably higher than this, as discussed in detail elsewhere.[38] At high temperatures (fast respiration rates) and in tissues and organs with few intercellular spaces, particularly in large apical meristems, respiration may become limited by oxygen transport. A further factor contributing to the apparently high oxygen requirements of some tissues may be the participation of terminal oxidases other than cytochrome oxidase, which have a low affinity for O_2 and do not contribute to oxidative phosphorylation.[64] However, they may be involved in the production of intermediates in carbohydrate metabolism that are essential for cell differentiation. Some early attempts to define limiting oxygen concentrations in soil or in solutions by aerating with defined gas mixtures should be viewed with caution because of uncertainty over the large impedance to diffusion in stationary moisture films of unknown and variable thickness. It is not surprising, therefore, that values of the limiting concentration of O_2 vary widely in the literature, and sometimes appear to be different for different root functions.

Despite the foregoing considerations, experiments using well stirred solution have sometimes demonstrated that very low concentrations of O_2 can be adequate. At oxygen partial pressures averaging 0.006 atm excised mustard roots took up oxygen at half the maximum rate.[37] Oxygen partial pressures of 0.01–0.02 atm gave near maximal rates of root growth, and uptake of ions and water in tobacco and soybean.[44, 105] Phosphate uptake by barley roots was maximal at partial pressures of 0.03 atm.[43] A recent discussion of limiting oxygen concentrations in relation to tissue respiration and theoretical models of root growth is given in Armstrong and Gaynard.[4]

The primary cause of inhibition of metabolism and of cell death in roots under oxygen stress is a matter for speculation. The generation of ATP in anaerobic respiratory pathways may be inadequate to maintain cell metabolism, for the concentration of ATP in plant cells suddenly subjected to anoxia can be halved in 15 min. Lack of oxygen, in common with the action of respiratory inhibitors, leads to rapid loss of membrane integrity with release of cell constituents to the medium[41] and an increased resistance to water flux.[76] The toxicity of the products of anaerobic metabolism are considered in a later section.

Waterlogging injury to shoots has been attributed to a variety of factors, reviewed in part by Kramer[61] and Grable.[34] These include: the accumulation of toxins, synthesized by micro-organisms in the anaerobic soil or released by damaged roots; desiccation, caused by decreased permeability of oxygen stressed roots to water; abnormal concentrations of hormones in aerial tissues, due both to water stress and to interference in delivery of hormones from roots to shoots; nutrient deficiency, arising from denitrification or from the failure of the damaged roots to transport ions to the shoots. Few investigators have attempted to analyse the relative contribution of each of the above factors to injury to shoots and to find whether they act concurrently.

TOXINS

Soil produced toxins

It is clear from experiments in nutrient solution that in some species many of the typical symptoms of waterlogging injury to roots and shoots can be induced merely by excluding oxygen from around the roots[84, 106, 107] but there are exceptions. Chlorosis and desiccation of leaves of pea plants produced by waterlogging of a compost rich in organic matter could *not* be reproduced in anaerobic solution culture, and unidentified toxins were implicated.[46] Whether the principal cause of damage to the shoots of plants grown in waterlogged *soil* arises more from the production of toxic substances in the soil than from the oxygen deficiency experienced by roots has long been questioned. When the free oxygen content of the soil solution becomes depleted, anaerobes respire via a sequence of redox components acting as electron acceptors. Some of these reduced components (NO_2^-, Mn^{2+} and S^-) and intermediates in microbial carbon metabolism such as the volatile fatty acids, phenolic acids and various hydrocarbons including ethylene, may accumulate in the soil solution to concentrations which could be harmful (i.e. toxic) to plants. Studies of phytotoxic compounds in soil have been reviewed.[71, 73, 80, 83, 90, 97] There is ample evidence that under some circumstances, damage to field crops is closely associated with accumulation of volatile fatty acids,[16, 18] hydrogen sulphide[2, 25] or aromatic compounds.[80, 102] Large concentrations of iron[55] or of manganese[35] in the leaves of plants in waterlogged soil have been associated with plant damage and death. Even in Padi rice, well adapted to oxygen deficiency, problems arising from inadequate control of Fe^{2+} in the soil solution, or the accumulation of soil toxins, are well recognized.[16, 83, 90]

It would be misleading, however, to ascribe the early effects of transient waterlogging principally to the effects of soil toxins. The accumulation of potentially toxic substances to concentrations injurious to well aerated plants may take days or weeks, depending on temperature, the availability of substrates and the relative rates of production and utilization by micro-organisms.[18] Relatively large quantities of readily metabolized organic substrates, originating from the incorporation of straw or fresh organic matter, are required for the production of volatile fatty acids in amounts leading to toxic concentrations.[16, 72] The presence of free nitrate in the soil water as in most agricultural soils, delays the onset of strongly reducing conditions.[83] Nitrite can accumulate in waterlogged soils as an intermediate in denitrification but in temperate climates it rarely attains concentrations (around 1 mM at pH 5.0) that are associated with toxicity.[67] The concentration of ethylene in waterlogged agricultural soils[93] can much exceed that reported to cause a marked reduction in the extension of barley and wheat roots in solution culture,[94] but the effect is readily reversible upon removal of ethylene. The role of ethylene is discussed in more detail below.

The movement of oxygen (diffusing from the aerial tissues) out of roots into the anaerobic medium produces a locally oxidizing atmosphere which may help to detoxify, so contributing to plant tolerance.[19, 36, 91] Rhizosphere micro-organisms may sometimes assist in detoxification, for example *Beggiatoa* on rice roots oxidizes hydrogen sulphide.[56] The long term persistence of phytotoxic compounds in waterlogged soils is most deleterious, since they may accumulate to toxic concentrations in shoots or may interfere in the growth of new adventitious roots adapted to low oxygen concentrations.

At one time CO_2 was thought to be of particular importance in relation to waterlogging damage because of the deleterious effects on plants when roots were treated in solution culture with gas mixtures containing CO_2 at partial pressures of 0.2 atm or more.[60, 61] It is now recognized that concentrations of CO_2 in the soil water rarely rise to those necessary to cause damage. There is evidence that smaller concentrations of CO_2 (partial pressure 0.05 atm) become inhibitory to ion uptake when oxygen concentrations are also low[39] but permeability to water is not affected.[39, 105]

Root produced toxins

When roots and other submerged organs are subjected to oxygen stress, anaerobic respiration may lead to the formation of acetaldehyde[9] and ethanol.[9, 21–23, 54, 103] It has been suggested that the accumulation of these compounds leads to death of root tissues[9, 21–23] and injury to shoots.[10, 31] Evidence has been presented for the hypothesis that ethanol accumulates in species intolerant to waterlogging:[21] with oxygen stress, the normal oxidation of phosphoenol-pyruvate (PEP) via the Krebs cycle is suppressed; PEP is then decarboxylated to acetaldehyde, which in turn is converted to ethanol by NADH and the inducible enzyme alcohol dehydrogenase (ADH). In herbaceous plants not adversely affected by prolonged waterlogging, there was no ADH induction or production of ethanol.[21] In a survey of 19 species, the relative increase in ADH activity in root extracts was correlated with intolerance of waterlogging[74] but in more recent work with rice and maize, ADH activity was not a reliable guide to the rate of production of ethanol under oxygen stress.[53, 103] In barley cultivars, ADH activity failed to correlate with waterlogging tolerance[104] and more ADH activity could be induced in rice than in waterlogging-intolerant barley.[104] Furthermore, oxygen deficiency results in the production of ethanol in species such as rice[7, 53] winter wheat and rye[9] and willow which are considered relatively tolerant of waterlogging.

An important consideration is that in studies with waterlogging-tolerant marsh plants, growth and survival would have been accompanied by the development of many more adventitious roots (with airspaces that allowed internal transport of oxygen from the aerial tissues) which might therefore be expected to exhibit different biochemical properties compared with the damaged roots of intolerant species.

The detection of ethanol seems to provide a rapid and sensitive indication of anoxia in plant tissues[9, 23] and it has been taken as evidence of partial anoxia in the meristems of well aerated rice seedlings[7] and maize roots.[23] But ethanol accumulation has not been identified conclusively as a factor in waterlogging damage to roots. Barley roots can tolerate 5 mM ethanol in the external solution without showing damage to cell membranes.[65] The proportion of wheat and rye seed that germinate is approximately halved at concentrations in the range 30–170 mM.[9, 96] Although ethanol was externally supplied, the results suggest that ethanol is not particularly toxic. The concentrations of ethanol measured in root extracts[23] and sap of tree species[22] would only rarely be high enough to be toxic. An appreciable fraction of the ethanol is lost from the roots to the surrounding medium, or is exuded through lenticels or transpired.[10, 66]

In detached roots of pear, plum and peach, ethanol production seems to be so regulated that it does not reach a toxic concentration.[89] However in tomato, the concentration of ethanol in xylem exudates rose to 8.7 mM after only 12 h waterlogging, and ethanol supplied at such concentrations in aerated nutrient solution to intact plants was shown to be harmful to shoot growth.[31] In interpreting these results it should be borne in mind that in intact, transpiring plants, the concentration of ethanol in the xylem sap would be greatly diluted. Subsequent work indicated that these relatively large concentrations were to be expected only when plants were subjected to high light intensity and soil temperatures up to 30°C.[10]

In *Prunus* species (apricot, cherry, peach and plum) hydrogen cyanide is released from cyanogenic glycosides in the roots damaged by oxygen deficiency and Rowe and Beardsell[89] implicate HCN as the factor leading to root death in these species.

WATER RELATIONS

One of the earliest effects of deficient aeration in soil or in solution culture is to increase the resistance of roots to the radial movement of water.[62, 76] In some species this results in a fall in the water potential, and wilting.[60, 62] In others, stomatal closure ensures the maintenance of turgor. Thus, in tomato[50] and in species of *Populus* and *Salix*[81] there was rapid closure of stomata, decreased transpiration, and no development of water stress in leaves[50, 81] even during continuous waterlogging of up to 37 days.[81] Stomatal closure caused by waterlogging can greatly slow the rate of photosynthesis.[87]

The changes in permeability of roots to water is complex. An initial decrease in permeability is followed typically by a marked rise, after between 6 and 24 h anoxia, which has been attributed to cell death and the irreversible destruction of cell membranes.[60, 62, 76] With prolonged exposure, the flow of water decreases again probably due to blockage of xylem vessels, but partial decomposition of roots finally removes this constraint, and root resistance to water flow becomes negligible.[62]

PLANT HORMONES

In recent years much evidence, mainly from studies with dicotyledons, has implicated plant hormones in the control of some of the early symptoms of waterlogging damage to shoots. Poor aeration of the rooting medium in sunflower inhibits the synthesis and translocation to shoots of cytokinins.[14] Retardation of stem or leaf elongation, and premature leaf senescence, can be partly prevented by spraying shoots with a synthetic cytokinin, 6-benzylaminopurine in tomato[85, 92] and barley.[29] Anoxia reduces the synthesis of gibberellins in roots of tomato[88] and spraying shoots with GA_3 improves stem extension compared with unsprayed plants. Gibberellins and cytokinins sprayed together partly offset the growth retardation (fresh and dry weight) of leaves and stems in tomato.[49, 92]

The concentrations of free abscisic acid (ABA) extracted from leaves of dwarf bean, tomato, wheat and rice, grown in vermiculite increase many-fold within a few hours when oxygen is excluded from around the roots.[42] The mechanism triggering this important response is unknown: the suggestion that it might be a transient water-stress in the leaves now seems less plausible in view of the low water potentials that must develop in excised wheat leaves before there is acceleration of ABA synthesis.[108] The release of ABA provides an explanation for the early closure of stomata; and the inhibition of shoot growth and early leaf senescence are possibly conditioned by increased levels of ABA in the stressed shoots.[78]

On the role of hormones in relation to waterlogging symptoms, ethylene has attracted much interest. Leaf chlorosis, suggesting premature senescence, has been correlated with increased internal concentrations of ethylene in aerial tissues of *Vicia faba* and tomato.[30] Evidence implicating endogenous ethylene as the factor promoting various symptoms of waterlogging damage in sunflower has also been presented,[57] based on experiments employing large concentrations of 'Ethrel' to release ethylene within the tissue. In no study has there been a clear demonstration that within the time scale of the appearance of leaf senescence, endogenous ethylene in shoots rises to concentrations that induce the symptom when applied externally. In *Vicia*, increase in ethylene concentration was associated with leaf water stress,[30] but in tomato, increases occurred without any appreciable change in leaf water potentials during waterlogging.[50]

Leaf epinasty in tomato results from unequal *promotion* of cell expansion in the petiole by ethylene and is a typical response to waterlogging in some species.[48] It may serve to reduce light interception by leaves and hence slow transpiration. Incidentally, ethylene also promotes cell expansion in stems of plants adapted to submerged conditions, and at low concentrations accelerates extension of rice roots.[94] In tomato, epinasty is induced equally by soil waterlogging and by deoxygenating nutrient solution, and the concentration of ethylene in the leaves increases to a level which causes epinasty when applied to the shoots of plants with well aerated roots.[48] Substances that antagonize the action of ethylene inhibit the epinastic response to waterlogging or to low concentrations of oxygen in nutrient solution.[11, 48] The presence of roots seems to be essential for the induction of ethylene in the shoots, suggesting that substances, as yet unidentified, are transported from the damaged roots to shoots and stimulate ethylene production in aerial tissues.[48, 50] Although C^{14}-labelled ethylene has been shown to move from the rooting medium into shoots in tomato and contribute to epinasty,[47] soil derived ethylene is not an essential component.

The role of ethylene in monocotyledons has received much less attention. When a concentration of 10 p.p.m. in air, comparable to that which can occur in waterlogged soil water[93] was continuously supplied to the roots of barley for up to 21 days, there were only minor reductions in the final dry weight of roots and shoots and uptake of inorganic ions.[24] The form of the seminal root system was modified by slower extension of the main roots (axes), but this was compensated for by their increased diameter, and by the stimulation of laterals. Despite the high solubility of ethylene in water and its likely transport into the shoots, no symptoms of leaf senescence or abnormal development were recorded.[24]

NUTRIENT STATUS

Absorption and translocation of inorganic ions by unadapted roots of intact plants grown in nutrient solution are well known to be severely inhibited when oxygen supplies are inadequate for aerobic respiration.[39, 44] The internal aeration even of adapted roots may be less than optimal for ion uptake.[54] The concentrations of the major nutrients in the shoots of soil grown plants thus decrease during transient waterlogging,[68, 95] ion uptake being affected more than dry matter synthesis.

The possibility that the lower nutrient status of the shoot might contribute to waterlogging injury has received scant attention. Recent work with barley in waterlogged soil where the involvement of soil toxins seemed unlikely, suggested that inhibition of ion uptake by oxygen stressed roots contributed to the promotion of leaf senescence and to retardation of shoot growth.[27, 28] After 2 days waterlogging there was little net accumulation by shoots of nitrate, phosphate and potassium (although the concentration of these ions in the soil water remained large). The average concentration of these elements in the shoots decreased immediately after the onset of waterlogging due to the continuing synthesis of dry matter. There was a redistribution of nitrogen from the older to the younger leaves, offsetting in part the lack of supply from the roots,[28] and this preceded the onset of chlorophyll degradation.

Although the redistribution of nitrogen is an aspect of leaf senescence, and is undoubtedly under hormonal control, it may be triggered by the failure of the roots to absorb nutrients from the deoxygenated soil. By supplying calcium nitrate, daily, to the surface of the waterlogged soil, no symptoms of waterlogging damage appeared in barley shoots.[29] The maintenance of a more favourable redox potential in the soil by nitrate did not seem to be involved. The effect was dependent on the narrow, superficial zone of aerated soil where root growth and function could continue: in completely anaerobic soil there was no benefit from calcium nitrate additions. Leaf chlorosis, similar to that induced by waterlogging, could be promoted in plants with roots in well aerated nutrient solution simply by omitting nitrate, transiently, from the culture solutions.[29]

These results suggest that heavy fertilizer dressings alleviate waterlogging damage to crops under field conditions not simply be replacing the nitrogen lost by microbial denitrification and immobilization, as is widely supposed.

ADAPTATIONS TO ANAEROBIC CONDITIONS

Adaptive mechanisms that confer some degree of tolerance to anaerobic conditions comprise, 1. the development of airspaces that enhance diffusion of oxygen into roots from the aerial tissues, or from other roots located in the oxygen enriched surface soil, 2. metabolic changes which allow cells to tolerate anaerobic respiration, 3. physiological mechanisms that slow the growth rate of the shoots, redistribute inorganic nutrients and minimize water loss until aeration is restored. This third aspect has been touched on in preceding sections.

Adventitious roots and internal aeration

In the young roots tissues of herbaceous mesophytes grown in well aerated media, intercellular spaces occupy less than 10% of the root volume. Tracer studies with $^{15}O_2$ have shown that oxygen can diffuse in the continuum of gas filled spaces from the aerial tissues into the roots.[8, 52] However, the rates are far from sufficient for the total oxygen requirement of the root system,[79, 86] other than in roots of only a few cm in length,[37] even with rice.[54, 86] With prolonged waterlogging, adventitious roots with well developed air spaces (aerenchyma) emerge from the base of the stem and grow into the anaerobic soil or solution. The airspaces usually develop in the cortex, but sometimes occur instead in the stele.[19] The response is widespread among herbaceous[3, 75, 101] and woody species, including trees,[19, 33, 82] and is particularly important to the success of wetland species including rice. It has long been assumed that these adventitious roots carry on the functions of the initial root system, since partial recovery of shoots of herbaceous plants from waterlogging damage coincides with the development of adventitious roots.[51, 60] However, there have been few studies of the ability of these roots to absorb ions and water or to synthesize hormones.

The development of aerenchyma undoubtedly fulfils a dual role: it greatly reduces resistance to internal diffusion of oxygen and it diminishes the volume of respiring tissue. Oxygen flux out of the roots of trees and herbaceous species in anaerobic media has been detected by the use of micro-electrodes and by the reoxidation of reduced dyes.[3, 19, 82] The entry of oxygen to the pore space system is by way of leaf stomata, and in woody species through lenticels on the base of the stem close to the water line and on submerged, aerated roots.[19, 82] There is conspicuous hypertrophy of lenticels on the stem base and on parts of the root system above the water table with waterlogging[81] which may assist oxygen entry.

The pattern of oxygen loss from roots is not uniform along their length. In pea roots lacking aerenchyma, loss is mainly from the basal zones with little from the apical 6 cm unless the respiratory demand of the tissue is reduced by cooling.[40] In rice with well developed air spaces, loss in the basal region is very slow so that rhizosphere oxygenation is confined to the apical regions until lateral roots break through the cortex.[3] In adapted roots of Lodgepole pine (*Pinus contorta*), oxygen leakage follows a similar pattern to that in rice, with loss also through submerged lenticels.[82] The role of knee roots, pneumatophores and adventitious roots in trees has been reviewed by Gill.[33] The aerenchymatous, negatively geotropic nodule roots in *Myrica gale* which emerge above the water surface, have recently been shown to function in the transport of O_2 to the submerged, nitrogen fixing nodules.[98]

The inference that radial loss of oxygen to the rhizosphere is an indication of adequate internal aeration, losses being apparently in excess of requirements for respiration, does not take account of the large resistance to the diffusion of dissolved oxygen from cell to cell in the apical meristem. Oxygen leakage may occur along the length of roots that are simultaneously synthesizing ethanol by anaerobic metabolism.[22] Cortical airspaces in growing roots, even in rice,[3] are not normally found within a few cm of the root tip. In experiments with pea roots (which do not readily form aerenchyma) grown in deoxygenated agar, roots were able to extend 20 cm or more in the oxygen-free medium. Diffusion of oxygen out of young adapted roots of Lodgepole pine has been demonstrated at a distance of 30 cm from an air source.[82] Eventually the length of the diffusion path from the aerial tissues, the internal porosity of the root, the consumption of oxygen *en route*, and leakage, limit the concentration maintained in the meristem, and root extension ceases. Theoretical calculations,[70] and an electrical analogue[5] have been used to estimate the importance of each of these factors in the control of root extension when oxygen is supplied by internal diffusion.

The ability to produce roots with a large volume of air filled pores under waterlogged conditions would be expected to confer some tolerance to extended periods of waterlogging. Water-damage was less severe in maize, sunflower and wheat *cv.* Pato than in tomato and wheat *cv.* Inia, and the differences were associated with the greater root porosity in the more tolerant species.[109] Correlations have also been found between waterlogging tolerance, root porosity and O_2 diffusion in *Pinus* and *Picea* species.[20, 82] However, in a comparison between four rice varieties that all produced roots of greater porosity during waterlogging, there was no simple relation between plant growth and root porosity,[26] a clear indication that internal O_2 diffusion is not the only modifying factor in plant responses to waterlogging.

Factors promoting the development of adventitious roots under anaerobic conditions are poorly understood, but it has been suggested that the accumulation of carbon assimilates and auxin at the base of the stem,[61] and increased concentrations of ethylene,[47] are implicated.

Although the diffusion of oxygen in internal airspaces can contribute to the growth or survival of roots in anaerobic media, Crawford[23] argues compellingly that the long-term *survival* of roots, particularly of trees that may undergo seasonal flooding to several metres depth, depends upon metabolic adaptations.

Metabolic adaptations to anaerobic conditions

In herbaceous species that tolerate waterlogging, there is evidence that under O_2 stress, PEP is used in the dark fixation of carbon dioxide, with the synthesis of oxaloacetate and hence malate,[21, 22, 74] thus avoiding the accumulation of ethanol in concentrations *presumed* to be toxic. By contrast, in non-tolerant species waterlogging stimulates the rate of glycolysis,[21] with accumulation of ethanol, as previously discussed. Similar contrasting responses have been found in the roots of trees.[23] In tolerant species, malic enzyme activity seems to be depressed, so blocking the conversion of malate to pyruvate and then to ethanol. Concentrations of malate approaching 4 mM have been measured in the sap of birch trees growing in wet soil.[22] It has been suggested that the oxygen debt in the roots is repaid by translocation of malate to the leaves, where it may be involved in further metabolism. Evidence of the accumulation of other relatively non-toxic end products of anaerobic metabolism such as lactate and glycerol has been presented or reviewed.[9, 22]

One difficulty to be resolved concerns the energy relations of cells synthesizing malate or glycerol anaerobically. In the synthesis of ethanol there is a net yield of 2 moles of ATP per mole of hexose respired, because of the step catalysed by pyruvate kinase. But when PEP is converted to malate there is no net yield of ATP from respiration if the step from PEP to oxaloacetate is catalysed by PEP carboxylase. On the other hand, catalysis by PEP carboxykinase would yield 2 moles of nucleoside triphosphate, but the role of the latter enzyme in anoxic roots is unknown. Recent work with rice seedlings[7] clearly indicates that they can derive their energy from the anaerobic production of ethanol, with only a minor synthesis of malate, and tolerate the ethanol which they accumulate.

In the absence of molecular oxygen an alternative pathway is required for the oxidation of NADH. Nitrate might act as an alternative electron acceptor to oxygen in roots in poorly aerated media, as it does in some anaerobes. This attractive hypothesis[6] is still largely based on observations that the growth of cereals is often improved more with nitrate as the sole nitrogen source, than with ammonium, under oxygen deficiency. More recently, increase in the activity of nitrate reductase in leaves and roots of waterlogging-tolerant species was found when roots were subjected to anaerobic conditions.[32, 63, 64] Increases have also been found in activity of glutamic dehydrogenase and lactic dehydrogenase,[63] with accompanying rise in concentrations of soluble amino acids.[32, 63] All the above enzymes would be expected to provide alternative pathways for reoxidation of NADH, although in the light of recent evidence[77] concerning pathways of amino acid synthesis, the role of glutamic dehydrogenase requires reconsideration.

Lambers and co-workers[64] have recently found that in waterlogging-tolerant *Senecio* species the major proportion of the oxygen uptake by roots is through cyanide insensitive terminal oxidases, with low oxygen affinity. With restricted oxygen supply, respiration via cytochrome oxidase, and oxidative phosphorylation, may still proceed at a rate comparable with well aerated roots, although rates of oxygen consumption are greatly slowed. Doubts remain, however, about a causal relationship between tolerance and the greater activity of enzymes that reoxidize NADH. As mentioned before, plants have often been subjected to anaerobic conditions for periods of 1–4 weeks before assay for enzyme activity. In tolerant species there is a strong possibility that a larger proportion of the root material would have had an improved internal aeration. It is therefore noteworthy that in studies with sterile seminal roots of barley rapidly subjected to oxygen stress in solution culture,[66] no enhancement was found in nitrate reductase activity, and there was no direct suppression by nitrate of ethanol synthesis, through diversion of NADH from the reduction of acetaldehyde to the reduction of nitrate. This lack of evidence of 'nitrate respiration' in anoxic barley roots seems to accord with observations that in waterlogged soil, additions of calcium nitrate to the seminal root system did not help to improve shoot growth or delay senescence when the entire root system was anaerobic.[29]

CONCLUSIONS

Numerous factors controlled by soil and plant properties contribute to the waterlogging syndrome in plants, and although damage cannot be attributed to any single mechanism, oxygen deficiency in the rooting medium is clearly the trigger. Deoxygenated nutrient solution closely simulates waterlogged soil, except where soil toxins accumulate to harmful concentrations within the duration of the experiment.

In wetland species, anatomical and biochemical adaptations are important, and the contribution of these mechanisms to the survival of less well adapted crop plants merits consideration. Some symptoms of waterlogging damage, such as early senescence of older leaves, stomatal closure and epinasty may be of adaptive significance.

Other than by drainage, or the provision of O_2 in some other way, there remains considerable scope for the partial alleviation of the various stresses suffered by shoots by manipulation of water relations, hormones, and mineral nutrition.

REFERENCES

1. ALEKSANDROVA, A.Z. and SKAZKIN, F.D. The effect of excess soil moisture on barley at various stages of development. *Dokl. Akad. Nauk. SSSR* **159** (1964) 203–205.

2. ALLAM, A.I. and HOLLIS, J.P. Sulphide inhibition of oxidases in rice roots. *Phytopathology* **62** (1972) 634–639.

3. ARMSTRONG, W. Radial oxygen losses from intact rice roots as affected by distance from the apex, respiration and waterlogging. *Physiologia. Pl.* **25** (1971) 192–197.

4. ARMSTRONG, W. and GAYNARD, T.J. The critical oxygen pressure for respiration in intact plants. *Physiologia. Pl.* **37** (1976) 200–206.

5. ARMSTRONG, W. and WRIGHT, E.J. An electrical analogue to simulate the oxygen relations of roots in anaerobic media. *Physiologia. Pl.* **36** (1976) 383–387.

6. ARNON, D.I. Ammonium and nitrate nitrogen nutrition of barley and at different seasons in relation to hydrogen -ion concentrations, manganese, copper and oxygen supply. *Soil Sci.* **44** (1937) 91–113.

7. AVADHANI, P.N., GREENWAY, H., LEFROY, P. and PRIOR, L. Alcoholic fermentation and malate metabolism in rice germinating at low oxygen concentrations. *Aust. J. Pl. Physiol.* **5** (1978) 15–25.

8. BARBER, D.A., EBERT, M. and EVANS, N.T.S. The movement of $^{15}O_2$ through barley and rice plants. *J. exp. Bot.* **13** (1962) 397–403.

9. BELETSKAYA, E.K. Changes in metabolism of winter crops during their adaptation to flooding. *Soviet Pl. Physiol.* **24** (1977) 750–756.

10. BOLTON, E.F. and ERICKSON, A.E. Ethanol concentration in tomato plants during soil flooding. *Agron. J.* **62** (1970) 220–224.

11. BRADFORD, K.J. and DILLEY, D.R. Effects of root anaerobiosis on ethylene production, epinasty and growth of tomato plants. *Pl. Physiol.* **61** (1978) 506–509.

12. BROWN, N.J., FOUNTAINE, E.R. and HOLDEN, M.R. The oxygen requirement of crop roots and soils under near field conditions. *J. agric. Sci.* **64** (1965) 195–203.

13. BUMBIERIS, M. The role of *Phytophthora crytogea* and waterlogging in the decline of *Pinus radiata. Aust. J. Pl. Physiol.* **24** (1976) 703–709.

14. BURROWS, W.J. and CARR, D.T. Effects of flooding the root system of sunflower plants on the cytokinin content in the xylem sap. *Physiologia. Pl.* **22** (1969) 1105–1112.

15. CANNELL, R.Q., GALES, K., SNAYDON, R.W. and SUHAIL, B.A. Effects of short-term waterlogging on the growth and yield of peas (*Pisum sativum* L.) *Ann. Appl. Biol.* (in press).

16. CHANDRASEKARAN, S. and YOSHIDA, T. Effects of organic acid transformations in submerged soils on growth of the rice plant. *Soil Sci. Pl. Nutr.* **19** (1973) 39–45.

17. CHAUDHARY, T.N., BHATNAGAR, V.K. and PRIHAR, S.S. Corn yield and nutrient uptake as affected by water-table depth and soil submergence. *Agron. J.* **67** (1975) 745–749.

18. CHO, D.Y. and PONNAMPERUMA, F.N. Influence of soil temperature on the chemical kinetics of flooded soils and the growth of rice. *Soil Sci.* **112** (1971) 184–194.

19. COUTTS, M.P. and ARMSTRONG, W. Role of oxygen transport in the tolerance of trees to waterlogging. In CANNELL, M.G.R. and LAST, F.T. (Eds) *Tree Physiology and Yield Improvement*, Academic Press, London (1976) pp. 361–385.

20. COUTTS, M.P. and PHILIPSON, J.J. Tolerance of tree roots to waterlogging. I. Survival of sitka spruce and lodgepole pine. *New Phytol.* **80** (1978) 63–69.

21. CRAWFORD, R.M.M. Alcohol dehydrogenase activity in relation to flooding tolerance in roots. *J. exp. Bot.* **18** (1967) 458–464.

22. CRAWFORD, R.M.M. Some metabolic aspects of ecology. *Trans. bot. Soc. Edinb.* **41** (1972) 309–322.

23. CRAWFORD, R.M.M. Tolerance of anoxia and the regulation of glycolysis in tree roots. In CANNELL, M.G.R. and LAST, F.T. (Eds) *Tree Physiology and Yield Improvement*, Academic Press, London (1976) pp. 387–401.

24. CROSSETT, R.J. and CAMPBELL, D.J. The effects of ethylene in the root environment upon the development of barley. *Pl. Soil* **42** (1975) 453–464.

25. CULBERT, D.L. and FORD, H.W. The use of a multi-celled apparatus for anaerobic studies of flooded root systems. *Hort. Sci.* **7** (1972) 29–31.

26. DAS, D.K. and JAT, R.L. Influence of three

soil-water regimes on root porosity and growth of four rice varieties. *Agron. J.* **69** (1977) 197–200.

27. DREW, M.C. and SISWORO, E.J. Early effects of flooding on nitrogen deficiency and leaf chlorosis in barley. *New Phytol.* **79** (1977) 567–571.

28. DREW, M.C. and SISWORO, E.J. The development of waterlogging damage in young barley plants in relation to plant nutrient status and changes in soil properties. *New Phytol.* **82** (1979) 301–314.

29. DREW, M.C., SISWORO, E.J. and SAKER, L.R. Alleviation of waterlogging damage to young barley plants by application of nitrate and a synthetic cytokinin, and comparison between the effects of waterlogging, nitrogen deficiency and root excision. *New Phytol.* **82** (1979) 315–329.

30. EL-BELTAGY, A.S. and HALL, M.A. Effect of water stress upon endogenous ethylene levels in *Vicia faba. New Phytol.* **73** (1974) 47–60.

31. FULTON, J.M. and ERICKSON, A.E. Relation between soil aeration and ethyl alcohol accumulation in xylem exudates of tomatoes. *Soil Sci. Soc. Am. Proc.* **28** (1964) 610–614.

32. GARCIA-NOVO, F. and CRAWFORD, R.M.M. Soil aeration, nitrate reduction and flooding tolerance in higher plants. *New Phytol.* **72** (1973) 1031–1039.

33. GILL, C.J. The flooding tolerance of woody species – a review. *For. Abstr.* **31** (1970) 671–688.

34. GRABLE, A.R. Soil aeration and plant growth. *Adv. Agron.* **18** (1966) 57–106.

35. GRAVEN, E.H., ATTOE, O.J. and SMITH, D. Effect of liming and flooding on Mn toxicity in alfalfa. *Soil Sci. Soc. Am. Proc.* **29** (1965) 702–706.

36. GREEN, M.S. and ETHERINGTON, J.R. Oxidation of ferrous iron by rice (*Oryza sativa* L.) roots: a mechanism for waterlogging tolerance? *J. exp. Bot.* **28** (1977) 678–690.

37. GREENWOOD, D.J. and GOODMAN, D. Studies on the supply of oxygen to the roots of mustard seedlings (*Sinapis alba*). *New Phytol.* **70** (1971) 85–96.

38. GRIFFIN, D.M. A theoretical study relating the concentration and diffusion of oxygen to the biology of organisms in soil. *New Phytol.* **67** (1968) 561–577.

39. HAMMOND, L.C., ALLAWAY, W.H. and LOOMIS, W.E. Effects of oxygen and carbon dioxide upon absorption of potassium by plants. *Pl. Physiol. (Lancaster)* **30** (1955) 155–161.

40. HEALY, M.T. and ARMSTRONG, W. The effectiveness of internal oxygen transport in a mesophyte, *Pisum sativum. Planta.* **103** (1972) 302–309.

41. HIATT, A.J. and LOW, R.H. Loss of organic acids, amino acids, K and Cl from barley roots treated anaerobically and with metabolic inhibitors. *Pl. Physiol. (Lancaster)* **42** (1967) 1731–1736.

42. HIRON, R.W.P. and WRIGHT, S.T.C. The role of endogenous abscisic acid in the response of plants to stress. *J. exp. Bot.* **24** (1973) 769–781.

43. HOPKINS, H.T. Absorption of ionic species of orthophosphate by barley roots: effects of 2,4-dinitrophenol and oxygen tension. *Pl. Physiol. (Lancaster)* **31** (1956) 155–161.

44. HOPKINS, H.T., SPECHT, A.W. and HENDRICKS, S.B. Growth and nutrient accumulation as controlled by oxygen supply to the plant roots. *Pl. Physiol (Lancaster)* **25** (1950) 193–209.

45. HUCK, M.G. Variation in tap root elongation rate as influenced by composition of the soil air. *Agron. J.* **62** (1970) 815–818.

46. JACKSON, M.B. Rapid injury to peas by soil waterlogging. *J. Sci. Fd Agric.* **30** (1979) 143–152.

47. JACKSON, M.B. and CAMPBELL, D.J. Movement of ethylene from roots to shoots, a factor in the responses of tomato plants to waterlogged soil conditions. *New Phytol.* **74** (1975) 397–406.

48. JACKSON, M.B. and CAMPBELL, D.J. Waterlogging and petiole epinasty in tomato: the role of ethylene and low oxygen. *New Phytol.* **76** (1976) 21–29.

49. JACKSON, M.B. and CAMPBELL, D.J. Effects of benzyladenine and gibberellic acid on the responses of tomato plants to anaerobic root environments and to ethylene. *New Phytol.* **82** (1979) 331–340.

50. JACKSON, M.B., GALES, K. and CAMPBELL, D.J. Effect of waterlogged soil conditions on the production of ethylene and on water relationships in tomato plants. *J. exp. Bot.* **29** (1978) 183–193.

51. JACKSON, W.T. The role of adventitious roots in recovery of shoots following flooding of the original root system. *Am. J. Bot.* **42** (1955) 816–819.

52. JENSEN, C.R., STOLZY, L.H. and LETEY, J. Tracer studies of oxygen diffusion through roots of barley, corn and rice. *Soil Sci.* **103** (1967) 23–29.

53. JOHN, C.D. and GREENWAY, H. Alcoholic fermentation and activity of some enzymes in rice roots under anaerobiosis. *Aust. J. Pl. Physiol.* **3** (1976) 325–336.

54. JOHN, C.D., LIMPUNUNTANA, V. and GREENWAY, H. Adaptation of rice to anaerobiosis. *Aust. J. Pl. Physiol.* **1** (1974) 13–20.

55. JONES, H.E. and ETHERINGTON, J.R. Comparative studies of plant growth and distribution in relation to waterlogging. I. The survival of *Erica cinerea* L. and *E. tetralix* L. and its apparent relationship to iron and manganese uptake in waterlogged soil. *J. Ecol.* **58** (1970) 487–496.

56. JOSHI, M.M. and HOLLIS, J.P. Interaction of *Beggiatoa* and rice plant: detoxification of hydrogen sulphide in the rice rhizosphere. *Science* **195** (1977) 179–180.

57. KAWASE, M. Role of ethylene in induction of flooding damage in sunflower. *Physiologia. Pl.* **31** (1974) 29–38.

58. KHAN, G. The occurrence of mycorrhizas in halophytes hydrophytes and xerophytes, and of *Endogone* spores in adjacent soils. *J. gen. Microbiol.* **81** (1974) 7–14.

59. KOZLOWSKI, T.T. Water supply and leaf shedding In KOZLOWSKI, T.T. (Ed) *Water Deficits and Plant Growth* Vol. 4, Academic Press, New York (1976) pp. 191–231.

60. KRAMER, P.J. Causes of injury to plants resulting from flooding of the soil. *Pl. Physiol. (Lancaster)* **26** (1951) 722–736.

61. KRAMER, P.J. *Plant and Soil Water Relationships*, McGraw-Hill, New York (1969).

62. KRAMER, P.J. and JACKSON, W.T. Causes of injury to flooded tobacco plants. *Pl. Physiol. (Lancaster)* **29** (1954) 241–245.

63. LAMBERS, H. Respiration and NADH-oxidation of the roots of flood-tolerant and flood-intolerant *Senecio* species as affected by anaerobiosis. *Physiologia. Pl.* **37** (1976) 117–122.

64. LAMBERS, H., STEINGRÖVER, E. and SMAKMAN, G. The significance of oxygen transport and of metabolic adaptation in flood-tolerance of *Senecio* species. *Physiologia. Pl.* **43** (1978) 277–281.

65. LEE, R.B. Effects of organic acids on the loss of ions from barley roots. *J. exp. Bot.* **28** (1977) 578–587.

66. LEE, R.B. Inorganic nitrogen metabolism in barley roots under poorly aerated conditions. *J. exp. Bot.* **29** (1978) 693–708.

67. LEE, R.B. The effect of nitrite on root growth of barley and maize. *New Phytol.* (in press).

68. LEYSHON, A.J. and SHEARD, R.W. Influence of short term flooding on the growth and plant nutrient composition of barley. *Can. J. Soil Sci.* **54** (1975) 463–473.

69. LONGMUIR, I.S. Respiration rate of bacteria as a function of oxygen concentrations. *Biochem. J.* **57** (1954) 81–87.

70. LUXMOORE, R.H. and STOLZY, L.H. Oxygen diffusion in the soil plant system. 6. A synopsis with commentary. *Agron. J.* **64** (1972) 725–729.

71. LYNCH, J.M. Products of soil micro-organisms in relation to plant growth. *CRC Crit. Rev. Microbiol.* **5** (1976) 67–107.

72. LYNCH, J.M. Production and phytotoxicity of acetic acid in anaerobic soils containing plant residues. *Soil Biol. Biochem.* **10** (1978) 131–135.

73. McCALLA, T.M. and HASKINS, F.A. Phytotoxic substances from soil micro-organisms and crop residues. *Bact. Rev.* **28** (1964) 181–207.

74. McMANMON, M. and CRAWFORD, R.M.M. A metabolic theory of flooding tolerance: the significance of enzyme distribution and behaviour. *New Phytol.* **70** (1971) 299–306.

75. McPHERSON, D.C. Cortical airspaces in the roots of *Zea mays* L. *New Phytol.* **38** (1939) 190–202.

76. MEES, G.C. and WEATHERLEY, P.E. The mechanism of water absorption by roots. 2. The role of hydrostatic pressure gradients across the cortex. *Proc. R. Soc. Lond. B.* **147** (1957) 381–391.

77. MIFLIN, B.J. and LEA, P.J. The pathway of nitrogen assimilation in plants. *Phytochemistry* **15** (1976) 873–885.

78. MILBORROW, B.V. The chemistry and physiology of abscisic acid. *A. Rev. Pl. Physiol.* **25** (1974) 259–307.

79. NURITDINOV, N. and VARTAPETYAN, B.B. Transport of oxygen from the overground parts into the roots of cotton. *Soviet. Pl. Physiol.* **23** (1976) 527–529.

80. PATRICK, Z.A. Phytotoxic substances associated with the decomposition in soil of plant residues. *Soil Sci.* **111** (1971) 13–18.

81. PEREIRA, J.S. and KOZLOWSKI, T.T. Variations among woody angiosperms in response to flooding. *Physiologia. Pl.* **41** (1977) 184–192.

82. PHILIPSON, J.J. and COUTTS, M.P. The tolerance of tree roots to waterlogging. 3. Oxygen transport in lodgepole pine and sitka spruce roots of primary structure. *New Phytol.* **80** (1978) 341–349.

83. PONNAMPERUMA, F.N. The chemistry of submerged soils. *Adv. Agron.* **24** (1972) 29–96.

84. PURVIS, A.C. and WILLIAMSON, R.E. Effects of flooding and gaseous composition of the root environment on growth of corn. *Agron. J.* **64** (1972) 674–678.

85. RAILTON, I.D. and REID, D.M. Effects of benzyladenine on the growth of waterlogged tomato plants. *Planta* **111** (1973) 261–266.

86. RAYMOND, P., BRUZAU, F. and PRADET, A. Etude du transport d oxygène des parties aériennes aux racines à l'aide d'un paramètre du métabolisme: la charge énergétique. *C.r. hebd. Séanc. Acad. Sci., Paris* **286** (1978) 1061–1063.

87. REGEHR, D.L., BAZZAZ, F.A. and BOGGESS, W.R. Photosynthesis, transpiration and leaf conductance of *Populus deltoides* in relation to flooding and drought. *Photosynthetica* **9** (1975) 52–61.

88. REID, D.M. and CROZIER, A. Effect of waterlogging on gibberellin content and growth of tomato plants. *J. exp. Bot.* **22** (1971) 39–48.

89. ROWE, R.N. and BEARDSELL, D.V. Waterlogging of fruit trees. *Horticultural Abstracts* **43** (1973) 533–548.

90. RUSSELL, E.W. *Soil conditions and plant growth.* (10th ed.) Longman, London (1973).

91. SANDERSON, P.L. and ARMSTRONG, W. Soil waterlogging, root rot and conifer windthrow: oxygen deficiency or phytotoxicity? *Pl. Soil.* **49** (1978) 185–190.

92. SELMAN, I.W. and SANDANAM, S. Growth responses of tomato plants in non-aerated water culture to foliar sprays of gibberellic acid and benzyladenine. *Ann. Bot.* **36** (1972) 837–848.

93. SMITH, K.A. and RESTALL, S.W.F. The occurrence of ethylene in anaerobic soil. *J. Soil Sci.* **22** (1971) 430–443.

94. SMITH, K.A. and ROBERTSON, P.D. Effect of ethylene on root extension of cereals. *Nature* **234** (1971) 148–149.

95. SHEARD, R.W. and LEYSHON, A.J. Short-term flooding of soil: its effect on the composition of gas and water phases of soil and on phosphorus uptake by corn. *Can. J. Soil Sci.* **56** (1976) 9–20.

96. SIEGEL, S.M. and HALPERN, L.A. The effect of branching at C-1 on the biological activity of alcohols. *Proc. natn. Acad. Sci. U.S.A.* **51** (1964) 765–768.

97. STEVENSON, F.J. Organic acids in soil In McLAREN, A.D. and PETERSON, G.H. (Eds) *Soil Biochemistry*, Vol. 1, Marcel Dekker, New York (1967) pp. 119–146.

98. TJEPKEMA, J. The role of oxygen diffusion from the shoots and nodule roots in nitrogen fixation by root nodules of *Myrica gale. Can. J. Bot.* **56** (1978) 1365–1371.

99. TRAN DANG HONG, MINCHIN, F.R. and SUMMERFIELD, R.J. Recovery of nodulated cowpea plants (*Vigna anguiculata* (L.) Walp.) from waterlogging during vegetative growth. *Pl. Soil* **48** (1977) 661–672.

100. VAN'T WOUDT, B.D. and HAGAN, R.M. Crop responses at excessively high soil moisture levels. In LUTHIN, J.N. (Ed) *Drainage of Agricultural Lands*, Amer. Soc. Agron., Madison (1957) pp. 514–578.

101. VARADE, S.B., STOLZY, L.H. and LETEY, J. Influence of temperature, light intensity, and aeration on growth and root porosity of wheat *Tritium aestivum. Agron. J.* **62** (1970) 505–507.

102. WANG, T.S.C., YANG, T.-K. and CHUANG, Z.-T. Soil phenolic acids as plant growth inhibitors. *Soil Sci.* **103** (1967) 239–246.

103. WIGNARAJAH, K. and GREENWAY, H. Effect of anaerobiosis on activities of alcohol dehydrogenase and pyruvate decarboxylase in roots of *Zea mays. New Phytol.* **77** (1976) 575–584.

104. WIGNARAJAH, K., GREENWAY, H. and JOHN, C.D. Effect of waterlogging on growth and activity of alcohol dehydrogenase in barley and rice. *New Phytol.* **77** (1976) 585–592.

105. WILLEY, C.R. Effect of short periods of

anaerobic and near-anaerobic conditions on water uptake by tobacco roots. *Agron. J.* **62** (1970) 224–229.

106. WILLIAMSON, R.E. Influence of gas mixtures on cell division and root elongation of broad bean, *Vicia faba* L. *Agron. J.* **60** (1968) 317–321.

107. WILLIAMSON, R.E. Effect of soil gas composition and flooding on growth of *Nicotiana tabacum* L. *Agron. J.* **62** (1970) 80–83.

108. WRIGHT, S.T.C. The relationship between leaf water potential (ψ leaf) and the levels of abscisic acid and ethylene in excised wheat leaves. *Planta* **134** (1977) 183–189.

109. YU, P.T., STOLZY, L.H. and LETEY, J. Survival of plants under prolonged flooded conditions. *Agron. J.* **61** (1969) 844–847.

ADDENDUM

Ethylene has been shown to promote the breakdown of cell walls in the hypocotyl of sunflower leading to air space development.[113] Likewise, increased concentrations of endogenously produced ethylene seem to be involved in promoting both the early emergence of adventitious roots and the lysigenous production of extensive cortical air spaces (aerenchyma) in them in response to flooding.[112] However, when solutions were made totally anaerobic by bubbling with nitrogen gas, there was an inhibition of ethylene production by the roots and no air space development. These findings are clearly at variance with the well-established view[75] that lysigenously developed air spaces are simply the consequence of cell death through anoxia. Entrapment of ethylene by unstirred water films acting as a barrier to outward diffusion may be the reason why internal concentrations of the gas become greater,[57] but it would be interesting to know whether biosynthesis is stimulated.

In tomato there is now evidence concerning the nature of the substance stimulating ethylene release, and causing petiole epinasty in shoots during waterlogging. The damaged roots release into the transpiration stream an ethylene precursor, aminocyclopropane carboxylic acid.[110] The compound accumulates in roots and is transported to the shoot in physiologically active concentrations before the rise in ethylene production and epinasty.

Air spaces in adventitious roots of maize can occupy 80 per cent of the mid cortex. Such degeneration of the cortex might be expected to greatly inhibit the radial transport of ions to the stele. However, no difference could be found between roots with and without air spaces in the uptake and transport of K^+ (Rb-86) in the presence of air.[111] Electron microprobe analyses of transverse sections suggested that cations could diffuse rapidly across the root via the wall residues of the degenerated cells to the endodermis, which remained undamaged.

Further doubts about Crawford's hypothesis on flooding tolerance[21,22] have been raised by a study of carbohydrate metabolism in the roots of wetland plants grown in non-aerated solutions.[114] No evidence could be found that flood tolerance was related to accumulation of malate rather than ethanol. In three marsh species, as in the non-tolerant pea, ethanol was a major product of fermentation during anaerobiosis with a smaller amount of lactate. No PEP carboxykinase activity could be found.

REFERENCES

110. BRADFORD, K.J. and YANG, S.F. Xylem transport of 1-aminocyclopropane-1-carboxylic acid, an ethylene precursor, in waterlogged tomato plants. *Plant Physiol.* (1979) (in press).

111. DREW, M.C., CHAMEL, A., GARREC, J.P. and FOURCY, A. Cortical air spaces (aerenchyma) in roots of corn subjected to oxygen stress: structure and influence on uptake and translocation of ^{86}Rb ions. *Plant Physiol.* (1980) (in press).

112. DREW, M.C., JACKSON, M.B. and GIFFARD, S. Ethylene-promoted adventitious rooting and development of cortical air spaces (aerenchyma) in roots may be adaptive responses to flooding in *Zea mays* L. *Planta* **147** (1979) 83–88.

113. KAWASE, M. Role of cellulase in aerenchyma development in sunflower. *Am. J. Bot.* **66** (1979) 183–190.

114. SMITH, A.M. and AP REES, T. Pathways of carbohydrate fermentation in the roots of marsh plants. *Planta* **146** (1979) 327–334.

WHAT FACTORS ARE INVOLVED IN THE ONSET OF RIPENING IN CLIMACTERIC FRUIT?

G.E. HOBSON

Glasshouse Crops Research Institute, Rustington, Littlehampton, West Sussex, BN16 3PU, England

INTRODUCTION

Fruits have been loosely categorised according to their respiratory behaviour late in development.[6, 45] Those that show a steady decline with time have been termed 'non-climacteric', exemplified by citrus fruits, grapes and strawberries. In contrast, other fruits display an upsurge in respiration more or less coincident with ripening and have been designated 'climacteric' fruits, illustrated by the behaviour of tomatoes, pome fruits and avocados. The importance of this division is lessened if the view is taken that truncation or expansion of certain phases of development brings the patterns much closer together.[61, 73] Two recent reviews have excellently described the current status of phytohormones in ripening;[39, 45] the object of this article is to summarise recent work concerning the candidate mechanisms that initiate and control ripening and to suggest how they might be interrelated. Most information is available from studies with climacteric fruits.

Although the edible portions of fruits develop from a wide range of botanically-discrete parts of the inflorescence,[13, 50] one of the things they have in common is that during growth and development the fruits act as sinks for various organic compounds.[13, 87]

Following maturation, defined in this account as the point when full development may be achieved independent of the parent plant, the balance of acids, sugars and aroma compounds changes rapidly as full ripeness is approached. Pigments and texture almost always alter drastically at the same time. The initiating and controlling steps in the sequence of developmental processes has long been sought,[61] and a number of methods have been developed whereby the ripening rate of fruit may be manipulated. Some of these have been exploited commercially, leading to enhanced storage life and quality of the product.

THE CRITERION OF MATURITY

Maturity is regarded in two ways in contemporary accounts of fruit ripening. On the one hand,[73] ripening is regarded as following on from maturity. Thus fruit removed from the parent plant when mature is, given suitable storage conditions, capable of independent progression to full ripeness, indistinguishable from fruit ripened on the plant. Alternatively, other accounts[19] consider ripening to be the final phase of maturation. Thus fully mature fruit are regarded as synonymous with fully ripe specimens. In this account the former definition is favoured as being less confusing.

The sequence of events undergone by a typical fruit would start with fertilisation and progress through cell division (growth), cell enlargement and maturity, ending with ripening and senescence. Senescence, normally thought of as an irreversible process,[66] is considered to start with the inception of the climacteric respiration rise or certainly not later than the occurrence of the respiration peak.

While maturity is relatively easy to define, it is difficult to identify in the life of a fruit.[29] In a standard environment, the time elapsing from fruit set to maturity of climacteric fruit is largely a genetically-determined attribute of the cultivar in question. However, manipulation of storage temperature, oxygen or ethylene concentrations as well as certain chemical treatments allows this phase of development to be drastically altered. Much of fruit technology is concerned with the transportation and distribution of mature fruit under conditions of inhibited ripening. This is followed in some instances with treatments to reverse this process so that the product then has sales appeal and can be seen to be about to ripen. It is indeed fortunate that most fruits are tolerant of this stop–go type of physiology and that the quality of the resulting fruit does not suffer overmuch.

GENERAL FACTORS IN FRUIT DEVELOPMENT

Because maturity in the life of a fruit is so difficult to pinpoint, most studies on the rate of fruit development follow the effect of various factors on the time elapsing between fruit set and the first externally-visible signs of ripening. Potassium nutrition has, for instance, an effect on the rate of tomato fruit development, with shortage of the element causing an acceleration.[5] Again, the removal of a fruit from the parent plant commonly substantially reduces the time between maturation and ripening.[13] Indeed, until the avocado, for example, is detached from the tree, there is considerable resistance to ripening. As a general rule, the time from the detachment of mature fruit to ripeness is inversely proportional to the time elapsing from flower anthesis to removal from the plant. In other words, fruit kept on the plant for relatively long periods ripen quickly once picked.[51] Furthermore, the relation between the days from fertilisation to ripeness for fruit from an indeterminate plant, such as the tomato, and the solar radiation integrals does not form a smooth curve as might be expected, but alters with the age of the parent plant (possibly affecting the growth rate of the fruit) and the position of the fruit on the truss.[14, 68]

ETHYLENE EFFECTS

It seems likely that all young fruit continuously produce small quantities of ethylene.[38] However, climacteric fruit in the presence of oxygen show a sharp increase in the production of the gas at a time more or less coincident with the beginning of the respiration rise. Non-climacteric fruit show no such increase.[46]

Ethylene is intimately involved in the pathway leading to normal ripening and delays in the onset of this process are possible if steps are taken to keep its level below threshold values. Subsequent ripening changes are determined by the availability of ethylene rather than by the onset of the respiratory rise.[57] On the other hand, climacteric species may be induced to ripen some time in advance of control fruit by treatment with the gas, and some of the changes associated with ripening occur even at low oxygen tensions.[76] As fruit approach full development, the exposure time required to induce ripening decreases.[42] In order to separate the effects of endogenous and exogenous sources of ethylene, propylene has been used in place of the latter.[46] Once ethylene production is induced, it is evolved in a manner suggesting an autocatalytic production mechanism.[11]

To stimulate a normal climacteric pattern in immature fruit, exposure to ethylene at about 10 ppm for at least 24 h is required.[38] The quality of fruit induced to ripen while still of small size is obviously adversely affected. There is much to be said for the view that immature tissue has a resistance to the effects of ethylene which progressively weakens towards full development.[84] It might therefore be justified to equate maturation with the assumption of a physiological state when resistance (in terms of a lag period) to moderate amounts of the gas has largely subsided.

Tomatoes from mutant lines containing genes that cause drastic alterations in the ripening behaviour of the fruit have recently been subjected to considerable investigation.[78] Two genes known as *non-ripening* and *ripening inhibitor* slow down ripening very considerably so that the fruit are transformed from climacteric to non-climacteric behaviour. In these instances there ceases to be a direct relation between the respiratory activity of the fruit and maximal ethylene production.[79] Work with the mutant genes has emphasised that some preconditioning of the climacteric-type fruit tissue must be achieved before ethylene production, respiratory increases and carotenoid synthesis can take place. This tissue sensitisation does not take place in mutant tomato strains which are, to all intents and purposes, non-climacteric. In fact, although young mutant fruit do produce small amounts of ethylene as would normally be expected,[79] the presence of catalytic amounts of ethylene or propylene does not induce ethylene production higher than this background level.[43]

The use of various inhibitors of the enzymic production of ethylene from methionine has led to the suggestion that some fruits, especially in the riper stages, utilise precursors other than methionine.[4] Laties and his collaborators have obtained evidence from $^{13}C/^{12}C$ ratio studies that ethylene comes from lipid sources, at least in the climacteric avocado.[35] No direct effect of ethylene on either the cytochrome or the alternative pathway of isolated sweet potato mitochondria has been found[1] but the hormone may attach itself to certain specific enzymes.[25]

NUCLEIC ACID TURNOVER

Analytical data from a number of fruits show that there is considerable variation in the change in total nucleic acids, total RNA and the ability to incorporate nucleotides during ripening. For instance, RNA has been found to increase in apple tissue,[37] remain constant in avocados and bananas[63, 86] and to fall in tomato tissue.[60] Perhaps of more fundamental importance is the rate of incorporation of RNA precursors, and this seems to be high at one or possibly two periods up to the climacteric peak in respiration.[33, 60, 63, 75] Climacteric fruit show a rapid decline in the capacity for the incorporation of nucleotides into RNA[60, 63] and this is largely substantiated by a decline in the rate of ribosome synthesis[33, 47] and the incorporation of amino acids into new proteins.[66]

PROTEIN CHANGES

No dramatic alterations in the total protein content occur as fruits ripen, although certain species may show significant trends during this period. Within an overall protein ceiling, there is evidence for a redirection of protein synthesis so that those enzymes primarily concerned with the multiplicity of changes associated with ripening are augmented at the expense of other proteins.[9, 26, 28, 30, 66] Ethylene does not appear to be implicated in these changes.[48]

Whereas a number of purely hydrolytic enzymes increase in activity during ripening,[66] others show a peak in activity at the point of fruit maturity. Most enzymes (perhaps all) exist in multimolecular forms. This has led to the suggestion that the assembly of enzymes immediately prior to ripening consists of specific proteins having high activity and showing a particular iso-enzymic distribution.[27] The tissue then moves into the climacteric phase with its attendant respiration rise, ethylene upsurge and selected enzyme changes.

In a more restricted sense, certain individual enzyme systems have been put forward as key participants in the ripening mechanism. The activation of phosphofructokinase by a changing environment of phosphate brought about by membrane permeability changes inside the cell has been seen as a cornerstone in the ripening mechanism of tomato fruit.[12] The same enzyme has been shown to exhibit altered kinetic properties coincident with the respiration rise in banana fruit,[67] but this change was not dependent on phosphate levels. It still has to be proven that the enzyme is a control point for climacteric fruit in general.

Solomos and Laties[70] considered that changes in cell membrane organisation was a more likely explanation for the ripening phenomenon than the involvement of specific protein synthesis. They suggested that there was a close interaction between fruit cell integrity and incipient ethylene synthesis in the preclimacteric fruit. In further work with avocado fruit, these same authors[71] suggested that glycolysis was activated during the climacteric respiration rise, with phosphofructokinase and pyruvatekinase acting as rate-determining steps. The stimulation of these two enzymes could be the result of modulators of their activity, probably inorganic ions. They went on to propose ethylene as the stimulatory chemical on glycolysis and aerobic respiration, through its action on cellular organisation and permeability. The presence of uncouplers of oxidative phosphorylation does not prevent the ripening of tissue concomitantly exposed to ethylene, either exogenously or endogenously applied. This suggested that substrate level phosphorylation was adequate to sustain normal changes that are undergone in ripening. Further information,[72] indicated that for ethylene to enhance fruit respiration, the presence of a functional cyanide-resistant electron path was necessary, but this is not substantiated in later reports.[77]

MEMBRANES

Alterations in the properties of membranes as an explanation for the ripening phenomenon has been well summarised by Sacher.[66] It seems clear that a certain amount of ultrastructural organisation is lost as ripening progresses,[3] with erosion of the middle lamella and cellulose fibrils[53] probably being a common occurrence. The cell walls in pears suffered loss of uronosyl, galactose-1-arabinose and arabinose residues, probably brought about by the action of pectolytic enzymes, synthesised at a time normally coincident with ripening,[34] especially since (in tomatoes at least) the various glycosidases appear to remain more or less constant in activity during ripening.[54] However, the glycosidases may act as 'wall modifying agents',[32] rendering the cell wall polysaccharides more susceptible to attack by hydrolases such as polygalacturonase.

A number of studies have lent support to the concept that cells become increasingly leaky during ripening, with the changes being apparent even before the onset of the climacteric respiration rise.[8, 55, 65] However, Brady and his associates[7, 8, 9] felt that the climacteric respiratory increase was dissociated from changes in total permeability. According to de Swardt and Rousseau,[74] the apparent free space showed two peaks, one prior to the respiration rise and the other coincidental with ripening. More probably the permeabilities of the tonoplast and plasmalemma do not change with ripening but the increasing activities of ions in the cytoplasm during this period may explain many of the previous observations.[85] However, the leakiness of other membranes such as those in sub-cellular particles in the cytoplasm may change during the ripening period.

The importance of membrane integrity in fruit physiology is emphasised by several lines of work. At the sub-cellular level, the chilling of certain fruits[59] is reflected in tissue damage coincident with the molecular re-ordering of membrane lipids and changes in their composition.[31] The retardation of senescence and ripening by calcium infiltration into the tissue in particular[88] may also be a membrane phenomenon through a reinforcement of the structural integrity of the cell walls. A number of physiological disorders of fruits can be reduced by applications of calcium through, it is thought, its stabilising the permeability properties of the membranes and strengthening the cell walls (see[69]).

PLANT GROWTH REGULATORS

Several reviews of the content and action of auxins, gibberellins, cytokinins and abscisic acid, with special relevance to fruit physiology have been undertaken.[15, 17, 39, 45] Since these compounds occur in fruit tissue they are presumed to play a part in the physiological behaviour patterns.

Within limits, ethylene and abscisic acid are regarded as senescence promoters, while auxins, gibberellins and cytokinins generally retard senescence.[10]

Contradictory results following the use of indoleacetic acid (IAA) or the synthetic auxin 2,4-D on various fruits may in many cases be explained in terms of the method of application and hence the distribution in the fruit, as well as the particular concentration used. In limited quantities and with adequate penetration, IAA has been shown to delay the ripening of bananas,[82] and pears.[22, 23] With avocados, Tingwa and Young[80] indicated that concentrations above about 10 μm brought about much more rapid ripening with augmented ethylene production, whereas concentrations below this level generally suppressed ethylene production and respiration. Dipping whole bananas in 2,4-D, most of which was shown to remain in the skin, stimulated ethylene production in the outer tissues which then set off ripening in the rest of the fruit.[83] Considerable experimentation[24, 64] has concerned the degradation of IAA by peroxidase and the role of the products as senescence inhibitors, but the importance of the mechanism is still in some doubt. The effects of IAA on ripening certainly depend strongly on the method by which the auxin is applied and the subsequent distribution in the fruit. While it is an attractive proposition that ripening in a number of fruits is associated with a fall in the level of auxin in the tissue,[22] the tomato may be an exception to this generalisation.[20]

Gibberellin retards both colouration and softening once ripening in the tomato has commenced,[2] allowing the respiration climacteric to proceed normally.[18] While there is incomplete antagonism between gibberellin and ethylene, it seems unlikely that this mechanism is important in the initiation of ripening. While cytokinins appear to affect the rate at which tomatoes ripen,[16, 81] rather than the time of onset of the event, on present evidence it is still difficult to ascribe a direct role in the ripening mechanism to these hormones.

Abscisic acid treatments certainly promote fruit ripening but the chemical does not appear to act through stimulated ethylene production.[49] Although abscisic acid is most concentrated in tomato fruit at the time of lycopene synthesis,[40] there is no convincing evidence that the acid is directly involved in ethylene synthesis, nor is its accumulation controlled by ethylene.

THE ACTIVATION OF THE RIPENING MECHANISM

Since the time elapsing between anthesis and ripening for climacteric fruits (assuming standard environmental conditions) is close to a constant, it is likely that this property is under genetic control. Hence a basic developmental mechanism involving transcription and translation at appropriate stages in the life of fruits is probably involved. A limited amount of experimental data on the effects of inhibitors of protein synthesis on fruit ripening is available (see[66]) and it is commonly found that further development is arrested when preclimacteric tissue is exposed to them.

As there is little evidence to suggest that the total protein content of fruits alters drastically immediately prior to ripening, it is more likely that the synthesis of enzymes required for ripening is at the expense of other proteins. Limited ethylene treatment of banana slices caused increased incorporation into a wide range of soluble proteins but did not accelerate ripening.[7] Hence ethylene does not appear to cause the immediate synthesis of enzymes specifically concerned with the onset of ripening (see[30]). The view can be held that for normal ripening to take place, enzymes must exert a certain level of activity distributed between an appropriate number of isoenzymic forms.[21, 28, 56, 58]

Certain treatments may upset the normal co-ordination of events during ripening. Gibberellin arrests the colour change in tomato fruit but not the climacteric rise. Hence these two events, normally coincident with ripening, are not immutably linked. Again there is evidence that the climacteric rise is independent of protein synthesis[62, 71] and its occurrence does not invariably lead to ripening.[44, 57] Nevertheless, for the production of fruit of unimpaired quality, the changes associated with ripening need to be integrated and complete; ethylene may well serve to co-ordinate these separate events.[52] A tomato is able to ripen when less than 40% of its total growth period has elapsed and the time to commencement of ripening is approximately halved by continuous treatment with ethylene or its analogues. Hence the ability to ripen is acquired early in the life of a fruit.[41]

One of the difficulties in putting together any scheme concerning the sequence of events leading to a fully ripe fruit (see Fig. 1) is to separate cause from effect. It is, perhaps, more satisfactory not to envisage a chain of events with one development dependent upon the completion of the previous one, but rather to suggest that a number of key processes take place simultaneously, each one with its own control mechanism but generally loosely co-ordinated one with another.

One of the early occurrences, coincidental with seed maturation, is of increasing sensitivity towards ethylene. As this sensitivity alters, perhaps by changes in the phytohormone balance, the basic ethylene level is sufficient for the production process to become autocatalytic. Some of the *ad hoc* treatments that have been found to affect some or all of the ripening changes associated with climacteric fruit are also listed on the Figure. It is clear that oxygen is necessary to sustain ripening; ethylene synthesis from methionine,[36] chlorophyll degradation and other changes[76] demand aerobic conditions. However, ethylene seems capable of stimulating CO_2 production even at low oxygen tensions, and has an effect on many, but not all, parameters of ripening.

The evidence that protein synthesis, although perhaps limited and highly selective, is involved in the ripening process is convincing and amino acid incorporation has been shown to continue until close to the point of maximum respiration. Permeability changes in the fruit cell are detectable some time before the onset of the climacteric rise but the factors controlling the leakiness of the major cell membranes are at present ill-defined. Indeed, the view has been taken that the tonoplast and plasmalemma do not undergo significant changes at any time during ripening and the increasing leakiness of cells is due to nothing more than ion activation.[85]

It is clear that some treatments to which fruit are subjected allow certain facets of ripening to continue while others are inhibited. This suggests that fruit development occurs along a number of separate pathways affecting a wide range of metabolic systems, and these are loosely linked and co-ordinated in time. Figure 1 is intended to convey this picture and to set out some of the main changes that accompany ripening. The exact point or points in the mechanism at which inhibitors or accelerators have their action is, in some cases, still far from clear. However, the control over whether a cell ripens or not is obviously a fine one and at least one of the factors involved is non-diffusible because there is a fairly sharp demarcation line between the differently coloured areas of tomato fruit that ripen unevenly.[28] If this personal view of the ripening mechanism, which has much in common with the emphasis that Vickery and Bruinsma[85] put on re-organisation rather than disintegration, stimulates thought (or even experiment) by any of the readers, its purpose will have been fulfilled.

Acknowledgements

The author is greatly indebted to Dr. W.B. McGlasson, in whose laboratory this review was largely written, and Professors G.G. Laties, J.A. Sacher and R.E. Young for their generous help with gathering material for this review.

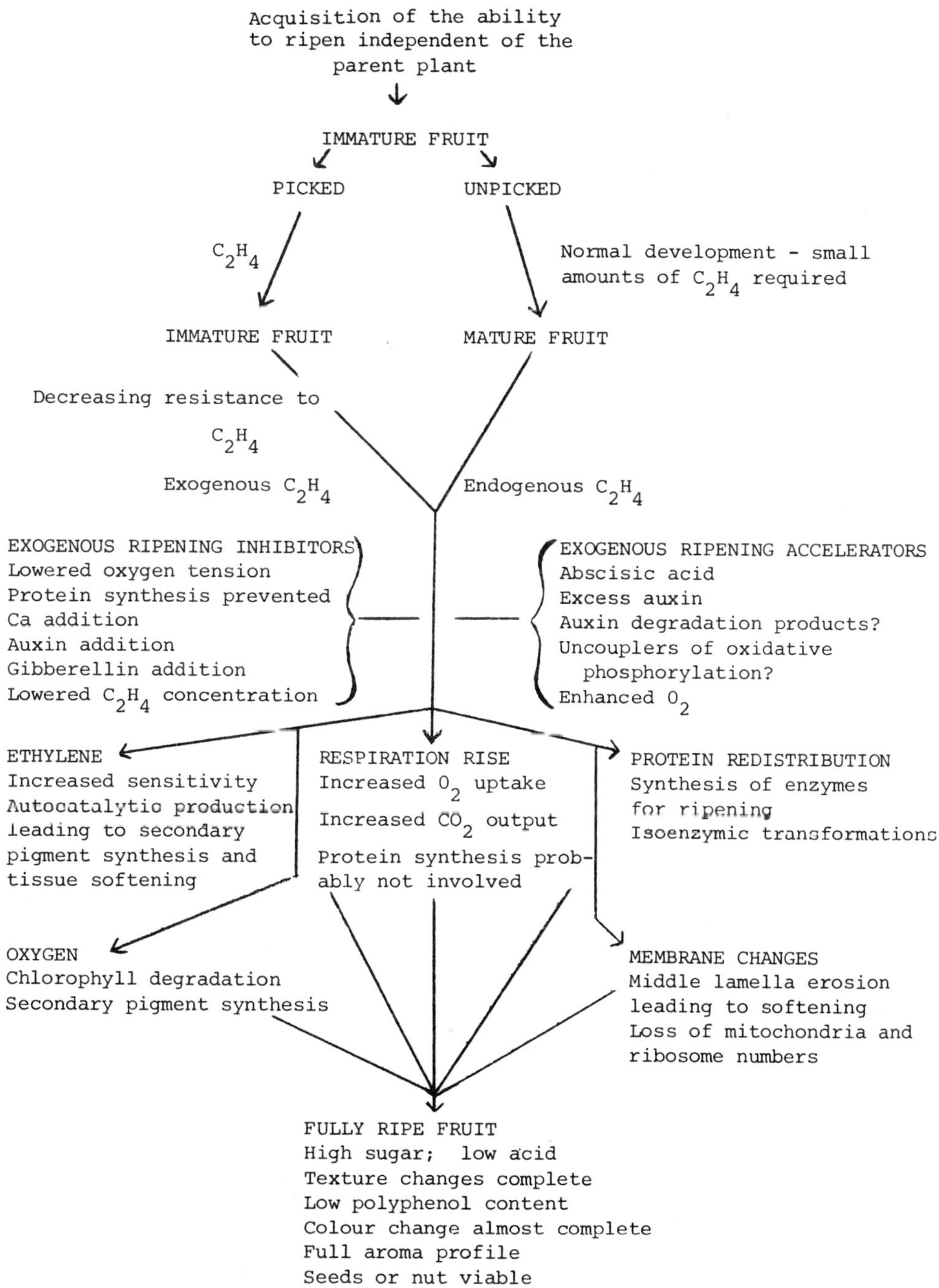

Fig. 1. Tentative mechanism for the ripening of climacteric fruit

REFERENCES

1. ARRON, G.P., DAY, D.A., GROVER, S.D. and LATIES, G.G. Effect of ethylene on the respiratory response of isolated sweet potato mitochondria. *Aust. J. Pl. Physiol.* **5** (1978) 239–248.

2. BABBITT, J.K., POWERS, M.J. and PATTERSON, M.E. Effects of growth regulators on cellulase, polygalacturonase, respiration, color and texture of ripening tomatoes. *J. Am. Soc. Hort. Sci.* **98** (1973) 77–81.

3. BAKER, J.E. Morphological changes during maturation and senescence. In ER. B. PANTASTICO (Ed) *Postharvest Physiology, Handling and Utilization of Tropical and Subtropical Fruits and Vegetables.* Avi Publishing, Westport, CT (1975) pp. 128–147.

4. BAKER, J.E., LIEBERMAN, M. and KUNISHI, A. Inhibition of ethylene production in tomato and avocado fruit slices by a rhizobitoxine analogue and free radical scavengers. *Pl. Physiol.* **57** (1976) Suppl. 501.

5. BESFORD, R.T. and MAW, G.A. Effect of potassium nutrition on tomato plant growth and fruit development. *Pl. and Soil* **42** (1975) 395–412.

6. BIALE, J.B. Respiration of fruits. In W. RUHLAND (Ed) *Encyclopedia of Plant Physiology.* Vol. 12/2. Springer, Berlin (1960) pp. 536–592.

7. BRADY, C.J. and O'CONNELL, P.B.H. On the significance of increased protein synthesis in ripening banana fruits. *Aust. J. Pl. Physiol.* **3** (1976) 301–310.

8. BRADY, C.J., O'CONNELL, P.B.H., SMYDZUK, J. and WADE, N.L. Permeability, sugar accumulation, and respiration rate in ripening banana fruits. *Aust. J. biol. Sci.* **23** (1970a) 1143–1152.

9. BRADY, C.J., PALMER, J.K., O'CONNELL, P.B.H. and SMILLIE, R.M. An increase in protein synthesis during ripening of banana fruit. *Phytochemistry* **9** (1970b) 1037–1047.

10. BRUINSMA, J., KNEGT, E. and VARGA, A. The role of growth regulating substances in fruit ripening. *Colloq. Int. du C.N.R.S.* No. 238 (1975), pp. 193–199.

11. BURG, S.P. and BURG, E.A. Ethylene action and the ripening of fruits. *Science* **148** (1965) 1190–1196.

12. CHALMERS, D.J. and ROWAN, K.S. The climacteric in ripening tomato fruit. *Pl. Physiol.* **48** (1971) 235–240.

13. COOMBE, B.G. The development of fleshy fruits. *Ann. Rev. Pl. Physiol.* **27** (1976) 507–528.

14. COOPER, A.J. Observations on the growth of the fruit on glasshouse tomato plants between March and September. *J. hort. Sci.* **34** (1959) 96–103.

15. CRANE, J.C. The role of hormones in fruit set and development. *Hortscience* **4** (1969) 108–111.

16. DAVEY, J.E. and STADEN, J. VAN. Endogenous cytokinins in the fruits of ripening and non-ripening tomatoes. *Pl. Sci. Lett.* **11** (1978) 359–364.

17. DILLEY, D.R. Hormonal control of fruit ripening. *Hortscience* **4** (1969) 111–114.

18. DOSTAL, H.C. and LEOPOLD, A.C. Gibberellin delays ripening of tomatoes. *Science* **158** (1967) 1579–1580.

19. DULL, G.G. The pineapple: general. In A.C. HULME (Ed). *The Biochemistry of Fruits and their Products*, Vol. 2. Academic Press, London (1971) pp. 303–324.

20. EL-BELTAGY, A.S., PATRICK, J.P., HEWETT, E.W. and HALL, M.A. Endogenous growth regulator levels in tomato fruits during development. *J. hort. Sci.* **51** (1976) 15–30.

21. FRENKEL, C. Involvement of peroxidase and indole-3-acetic acid oxidase isozymes from pear, tomato and blueberry fruit in ripening. *Pl. Physiol.* **49** (1972) 757–763.

22. FRENKEL, C. Role of oxidative metabolism in the regulation of fruit ripening. *Colloq. Int. du C.N.R.S.* No. 238 (1975) pp. 201–209.

23. FRENKEL, C. and DYKE, R. Auxin inhibition of ripening. *Pl. Physiol.* **51** (1973) 6–9.

24. FRENKEL, C., HADDON, V.R. and SMALLHEER, J.M. Promotion of softening and ethylene synthesis in Bartlett pears by 3-methylene oxindole. *Pl. Physiol.* **56** (1975) 647–649.

25. GERTMAN, E. and FUCHS, Y. Changes in pectinmethylesterase activity caused by ethylene at different temperatures. *Pl. Cell Physiol., Tokyo* **15** (1974) 501–505.

26. HOBSON, G.E. Electrophoretic investigation of enzymes from developing *Lycopersicon esculentum* fruit. *Phytochemistry* **13** (1974) 1383–1390.

27. HOBSON, G.E. Protein redistribution and

tomato fruit ripening. *Colloq. Int. du C.N.R.S.* No. 238 (1975) pp. 265–269.

28. HOBSON, G.E. and DAVIES, J.N. Protein and enzyme changes in tomato fruit in relation to blotchy ripening and potassium nutrition. *J. Sci. Fd Agric.* **27** (1976) 15–22.

29. HULME, A.C. The Mango. In A.C. HULME (Ed). *The Biochemistry of Fruits and their Products*, Vol. 2. Academic Press, London (1971) pp. 233–254.

30. HULME, A.C. The proteins of fruits: their involvement as enzymes in ripening. A review. *J. Fd Technol.* **7** (1972) 343–371.

31. KANE, O., MARCELLIN, P. and MAZLIAK, P. Incidence of ripening and chilling injury on the oxidative activities and fatty acid compositions of the mitochondria from mango fruits. *Pl. Physiol.* **61** (1978) 634–638.

32. KARR, A.L. and ALBERSHEIM, P. Polysaccharide-degrading enzymes are unable to attack plant cell walls without prior action by a 'wall-modifying enzyme'. *Pl. Physiol.* **46** (1970) 69–80.

33. KU, L.K. and ROMANI, R.J. The ribosomes of pear fruit. *Pl. Physiol.* **45** (1970) 401–407.

34. LABAVITCH, J. and AHMED, A.E. Cell wall metabolism in ripening fruit. *Pl. Physiol.* **61** (1978) Suppl. 116.

35. LATIES, G.G. Private communication (1978).

36. LIEBERMAN, M., KUNISHI, A., MAPSON, L.W. and WARDALE, D.A. Stimulation of ethylene production in apple tissue slices by methionine. *Pl. Physiol.* **41** (1966) 376–382.

37. LOONEY, N.E. and PATTERSON, M.E. Changes in total ribonucleic acid during the climacteric phase in yellow transparent apples. *Phytochemistry* **6** (1967) 1517–1520.

38. McGLASSON, W.B. The ethylene factor. In A.C. HULME (Ed) *The Biochemistry of Fruits and their Products*, Vol. 1, Academic Press, London (1970) pp. 475–519.

39. McGLASSON, W.B. Role of hormones in ripening and senescence. In H.O. HULTIN (Ed). *Postharvest Biology and Biotechnology.* Avi Publishing, Westport, CT., U.S.A. (in press).

40. McGLASSON, W.B. and ADATO, I. Changes in the concentrations of abscisic acid in fruits of normal and *Nr, rin* and *nor* mutant tomatoes during growth, maturation and senescence. *Aust. J. Pl. Physiol.* **3** (1976) 809–817.

41. McGLASSON, W.B. and ADATO, I. Relationship between the capacity to ripen and ontogeny in tomato fruits. *Aust. J. Pl. Physiol.* **4** (1977) 451–458.

42. McGLASSON, W.B. and PRATT, H.K. Effects of ethylene on Canteloupe fruits harvested at various ages. *Pl. Physiol.* **39** (1964) 120–127.

43. McGLASSON, W.B., DOSTAL, H.C. and TIGCHELAAR, E.C. Comparison of propylene-induced responses of immature fruit of normal and *rin* mutant tomatoes. *Pl. Physiol.* **55** (1975) 218–222.

44. McGLASSON, W.B., PALMER, J.K., VENDRELL, M. and BRADY, C.J. Metabolic studies with banana fruit slices. II. Effect of inhibitors on respiration, ethylene production and ripening. *Aust. J. Biol. Sci.* **24** (1971) 1103–1114.

45. McGLASSON, W.B., WADE, N.L. and ADATO, I. Phytohormones and Fruit Ripening. In D.S. LETHAM, P.B. GOODWIN and T.J.B. HIGGINS (Eds) *Phytohormones and Related Compounds – a Comprehensive Treatise.* A.S.P. Biological and Medical Press, B.V. Amsterdam (1978) Vol. 2 (in press).

46. McMURCHIE, E.J., McGLASSON, W.B. and EAKS, I.L. Treatment of fruit with propylene gives information about the biogenesis of ethylene. *Nature, Lond.* **237** (1972) 235–236.

47. MAREI, N. and ROMANI, R. Ribosomes from fig fruits: physical properties and the occurrence of transient dimers. *Biochim. biophys. Acta* **247** (1971) 280–290.

48. MATTOO, A.K. and VICKERY, R.S. Subcellular distributions of isoenzymes in fruits of a normal cultivar of tomato and of the *rin* mutant at two stages of development. *Pl. Physiol.* **60** (1977) 496–498.

49. MIZRAHI, Y., DOSTAL, H.C., McGLASSON, W.B. and CHERRY, J.H. Effects of abscisic acid and benzyladenine on fruits of normal and *rin* mutant tomatoes. *Pl. Physiol.* **56** (1975) 544–546.

50. MONSELISE, S. A comparative evaluation of the effects of applied regulators and other factors on maturation and ripening of orange and apple fruits. *Colloq. Int. du C.N.R.S.* No. 238 (1975) pp. 97–104.

51. PALMER, J.K. The banana. In A.C. HULME (Ed.) *The Biochemistry of Fruits and their Products*, Vol. 2. Academic Press, London (1971) pp. 65–105.

52. PASSAM, H.C. and BIRD, M.C. The respiratory activity of Honeydew melons during the climacteric. *J. exp. Bot.* **29** (1978) 325–333.

53. PESIS, E., FUCHS, Y. and ZAUBERMAN, G. Cellulase activity and fruit softening in avocado. *Pl. Physiol.* **61** (1978) 416–419.

54. PHARR, D.M., SOX, H.N. and NESBITT, W.B. Cell wall-bound nitroglycosidases of tomato fruits. *J. Am. Soc. Hort. Sci.* **101** (1976) 397–400.

55. POOVAIAH, B.W., MIZRAHI, Y., DOSTAL, H.C., CHERRY, J.H. and LEOPOLD, A.C. Water permeability during tomato fruit development in normal and *rin* non-ripening mutant. *Pl. Physiol.* **56** (1975) 813–815.

56. PRESSEY, R., AVANTS, J.K. and DULL, G.G. Difference in polygalacturonase composition of clingstone and freestone peaches. *Abs. XXth Int. Hort. Congr., Sydney* (1978) 1096.

57. QUAZI, M.H. and FREEBAIRN, H.T. The influence of ethylene, oxygen, and carbon dioxide on the ripening of bananas. *Bot. Gaz.* **131** (1970) 5–14.

58. RACUSEN, D. and FOOTE, M. Peroxidase isozymes in bean leaves by preparative disc electrophoresis. *Can. J. Bot.* **44** (1966) 1633–1638.

59. RAISON, J.K. Temperature-induced phase changes in membrane lipids and their influence on metabolic regulation. *Symp. Soc. exp. Biol.* **27** (1973) 485–512.

60. RATTANAPANONE, N., GRIERSON, D. and STEIN, M. Ribonucleic acid metabolism during the development and ripening of tomato fruits. *Phytochemistry* **16** (1977) 629–633.

61. RHODES, M.J.C. The climacteric and ripening of fruits. In A.C. HULME (Ed) *The Biochemistry of Fruits and their Products*, Vol. 1. Academic Press, London (1970) pp. 521–533.

62. RICHMOND, A. and BIALE, J.B. Protein synthesis in avocado fruit tissue. *Archs. Biochem. Biophys.* **115** (1966) 211–214.

63. RICHMOND, A. and BIALE, J.B. Protein and nucleic acid metabolism in fruits II. RNA synthesis during the respiratory rise of the avocado. *Biochim. Biophys. Acta* **138** (1967) 625–627.

64. RUDD, T., ARANJO, P. and FRENKEL, C. Onset of ripening in pears in relation to the auxin degrading function of peroxidase. *Pl. Physiol.* **61** (1978) Suppl. 140.

65. SACHER, J.A. Permeability; RNA and protein turnover during aging of fruit and leaf tissues. *Symp. Soc. exp. Biol.* **21** (1967) 269–303.

66. SACHER, J.A. Senescence and postharvest physiology. *A. Rev. Pl. Physiol.* **24** (1973) 197–224.

67. SALMINEN, S.O. and YOUNG, R.E. The control properties of phosphofructokinase in relation to the respiratory climacteric in banana fruit. *Pl. Physiol.* **55** (1975) 45–50.

68. SHORT, T.H., BAUERLE, W.L. and CURRY, R.B. A preliminary report on predicting maturity data and yield of greenhouse tomatoes. *Res. Summary Ohio Agric. Res. and Dev. Centre*, No. 82 (1975), pp. 9–11.

69. SIMON, E.W. The symptoms of calcium deficiency in plants. *New Phytol.* **80** (1978) 1–16.

70. SOLOMOS, T. and LATIES, G.G. Cellular organisation and fruit ripening. *Nature, Lond.* **245** (1973) 390–392.

71. SOLOMOS, T. and LATIES, G.G. Similarities between the actions of ethylene and cyanide in initiating the climacteric and ripening of avocados. *Pl. Physiol.* **54** (1974) 506–511.

72. SOLOMOS, T. and LATIES, G.G. Effects of cyanide and ethylene on the respiration of cyanide-sensitive and cyanide-resistant plant tissues. *Pl. Physiol.* **58** (1976) 47–50.

73. SPENCER, M. Fruit ripening. In J. BONNER and J.E. VARNER (Eds) *Plant Biochemistry.* 1st Edition. Academic Press, London (1965) pp. 793–825.

74. SWARDT, G.H. de and ROUSSEAU, G.G. Relationships between changes in membrane permeability and the respiration climacteric in pericarp tissue of tomatoes. *Planta* **112** (1973) 83–86.

75. SWARDT, G.H. de, SWANEPOEL, J.H. and DUVENAGE, A.J. Relationships between changes in ribosomal RNA and total protein synthesis, and the respiration climacteric in pericarp tissues of tomatoes. *Z. Pflanzenphysiol.* **70** (1973) 358–363.

76. STREIF, J. and BANGERTH, F. The effect of different partial pressures of oxygen and ethylene on the ripening of tomato fruits. *Scient. Hort.* **5** (1976) 227–237.

77. THEOLOGIS, A. and LATIES, G.G. Operation of the cyanide resistant path at various stages of ripening. *Pl. Physiol.* **61** (1978) Suppl. 142.

78. TIGCHELAAR, E.C., McGLASSON, W.B. and BUESCHER, R.W. Genetic regulation of tomato fruit ripening. *Hortscience* **13** (1978) 508–513.

79. TIGCHELAAR, E.C., McGLASSON, W.B. and FRANKLIN, M.J. Natural and ethephon-stimulated ripening of F_1 hybrids of the ripening inhibitor (*rin*) and non-ripening (*nor*) mutants of tomato (*Lycopersicon esculentum* Mill). *Aust. J. Pl. Physiol.* **5** (1978) 449–456.

80. TINGWA, P.O. and YOUNG, R.E. The effect of indole-3-acetic acid and other growth regulators on the ripening of avocado fruits. *Pl. Physiol.* **55** (1975) 937–940.

81. VARGA, A. and BRUINSMA, J. The growth and ripening of tomato fruits at different levels of endogenous cytokinins. *J. Hort. Sci.* **49** (1974) 135–142.

82. VENDRELL, M. Reversion of senescence: effects of 2,4-dichlorophenoxyacetic acid and indoleacetic acid on respiration, ethylene production, and ripening of banana fruit slices. *Aust. J. biol. Sci.* **22** (1969) 601–610.

83. VENDRELL, M. Relationship between internal distribution of exogenous auxins and accelerated ripening of banana fruit. *Aust. J. biol. Sci.* **23** (1970) 1133–1142.

84. VENDRELL, M. and McGLASSON, W.B. Inhibition of ethylene production in banana fruit tissue by ethylene treatment. *Aust. J. Biol. Sci.* **24** (1971) 885–895.

85. VICKERY, R.S. and BRUINSMA, J. Compartments and permeability for potassium in developing fruits of tomato (*Lycopersicon esculentum* Mill). *J. exp. Bot.* **24** (1973) 1261–1270.

86. WADE, N.L. O'CONNELL, P.B.H. and BRADY, C.J. Content of RNA and protein of the ripening banana. *Phytochemistry* **11** (1972) 975–979.

87. WALKER, A.J., HO, L.C. and BAKER, D.A. Carbon translocation in the tomato: pathways of carbon metabolism in the fruit. *Ann. Bot.* **42** (1978) 901–910.

88. WILLS, R.B.H., TIRMAZI, S.I.H. and SCOTT, K.J. Use of calcium to delay ripening of tomatoes. *Hortscience* **12** (1977) 550–551.

ADDENDUM

Since this Commentary was completed, a number of articles have appeared that amplify our knowledge of certain aspects of ripening behaviour. A series of electron micrographs published by Ben-Arie and her associates[89] of tissue changes during the ripening of apples and pears showed that although the middle lamellae were progressively eroded, intercellular plasmodesmata persisted during this process. Similar effects were produced by the application of polygalacturonase (PG) with apple tissue, and PG and cellulase with pear tissue, emphasizing again the important roles of these enzymes in ripening.[78] The separation of two isoenzymic forms of PG in normal ripening tomatoes was reported by Grierson and his collaborators,[90] whereas a slow-ripening mutant (*Never ripe*) was found to contain only one form. In studies on changes in tomato cell walls, Gross and Wallner[91] deduced that PG selectively attacks parts of the cell wall containing few neutral sugar residues. The concomitant loss of galactose and arabinose during ripening was caused neither by PG nor B-galactosidase action. Endo-pectolytic enzymes of fungal origin released galacturonides, neutral sugars and proteins[92] as soluble products during the maceration of potato tuber. The protein release under the influence of pectolytic enzymes may be an important facet of the ripening mechanism. An excellent review of ethylene formation and action has been brought together by Lieberman.[93] In it the part played by 1-aminocyclopropane-1-carboxylic acid as an intermediate in the conversion of methionine via S-adenosylmethionine is reported, based on the elucidation of the pathway by Yang and his co-workers.[94]

REFERENCES

89. BEN-ARIE, R., KISLEV, N. and FRENKEL, C. Ultrastructural changes in the cell walls of ripening apple and pear fruit. *Pl. Physiol.* **64** (1979) 197–202.

90. GRIERSON, D., TUCKER, G. and ROBERTSON, N. The regulation of gene expression during the ripening of tomato fruits. *Proc. 7th Long Ashton Symposium*, 1979, Academic Press, London (in press).

91. GROSS, K.C. and WALLNER, S.J. Degradation of cell wall polysaccharides during tomato fruit ripening. *Pl. Physiol.* **63** (1979) 117–120.

92. ISHII, S. Analysis of the components released from potato tuber tissues during maceration by pectolytic enzymes. *Pl. Physiol.* **62** (1979) 586–589.

93. LIEBERMAN, M. Biosynthesis and action of ethylene. *Ann. Rev. Pl. Physiol.* **30** (1979) 533–591.

94. CAMERON, A.C., FENTON, C.A.L., YU, Y., ADAMS, D.O. and YANG, S.F. *HortScience* **14** (1979) 178–180.

REGULATION OF PHOTORESPIRATION*

RATHNAM-CHAGUTURU†‡ and RAYMOND CHOLLET§

‡Dow Chemical, U.S.A.
Central Research – New England Laboratory
P.O. Box 400, Wayland, MA 01778 (U.S.A.)

§University of Nebraska
Laboratory of Agricultural Biochemistry
Lincoln, NE 68583 (U.S.A.)

INTRODUCTION

Atmospheric O_2 is a potent reversible inhibitor of photosynthesis in leaves of higher plants which fix CO_2 exclusively via the Calvin cycle, the so-called C_3 species. The inhibitory effect of O_2 on C_3 photosynthesis consists of two related components:[20, 24, 27, 102] (a) a direct inhibition of photosynthesis by O_2; and (b) a reduction in net photosynthesis by an O_2-stimulated formation of glycolic acid and the subsequent oxidation of glycolate, in part, to CO_2 by a process termed photorespiration. Therefore, photosynthesis in C_3 plants under atmospheric conditions (21% O_2, 330 μl/l CO_2) is a composite of two integrated pathways of carbon metabolism. The two cyclic pathways, the C_3 photosynthetic carbon reduction (PCR) cycle and the C_2 photorespiratory carbon oxidation (PCO) cycle, are initiated by the carboxylation and oxygenation, respectively, of the primary CO_2 acceptor ribulose 1,5-bisphosphate (RuBP) by the bifunctional enzyme RuPB carboxylase/oxygenase.[24, 68, 110] The relative flux of carbon through these two pathways is determined by the prevailing CO_2 and O_2 compete for the same catalytic site and behave in a mutually antagonistic manner.[16a, 24, 27, 68, 110] A variety of recent experiments with photosynthetic bacteria,[67, 109] algae,[67] intact leaves[13, 106] and isolated intact chloroplasts[13, 58] indicate that at least 90% of the glycolate formed photosynthetically is produced by oxygenation of RuBP to P-glycolate involving this bifunctional enzyme. Oxygen interacts with the PCO cycle not only in the formation of P-glycolate in the chloroplast, but also in the subsequent oxidative metabolism of glycolate in the leaf peroxisomes and mitochondria which ultimately results in the release of previously fixed carbon and nitrogen as CO_2 and NH_3, respectively, during the conversion of glycine to serine in the mitochondrion.[13, 56, 68, 110] Since all major crop species except corn, sorghum, sugarcane and pearl millet are C_3 plants, control of photorespiration and the associated O_2 inhibition of C_3 photosynthesis represents an enormous potential for increasing net photosynthesis and thus presumably agricultural productivity.

This review describes the "regulatory" mechanisms operating in cyanobacteria, algae and nonsucculent higher plants with reduced photorespiration and analyzes the potential for chemical or genetic control of photorespiration in C_3 plants. Recent accounts of the biochemistry of the C_2 PCO cycle have appeared elsewhere.[3, 13, 24, 68, 102, 110]

PLANTS WITH REDUCED PHOTORESPIRATION

C_4 plants

In contrast to C_3 species, the leaves of C_4 plants show little or no enhancement of net photosynthesis at low O_2, do not evolve CO_2 in the light and have an O_2-insensitive CO_2 com-

*Published as Paper No. 5977, Journal Series, Nebraska Agricultural Experiment Station.
†To whom all correspondence should be addressed.

pensation point (Γ) close to zero.[20, 94, 97] Similarly, the results from long-term growth studies indicate that dry-matter accumulation in C_4 plants is essentially unaffected by either CO_2 fertilization or O_2 levels of up to 40%.[47, 85] This apparent lack of photorespiration in C_4 species has been attributed to the presence of a specialized "Kranz" or wreath-like leaf anatomy and a complex biochemical pathway for net CO_2 uptake.[23, 27, 42, 89a, 94] The leaf vascular tissue is surrounded by a concentric layer of large bundle sheath cells containing numerous mitochondria, peroxisomes and starch-filled chloroplasts, and this layer is in turn surrounded by one or more rings of mesophyll cells which also contain numerous chloroplasts. In general, the mesophyll and bundle sheath cells are situated such that each mesophyll cell is separated, at most, by one cell from the nearest bundle sheath cell. In addition, numerous plasmodesmata traverse the mesophyll-bundle sheath interfacial cell wall, providing a direct means for symplastic transport between the two photosynthetic cell types.[81, 94]

The biochemical mechanism for net CO_2 fixation in leaves of C_4 Plants involves two complete, interdependent pathways of photosynthetic carbon metabolism, the C_4 cycle and the conventional C_3 PCR cycle.[42, 89a, 94, 97] The C_4 pathway, operating as an appendage to the C_3 cycle, serves as a mechanism for initially fixing atmospheric CO_2 and ultimately concentrating it at the site of RuBP carboxylase. Atmospheric CO_2 is initially fixed in the mesophyll by phosphoenolpyruvate (PEP) carboxylase, with the resulting oxaloacetate being predominantly reduced to malate or aminated to aspartate via NADP-malate dehydrogenase and aspartate aminotransferase, respectively. These latter C_4 acids occur in large pools compared with oxaloacetate and are transferred to the bundle sheath cells probably by simple diffusion through plasmodesmata and the cell symplasm.[42, 81, 94] In the bundle sheath, the C_4 acids are enzymically decarboxylated at the C-4 carboxyl position and the resulting CO_2 is refixed and metabolized via RuBP carboxylase and the C_3 cycle which are exclusively localized in this cell type.[42, 43, 89a, 94] The precise metabolic steps involved in C_4 acid decarboxylation in the bundle sheath vary between different C_4 species, leading to the recent classification of C_4 plants into three subgroups based on the major C_4 acid decarboxylase present: NADP-malic enzyme-type, PEP carboxykinase-type and NAD-malic enzyme-type.[42, 89a, 94] The C_3 acid remaining after C_4 acid decarboxylation in the bundle sheath (ultimately pyruvate or alanine) is transported back to the mesophyll where it serves as a precursor of the initial carboxylation substrate PEP.[42, 89a, 94] This regenerative phase of the C_4 cycle is catalyzed by the mesophyll chloroplast enzyme pyruvate, Pi dikinase, the most diagnostic marker enzyme for C_4 photosynthesis in nonsucculent higher plants.[42, 94]

It is evident that the C_4 pathway of photosynthesis is characterized by having two sequential, but spatially-compartmented, carboxylations. The first, utilizing atmospheric CO_2, is catalyzed by PEP carboxylase in the mesophyll cytoplasm. The second, utilizing (probably exclusively) CO_2 generated internally by the various C_4 acid decarboxylation reactions, is catalyzed by RuBP carboxylase in the bundle sheath chloroplasts. The O_2-insensitive PEP carboxylase reaction[16a] in the mesophyll functions as a very efficient trap for fixation of CO_2 from the atmosphere. Moreover, the C_4 cycle, comprising C_4 acid synthesis, C_4 acid decarboxylation and PEP regeneration, serves as a biochemical CO_2 pump, increasing pCO_2 at the site of RuBP carboxylase/oxygenase in the bundle sheath above that in free equilibrium with the leaf.[42, 89, 89a, 94] From intact leaf CO_2 exchange studies[35] and measurements of the size of the intermediate $^{14}CO_2$ pool developed during steady-state C_4 photosynthesis in $^{14}CO_2$,[41] it has been estimated that the concentration of free CO_2 developed in the bundle sheath cells under physiological conditions of pCO_2 and pO_2 is at least seven-times that resulting from simple equilibration with CO_2 in air (~8 μM).[42, 94] This elevated level of CO_2 (>60 μM) is in marked contrast to the situation in C_3 species in which the pCO_2 at the site of RuBP carboxylase (~6 μM) is in more direct equilibrium with the external atmosphere.

With intact leaves of C_4 plants at physiological levels of CO_2 and O_2 there is no evidence for any CO_2 evolution or photorespiration with properties similar to those in C_3 species.[20] However, C_4 leaf tissue can metabolize exogenous glycolate or glycine to CO_2 *in vitro* via an O_2-sensitive route,[23, 90] an observation consistent with the presence of peroxisomes, mitochondria and photorespiratory enzyme activity in the tissue.[23, 89a, 94] Although the frequency of peroxisomes and the activities of photorespiratory enzymes, all of which are primarily localized in the bundle sheath, are generally lower than those observed in C_3 plants, there is little doubt that C_4 species have the *potential* for photo-synthesizing and metabolizing P-glycolate to CO_2 in the bundle sheath via the C_2 PCO cycle.[23, 27, 89a, 90, 94] The most likely explanation for the apparent lack of photorespiration and O_2 inhibition of C_4 photosynthesis involves the C_4 cycle-mediated CO_2 pump.[23, 27, 42, 89a, 94] As discussed previously, O_2 inhibits RuBP carboxylase activity and can substitute for CO_2 in the reaction to yield P-glycolate and P-glycerate. Furthermore, the O_2 effects on RuBP carboxylase are competitively inhibited by CO_2. At atmospheric levels of O_2, photosynthetic CO_2 fixation by the C_4 cycle in the mesophyll cell layer proceeds unhindered,[16a, 23, 89] (also see ref. 94) leading to increased pCO_2 in the bundle sheath. Since the intracellular pO_2 during active photosynthesis presumably remains unchanged,[108] the elevated CO_2/O_2 ratio in the bundle sheath would allow CO_2 to compete more effectively with O_2 for RuBP carboxylase during bundle sheath photosynthesis, thus reducing O_2 inhibition of net CO_2 uptake in leaves of C_4 plants. Similarly, an increased level of CO_2 in the bundle sheath would reduce the RuBP oxygenase-catalyzed formation of P-glycolate from RuBP, thereby decreasing the amount of glycolate available for photorespiratory oxidation to CO_2.

Supportive evidence for this C_4 cycle-mediated CO_2 concentrating mechanism is provided by a variety of studies,[23, 94] including recent experiments with C_4 leaf slices.[89, 92] Photosynthesis by control leaf slices having a functional C_4 cycle is, as expected, not inhibited by 21% O_2. However, $^{14}CO_2$ fixation by C_4 leaf slices pretreated with enzyme inhibitors to prevent the operation of the C_4 cycle (e.g. inhibitors of PEP carboxylase or the C_4 acid decarboxylases) is markedly inhibited by 21% O_2 at low pCO_2. This inhibitory effect of O_2 is, however, overcome by increased pCO_2. Since the only difference between the control and inhibitor-treated leaf slices is presumably the presence of a functional C_4 cycle in the former, it appears that the C_4 pathway regulates photo respiration in leaves of C_4 plants by increasing the pCO_2 in the bundle sheath at the site of RuBP carboxylase/oxygenase. The observation that the quantum yield of C_4 photosynthesis (i.e. CO_2 fixed per absorbed quantum) is insensitive to changes in pO_2, pCO_2 and temperature[35] (environmental factors known to perturb photorespiration in C_3 plants[27]) demonstrates conclusively that photorespiration does not occur in C_4 plants *in vivo.*[20]

Although a near-zero CO_2 compensation point in air and the absence of a detectable release of CO_2 in the light are usually regarded as diagnostic gas-exchange features of C_4 species, several recent developmental studies suggest that these characteristics may be altered during leaf ontogeny. Kennedy's group[53, 113] has reported that senescent C_4 leaf tissue has relative rates of photorespiration approaching those in C_3 plants based on Γ and the light/dark $^{14}CO_2$ efflux assay of photorespiration. However, the validity of the photorespiratory values obtained by this latter technique is questionable unless the results are confirmed with independent measures of photorespiratory CO_2 exchange, such as an O_2-sensitive CO_2 compensation point, the percent O_2 inhibition of photosynthesis or the rate of total CO_2 evolution (μmol CO_2 evolved) in the light.[21, 25, 27] Along these lines, Williams and Kennedy[113] reported that senescence in corn does *not* result in a C_3-type O_2 sensitivity of net photosynthesis or Γ which would be predicted for an actively photorespiring leaf. It thus appears that these and similar reports[33] claiming increased photorespiration with increasing leaf age in C_4 plants are experimentally deficient. It is likely that the finite CO_2 compensation points observed with such tissue at 21% O_2 (from 22 to 43 μl/l) are due to a combination of dark respiration and a senescent photosynthetic apparatus rather than photorespiration *per se.*

C_3-C_4 intermediate plants

From the foregoing discussion, it is evident that the leaves of species with C_4 photosynthesis possess a full syndrome of anatomical, physiological and biochemical characteristics that are clearly distinguished from those present in C_3 plants. Although all available evidence indicates that C_4 plants have evolved from C_3 species and that this has occurred independently many times during evolution,[16] until recently no intermediate species had been detected. Attempts to create hybrids between related C_3 and C_4 plants using conventional breeding techniques have met some success with *Atriplex* species.[16] The F_2 and F_3 generation hybrids possessed several anatomical and biochemical characteristics intermediate between the C_4 and C_3 parents, but these features were either not properly compartmented or causally coordinated to lead to a reduction in photorespiration and the associated O_2 inhibition of photosynthesis.[16] This requirement for a complete coordination of the anatomical and biochemical properties of the leaf will make it exceedingly difficult, if not impossible, to introduce the C_4 pathway *per se* into C_3 plants via genetic manipulation.

Recent attempts to find naturally occurring intermediate species have met with some success. Three species of the *Laxa* group of the genus *Panicum* (*P. milioides, P. decipiens* and *P. schenckii*) have been positively identified as being intermediate between C_3 and C_4 plants with respect to leaf anatomy and photorespiration estimated by the O_2 sensitivity of net photosynthesis and Γ.[18, 19, 51, 52, 74, 75, 85, 94] Of these, only *P. milioides* has been extensively characterized with respect to photosynthetic carbon metabolism.[94] Based on comparative studies with intact leaves, thin leaf slices and isolated mesophyll and bundle sheath cell types, Rathnam and Chollet[91-94] have recently proposed that CO_2 fixation in *P. milioides* is mediated via two photosynthetic pathways, a limited, but functional, C_4 pathway and the conventional C_3 cycle. The mechanism of CO_2 fixation by the limited C_4 cycle is basically similar to that in NAD-malic enzyme-type C_4 plants[42, 89a, 94] with respect to the sequence and intercellular compartmentation of reactions leading to the synthesis and subsequent decarboxylation of malate and aspartate, the refixation of the released CO_2, and the regeneration of PEP, the initial CO_2 acceptor. However, it has yet to be determined whether the limited C_4-like PEP carboxylation-regeneration system is present in *all* mesophyll cells or only in the radially-arranged cells immediately adjacent to the chloroplast-containing bundle sheath.[93, 94] Although RuBP carboxylase and a functional C_3 cycle are present in *both* cell types,[43, 91, 93] the likely role of the limited C_4 cycle is to concentrate CO_2 at the site of bundle sheath, but not mesophyll, RuBP carboxylase. This interpretation is consistent with several lines of experimental evidence which indicate that NAD-malic enzyme, the only C_4 acid decarboxylase detected in this[92] and the two other intermediate *Panicum* species (Rathnam and Chollet, unpublished results), is exclusively localized in the bundle sheath.[91, 93] It thus appears that reduced photorespiration in this C_3-C_4 intermediate species is due to a limited degree of NAD-malic enzyme-type C_4 photosynthesis permitting an increase in pCO_2 at the site of bundle sheath, but not mesophyll, RuBP carboxylase/oxygenase. The reduced rate of glycolate formation relative to photosynthesis recently reported for *P. milioides*[104] is consistent with this proposal. Similarly, preliminary experiments with thin leaf slices of *P. decipiens* and *P. schenckii* suggest that a limited C_4-like CO_2 pump (mediated by a PEP carboxylase/NAD-malic enzyme reaction sequence similar to that in *P. milioides*) is also responsible for the reduced photorespiration and O_2 sensitivity of net photosynthesis in these related intermediate species (Rathnam and Chollet, unpublished results).

Earlier attempts by other laboratories to demonstrate a limited potential for C_4 photosynthesis in *P. milioides* met with little success. Leaf extracts of this C_3–C_4 intermediate species were reported to have, at most, only slightly higher activities of the three C_4 acid decarboxylases in comparison to C_3 plants,[61, 75] although the activity of pyruvate, Pi dikinase was not determined. In addition, the initial studies with mesophyll protoplasts and bundle sheath strands isolated by standard enzymic procedures developed for C_4 leaf tissue did not show any clear compartmentation of PEP carboxylase, RuBP carboxylase or C_4 acid decarboxylase activity between the two cell types,[61] an important biochemical feature associated invariably with C_4 photosynthesis. These findings, together with the results from $^{14}CO_2$-pulse and pulse-chase labeling experiments with intact leaves,[51, 55] prompted the conclusion that there was *no* C_4 photosynthesis in *P. milioides* despite the presence of C_4-like leaf anatomy. However, re-evaluation of these earlier studies suggests some possible deficiencies in experimental protocol. For example, with respect to the initial intercellular compartmentation studies,[61] it has recently been demonstrated that the conventional enzyme digestion system and sequential filtration protocol that has been successfully employed for isolating pure leaf cell types from C_4 plants yields cross-contaminated mesophyll and bundle sheath preparations from *P. milioides*[93] (J.C. Servaites and B.J. Reger, personal communications). In addition, the procedure used previously to assay for NAD-malic enzyme activity[75] would have probably not activated this Mn^{2+}-dependent enzyme (B.J. Reger, personal communication). Similarly, it is likely that the earlier $^{14}CO_2$ pulse and pulse-chase photosynthesis experiments[51, 55] failed to detect significant labeling of malate and aspartate in *P. milioides* due to the high levels of total CO_2 employed in these studies (425–710 μl/l).[93, 94]

Although *P. milioides* is the only C_3–C_4 intermediate species that has been characterized in some detail with respect to its photosynthetic carbon metabolism, several other reports have appeared describing additional higher plants with one or more intermediate characteristics. Kennedy and Laetsch[54] reported that *Mollugo verticillata* is a C_3–C_4 intermediate species based on the initial products of photosynthesis in $^{14}CO_2$, leaf anatomy and the relative rate of photorespiration determined by the light/dark $^{14}CO_2$ efflux assay. More recently, four ecotypes of *M. verticillata* have been described with differences in the percent of label in C_4 acids, Γ and the percent O_2 inhibition of photosynthesis.[100] Although all four ecotypes, namely Kansas, Iowa, Mexico and Massachusetts, exhibit reduced O_2 inhibition of photosynthesis (11–17% inhibition by 21% O_2), only the Kansas ecotype has an intermediate CO_2 compensation point of 25 μl/l, the other three being similar to C_3 species. Since reduced O_2 sensitivity of photosynthesis would be expected to be associated with a lower Γ (as in the three intermediate C_3–C_4 *Panicum* species[74, 75]), the lack of such a correlation in the Iowa, Mexico and Massachusetts ecotypes is puzzling. Moreover, the same researchers have recently reported differences in C_4 cycle enzyme activity (including PEP carboxylase, NAD-malic enzyme and aspartate-and alanine-aminotransferase) between the four ecotypes[101] and suggested that the observed variation in enzyme activity is well correlated with the previously reported ecotypic differences in $^{14}CO_2$-labeling patterns and photosynthetic/photorespiratory CO_2 exchange.[100] However, the activity of PEP carboxylase in the four populations (9–14 μmol/mg Ch1.h or 19–13 μmol/g fresh weight.h) is similar to that in C_3 plants and either varied little between the ecotypes (on a chlorophyll basis) or was 55% greater in the Kansas population compared to the other three (on a fresh weight basis). In contrast, the previous $^{14}CO_2$-labeling studies indicated that the percent of the total ^{14}C fixed entering the C_4 acid pool during 3 s pulse photosynthesis was 2- to 3-times greater in the Kansas *and* Iowa ecotypes compared to the Mexico and Massachusetts populations. From a related study, Raghavendra *et al.*[86] have reported the simultaneous occurrence of C_3 and C_4 photosynthesis in *M. nudicaulis* with the older leaves being C_4, the young leaves C_3, and the medium-aged leaves intermediate between C_3 and C_4. The transition from the C_3 cycle in young leaves to the C_4 pathway in old leaves was manifest in at least five features — leaf anatomy, initial photosynthetic products, rates of photosynthesis and photorespiration (based on the light/dark ^{14}C-assay) and the degree of starch-staining in the bundle sheath. However, we have been unable to confirm this reported variation of leaf anatomy in *M. nudicaulis* in that under our growth conditions (800 $\mu E/m^2.s$, 16-h photoperiod, 21°C day/16°C night) all the leaves exhibit typical C_4 Kranz anatomy, irrespective of their position along the stem (Rathnam and Chollet, unpublished results). In addition, the relative photorespiratory values obtained with the light/dark ^{14}C-assay employed in these *Mollugo* studies are of questionable validity.[21, 25, 27] Similarly, incorporation of label into C_4 acids during short-term photosynthesis in $^{14}CO_2$ is not a valid indication of C_4 photosynthesis *per se* unless evidence is also provided for the transfer of radiocarbon from the C_4 acids to PCR cycle intermediates during a chase in $^{12}CO_2$–air.[94, 97] It thus appears that many aspects of the recent *Mollugo* studies are either experimentally deficient or internally inconsistent, thus weakening any case for reduced photorespiration or C_3–C_4 intermediacy in these plants.

Recently, several reports have appeared describing reduced photorespiration in *Moricandia arvensis*, a crucifer. Moss' group[34, 60] observed that *M. arvensis* exhibited intermediacy with respect to the relative rate of photorespiration (estimated by Γ and the rate of CO_2 efflux into CO_2-free air in the light) and the O_2 inhibition of net photosynthesis compared to representative C_3 and C_4 plants, including several C_3 species of the Cruciferae family. From similar studies, Apel's group[4, 5] (P. Apel. personal communication) has reported that this and a related species, *M. spinosa*, exhibit C_3–C_4 intermediacy with respect to the CO_2 compensation point and O_2 sensitivity of Γ. These results are strikingly similar to related CO_2 exchange studies by several groups with the intermediate C_3–C_4 *Panicum* species[52, 75, 85] and indicate that photosynthesis in *M. arvensis* is less sensitive to O_2 than is photosynthesis in typical C_3 plants, and that photorespiration, relative to photosynthesis, is reduced. In addition to these physiological similarities between this crucifer and the intermediate *Panicum* species, Apel and Ohle[4] and Rathnam and Chollet (unpublished results) have also observed a sheath of chloroplast-containing cells around the leaf vascular bundles of *M. arvensis.* Similarly, recent biochemical studies from Apel's laboratory[11] indicate that the PEP carboxylase activity in crude leaf extracts prepared from this species (80 μmol/mg Chl.h) is about twice that in a related C_3 species, *M. foetida.* Thus, reasonable evidence has been provided to indicate that *M. arvensis* is intermediate between C_3 and C_4 plants with respect to photorespiration and the associated O_2 inhibition of photosynthesis. Although the mechanism(s) by which photorespiration has been reduced remains to be elucidated, it is tempting to speculate that a limited C_4-like CO_2 concentrating mechanism is also responsible for the reduced photorespiration and O_2 sensitivity of net photosynthesis in *M. arvensis*, as in the intermediate C_3–C_4 *Panicum* species.[93, 94]

Cyanobacteria and algae

The mechanism of glycolate biosynthesis in a variety of cyanobacteria and algae appears identical to that in higher plants. All microbial and algal RuBP carboxylases examined to date also catalyze the oxygenase reaction[1, 70] and the activity of this enzyme and P-glycolate phosphatase are sufficient to account for the *in vivo* rates of glycolate formation.[67, 109] Following the synthesis of glycolate via the oxygenase reaction, the metabolic fate of this photorespiratory substrate markedly differs from that in higher plants. In a variety of cyanobacteria and algae glycolate is largely excreted into the surrounding medium.[110] However, in most cases this excretion is enhanced in the presence of α-hydroxypyridinemethanesulfonate (α-HPMS), suggesting that these organisms are also capable of further metabolizing this photorespiratory substrate via a glycolate oxidizing enzyme presumably glycolate dehydrogenase.[48, 110] Despite the potential to photosynthesize and excrete glycolate or metabolize it, in part, to CO_2, photorespiration *per se* (estimated by Γ, the inhibition of photosynthesis by up to 50% O_2 and light-enhanced CO_2 release) has not been detected in a variety of cyanobacteria and "classical" C_3 algae (e.g. *Chlorella* and *Chlamydomonas*) adapted to air levels of CO_2.[15, 22, 66, 72] Several workers have proposed, based mainly on enzyme activity and inhibitor experiments and the initial products of $^{14}CO_2$ fixation, that the cyanobacteria possess the C_4 pathway of photosynthesis (see ref. 30). However, this notion is inconsistent with the detailed *in vivo* ^{14}C-labeling studies by Bassham's and Miyachi's groups[72, 82] which indicate that these microorganisms, like the algae, fix CO_2 exclusively via the C_3 PCR cycle. More recently, it has become evident that several cyanobacteria and algae when adapted to a low CO_2 environment (~0.03% CO_2) have a much higher photosynthetic affinity for CO_2 than cells grown at elevated CO_2 (2–5%).[12, 36, 96] The apparent *Km* (CO_2) calculated for high CO_2-adapted cells at pH 7.0 (~15 μM) is similar to that of C_3 species of higher plants. The greater CO_2 affinity of the low CO_2 -grown cells (~0.8 μM) is, however, not a consequence of a change in the biochemical pathway of photosynthetic CO_2 fixation[12, 96] or in the kinetic properties of RuBP carboxylase/oxygenase.[12, 45] It has been suggested from concurrent measurements of the internal and external pH and HCO_3^- concentration that several cyanobacteria and algae adapted to growth on limiting carbon have a mechanism for concentrating inorganic carbon from the surrounding medium.[7, 8, 96] This CO_2 concentrating mechanism apparently involves an active HCO_3^- uptake system[8, 71, 96] and its induction is accompanied by a coordinate induction of carbonic anhydrase activity.[36, 45, 48] Thus, like C_4 plants and the intermediate C_3–C_4 *Panicum* species, these organisms appear to control photorespiration by a metabolic CO_2 concentrating mechanism which effectively suppresses the deleterious effects of O_2 on C_3 photosynthesis at the level of RuBP carboxylase/oxygenase. Although this system for concentrating inorganic carbon appears superficially much less complicated than the higher plant C_4 acid mechanism which involves changes in leaf anatomy, photosynthetic enzymes and the inter- and intra-cellular compartmentation of component reactions for achieving the same result, this generalization must await more detailed characterization of the active HCO_3^- influx system.

CHEMICAL CONTROL OF PHOTORESPIRATION

Specific chemical inhibition of the various enzymic reactions associated with the C_2 PCO cycle has frequently been cited[27, 77, 115, 116] as a ration approach to increase net photosynthesis and thus presumably agricultural productivity in major C_3 agronomic crops.

Inhibitors of glycolate biosynthesis

Zelitch and his coworkers[115, 116] have reported that glycidate (2,3-epoxypropionate), an epoxide analogue of glycolate and glyoxylate, is a potent inhibitor of the peroxisomal enzyme glutamate-glyoxylate aminotransferase. When illuminated tobacco leaf discs are incubated with the epoxide the pools of glyoxylate and glutamate increase as a result of the inhibition of the aminotransferase, which in turn presumably inhibits glycolate formation in the chloroplast by an undefined feedback mechanism.[115, 116] Although Zelitch's group claims that net photosynthesis is increased as a result of a decrease in photorespiration in glycidate-, glutamate- or glyoxylate-treated leaf discs, these observations have not been confirmed despite repeated attempts in other laboratories.[10, 23a, 25, 46, 62, 114]

Inhibitors of P-glycolate phosphatase and glycolate metabolism

P-glycolate phosphatase is inhibited by fluoride,[64] ribose 5-P[29] and certain intermediates of the PCO cycle,[9] suggesting that regulation of this chloroplastic PCO cycle enzyme is feasible.

However, P-glycolate, which accumulates when the phosphatase is inhibited by chemical[64] or genetic means,[106] is a very potent inhibitor of triose-P isomerase.[50] This would lead to a reduction in the pool size of RuBP in the C_3 PCR cycle and thus inhibit net photosynthesis.[64, 106]

Chemical inhibition of the peroxisomal enzyme glycolate oxidase has long been considered a rational way of designing a plant growth regulator.[31] Zelitch discovered that aldehyde-bisulfite addition compounds (salts of α-hydroxysulfonates) are effective inhibitors of this flavoprotein.[115] Several series of substituted glycolic, oxyacetic and glyoxylic acids and acetylenic compounds, all related to the natural substrate of the enzyme, have been synthesized and tested for their effect on glycolate oxidase.[31, 49, 88] Of these, only α-HPMS and 2-hydroxy-3-butynoate (HBA) have been examined in some detail with respect to photorespiration and photosynthesis. In the presence of these compounds glycolate accumulates in illuminated leaves or leaf discs, but there is no sustained increase in net CO_2 uptake;[49, 115] in fact, in most cases photosynthesis is inhibited.[38, 104, 114] Related experiments with C_3 leaf segments[38, 62] and isolated leaf mesophyll cells[38, 103] and protoplasts[95] suggest that when glycolate oxidation is inhibited by HBA or α-HPMS under conditions of active photorespiration, carbon diverted from the PCR cycle into the PCO cycle accumulates as glycolate with little recycling back into the C_3 cycle. This presumably leads to an eventual depletion of PCR cycle intermediates, particularly RuBP, and a concomitant inhibition of net photosynthesis. The observations that the inhibitory effect of HBA on isolated leaf protoplast or cell photosynthesis is relieved by exogenous ribose 5-P[95] (a C_3 cycle precursor of RuBP) or high pCO_2 or low pO_2[95, 103] (conditions in which P-glycolate synthesis would largely be prevented) support this notion and indicate that this compound does not inhibit photosynthesis *per se.* Essentially identical results have been observed when the subsequent metabolism of glycine in the mitochondrion is inhibited by isonicotinic acid hydrazide,[38, 62, 103] a potent inhibitor of the glycine decarboxylation reaction. Thus, inhibition of the subsequent metabolism of P-glycolate in the chloroplast, peroxisome or mitochondrion at atmospheric levels of CO_2 and O_2 is severely deleterious to net photosynthesis and it is essential that P-glycolate, once synthesized, be metabolized through the complete PCO cycle if normal photosynthetic rates are to be maintained. This indicates that increasing photosynthetic productivity by inhibition of P-glycolate metabolism is *not* feasible and that the search for chemical control of photorespiration must focus on reducing or preventing the initial diversion of fixed carbon from the C_3 PCR cycle into the C_2 PCO cycle. Since these two competing cycles are linked directly through RuBP carboxylase/oxygenase,[3, 24, 68] this goal must be achieved by alteration of the kinetic properties of this bifunctional enzyme so that the rate of carboxylation is increased relative to the rate of oxygenation.[77]

Differential regulation of RuBP carboxylase/oxygenase activities

Several detailed studies of the effects of various chloroplast metabolites on RuBP carboxylase/oxygenase activity have failed to identify an effector which differentially regulates the two competing reactions.[26, 40, 111] Although numerous other factors (e.g. leaf age, developmental stage of tomato fruits, type of enzyme preparation, glycidate, iodoacetamide, NH_2OH, pH, temperature, Me^{2+}, etc.) have been reported to alter the carboxylase:oxygenase ratio (literature not reviewed), most of these claims have not been verified despite repeated attempts in other laboratories.[3] The discrepancy in most instances can largely be attributed to assay artifacts resulting from the complex enzymic properties of RuBP carboxylase/oxygenase with respect to CO_2, Mg^{2+}, pH, and order of addition effects and the rapid rate of enzyme deactivation at low pCO_2.[1, 3] Much attention has recently been focused on two independent reports which indicate that hydroxylamine increases the ratio of carboxylase to oxygenase activity. Bhagwat *et al.*[14] reported that NH_2OH

specifically inhibits RuBP oxygenase activity while having no effect on the carboxylase reaction. Okabe *et al.*[78] observed that this reagent causes a marked activation of carboxylase activity with a concomitant inhibition of the oxygenase reaction. More recently, however, it has been shown[17] that NH_2OH inhibits both activities of the CO_2/Mg^{2+}-activated enzyme by a reversible direct inhibition as well as through reaction with RuBP to deplete this substrate.

Andrews and Lorimer[3] have suggested that oxygenation of RuBP and thus photorespiration are the unavoidable consequence of the active site chemistry of RuBP carboxylase and that the two competing activities are unlikely to be differentially regulated. However, temperature[6, 63, 77] and various transition metals (Mn^{2+}, Co^{2+}, Ni^{2+})[28, 99, 112] (Rejda and Chollet, unpublished results) have been conclusively shown to affect differentially the two activities *in vitro*. Furthermore, large differences in the size, shape and charge distribution in CO_2 and O_2 provide some optimism that the binding of the two gaseous substrates can be differentially affected.[77] Since the partial charges of the carbon and oxygen atoms in CO_2 are +0.22 and −0.11, respectively, whereas the atoms of molecular oxygen have no partial charge, Ogren[77] has suggested that ionic or structural alterations at the active site of RuBP carboxylase are more likely to influence CO_2 binding than that of O_2. The differential effect of reaction temperature[6, 63] on the *Km* (CO_2) (i.e. increases with increasing temperature) and *Km* (O_2) (i.e. essentially no effect) may demonstrate this point. Although few in number, the well-documented *in vitro* effects of Me^{2+} and temperature on the carboxylase and oxygenase reactions indicate clearly that it is possible to differentially regulate the two competing activities of this bifunctional enzyme.

GENETIC CONTROL OF PHOTORESPIRATION

Mutational approaches

Besides attempting to control photorespiratory CO_2 loss by chemical means, there has also been increasing interest in recent years to regulate photorespiration within a given C_3 species by the use of pedigree selection and genetic manipulation.[116] Extensive screening of natural and mutant populations of several important C_3 crops (e.g. soybean, wheat, barley, oats) for an altered relationship between photosynthesis and photorespiration has met with little success.[65, 69, 76] Thus it appears that these two competing processes are integrally and tightly coupled. Genetic variation in photorespiration, relative to photosynthesis, has, however, been reported by Zelitch and Day.[117, 118] The pleiotropic Su/su tobacco mutant grows slowly and presumably has high rates of photorespiration compared to its wild-type sibling.[117] Similarly, these workers concluded that a decrease in photorespiration was associated with an increase in net photosynthesis in several pedigree tobacco selections.[118] Both of these reports were based solely on the light/dark $^{14}CO_2$ efflux assay. As discussed previously, this photorespiration assay is of questionable validity, so confirmatory evidence based on other indicators of photorespiration (e.g. O_2 inhibition of net photosynthesis, Γ) is necessary before these conclusions can be accepted. Along these lines, Okabe *et al.*[80] have recently described a new yellow tobacco mutant which is genetically related to the yellow-green Su/su mutant. In contrast to the earlier report,[117] the new mutant Su/su var. Aurea displays the same photorespiratory activity as the wild-type based on determinations of Γ and the O_2 inhibition of net photosynthesis.[80]

Current concepts of photorespiratory carbon metabolism in higher plants have been derived almost entirely from the results of biochemical and physiological investigations. Very recently, Somerville and Ogren[106, 107] have initiated a novel genetic approach to reexamine these concepts by analysis of mutants of the C_3 plant *Arabidopsis thaliana* defective in photorespiration. Several properties of the first two classes of such mutants, one deficient in P-glycolate phosphatase activity[106] and the other devoid of serine-glyoxylate aminotransferase activity,[107] have direct bearing on the continuing controversies regarding the source of photorespiratory glycolate and CO_2 and the potential for increasing net photosynthesis by inhibiting the subsequent metabolism of P-glycolate. The pattern of $^{14}CO_2$ incorporation in the wild-type in the presence of the glycolate oxidase inhibitor HBA is consistent with similar studies with other C_3 species in that a large proportion of the label is found in glycolate, associated with a corresponding decrease in serine and glycine. In contrast, the P-glycolate phosphatase mutant does not accumulate significant amounts of labeled glycolate, glycine or serine in either the presence or absence of HBA, but rather P-glycolate builds up.[106] These *in vivo* observations, together with the low level of CO_2 evolution in the light and the absence of a CO_2 post-illumination burst in the mutant, indicate conclusively that P-glycolate is the precursor of glycolate, glycine, serine and photorespiratory CO_2. More noteworthy, analyses of both mutant classes with respect to net CO_2 uptake[106, 107] indicate that inhibition of the subsequent metabolism of P-glycolate in the chloroplast (i.e. the phosphatase mutant) or peroxisome (i.e. the aminotransferase mutants) is severely deleterious to net photosynthesis at atmospheric levels of O_2 and CO_2. These results are strikingly similar to the observations from the recent chemical inhibition studies described above and indicate conclusively that the search for genetic or chemical control of photorespiration must focus on reducing or preventing the intial diversion of fixed carbon from the C_3 PCR cycle into the C_2 PCO cycle, i.e. the oxygenase reaction. This is precisely what Nature has accomplished in the cyanobacteria, algae and C_4 plants by way of the various CO_2 concentrating mechanisms.

Attempts to genetically manipulate RuBP carboxylase/oxygenase in favor of carboxylation and thus net photosynthesis have met with little success. Kung's group has recently reported[98] that manipulation of the polypeptide composition of the catalytic and non-catalytic subunits of the holoenzyme has little effect on the carboxylase:oxygenase ratio. Similarly, mutations altering the catalytic activity of the higher plant-type enzyme in the chemolithotrophic bacterium *Alcaligenes eutrophus* quantitatively affect both activities identically.[2] Although it has been reported that the yellow-green Su/su tobacco mutant possesses an enzyme with altered kinetic properties with respect to CO_2 and O_2,[79] these observations have not been confirmed.[57] The availability of *Arabidopsis* mutants with defects in chloroplastic and peroxisomal enzymes of the C_2 PCO cycle has led to the construction of double mutants.[107] The reversion frequency of such a double mutant back to wild-type is on the order of 1×10^{-9}. Mutants of this type, in which photorespiration is changed from a merely wasteful process into one which is lethal at atmospheric levels of O_2 and CO_2, may permit the direct selection of secondary mutations which reduce or eliminate RuBP oxygenase activity since reversion by any other mechanism has been essentially eliminated.[107]

In a related study, Zelitch's group[116] is using tissue culture techniques for screening large populations of haploid plant cells for resistance to several photorespiratory inhibitors. It is hoped that the clonal populations regenerated from the inhibitor-resistant cells will represent specific biochemical mutations blocking photorespiration.

Ploidy effects

Increases in ploidy have been shown to quantitatively alter a number of photosynthetic-related properties in C_3 plants, including RuBP carboxylase protein levels and activity.[87, 105]

Recently, Garrett reported[37] a ploidy effect on the *Km* (CO_2) of RuBP carboxylase purified from various cultivars of perennial ryegrass, including an isogenic diploid and tetraploid line. Whereas the *Km* (CO_2) (carboxylase) and *Ki* (CO_2) (oxygenase) for the CO_2/Mg^{2+}-activated enzyme from the diploid cultivars were about twice those from the tetraploid lines, the kinetic constants for O_2 were identical for the two groups. Correlative leaf CO_2 exchange studies[37] indicated that photorespiration, estimated by Γ and the CO_2 post-illumination burst, was also reduced in the tetraploid cultivar. Various aspects of photosynthetic and photorespiratory carbon metabolism in thin leaf slices, mesophyll protoplasts and chloroplasts isolated from the isogenic diploid and tetraploid lines have also been examined.[95] The percent O_2 inhibition of photosynthesis in leaf slices and protoplasts from the tetraploid cultivar was less than that of the diploid line at both 21 and 49% O_2. Kinetic studies revealed that the *Km* (CO_2) for photosynthesis by the diploid protoplasts was about twice that of the tetraploid line. In contrast, the *Ki* (O_2) for protoplast photosynthesis was similar in both cultivars, as was the potential for oxidizing glycolate and glycine to CO_2 via the PCO cycle. Although the maximal rates of glycolate accumulation by the isolated protoplasts in the presence of 21% O_2 and HBA were similar in the two cultivars, the percent of total fixed ^{14}C entering the [^{14}C] glycolate pool and the ratio of the rate of [^{14}C] glycolate formation to $^{14}CO_2$ fixation at 21% O_2 and low pCO_2 were about two-times greater in protoplasts and intact chloroplasts isolated form the diploid line compared to the tetraploid. These results thus indicate that a significant difference exists between the isogenic diploid and tetraploid cultivars with respect to the rate of photorespiratory glycolate formation relative to photosynthesis, presumably reflecting the presence of RuBP carboxylase/oxygenase with dissimilar kinetic properties.[37] However, it remains to be established at the molecular level how a doubling of ploidy has brought about such a *qualitative* change in this bifunctional enzyme.

ROLE OF PHOTORESPIRATION

Why the apparently wasteful process of photorespiration exists in plants and what physiological role, if any, the PCO cycle has in plant growth continue as puzzling enigmas. It has been suggested that the oxygenation of RuBP is an unavoidable consequence of the active site chemistry of RuBP carboxylase[3] and that the photorespiratory conversion of P-glycolate back to 3-P-glycerate serves to prevent accumulation of this potent inhibitor of triose-P isomerase in the chloroplast and to salvage a major portion of the carbon initially diverted from the PCR cycle.[13, 68] Similarly, it has been claimed that the oxidative metabolism of glycolate via the C_2 PCO cycle fuels the production of ATP during the conversion of glycine to serine in the mitochondrion which yields NADH.[73] *In vivo*, however, there would be no net energy production since ATP and reducing equivalents must be reinvested for the conversion of hydroxypyruvate to 3-P-glycerate[110] and for the refixation of NH_3 released during the formation of serine and CO_2 from glycine in the mitochondrion.[56] indeed, calculations indicate that evolution of one mol of CO_2 by photorespiration requires about twice as much energy as the fixation of an equivalent amount of CO_2 via the C_3 PCR cycle.[44]

Most recently, it has been suggested that the internal generation of CO_2 and the orderly dissipation of photochemical energy by the ATP and NAD(P)H requirements of the integrated PCR and PCO cycles serve to protect the photosynthetic apparatus against photooxidative damage when leaves of higher plants are deprived of external CO_2 in the light (e.g. by daytime stomatal closure during periods of environmental stress).[44] This hypothesis is supported by experiments which show that when intact C_3 leaves are preilluminated in the absence of external CO_2 and at low pO_2 (i.e.

no internal source of CO_2 via photorespiration), the capacity for subsequent CO_2 assimilation at saturating light and its efficiency at low light intensities (i.e. the apparent quantum yield) are inhibited.[32, 83, 84] The reduced apparent quantum yield in intact leaves is reflected in a marked reduction of photosystem II activity of chloroplast thylakoids isolated from the pretreated tissue.[84] Although these and related studies[59] support the notion that CO_2 recycling through the integrated PCO and PCR cycles may be one means of effectively dissipating photochemical energy when atmospheric CO_2 supply to illuminated leaves is limited,[44] they are not unequivocal. For example, it remains to be established under what environmental conditions this presumed protective mechanism might be essential for continued, long-term plant growth since clearly the extreme experimental conditions used in these photoinhibition studies (CO_2-free 1% O_2, i.e. the *total* absence of external CO_2 and internal photorespiratory CO_2) would likely never be experienced in the field. Even under these extreme conditions, the photoinhibitory effects on photosynthesis in fully expanded leaves are almost completely reversed within a 24- to 48-h recovery period at atmospheric levels of O_2 and CO_2.[84] Similarly, the observation that photoinhibition of isolated C_3 leaf cell photosynthesis is *increased* markedly when the cells are preilluminated at air levels of O_2 (rather than at low pO_2) in the absence of CO_2[59] is inconsistent with the notion that the internal generation and recycling of photorespiratory CO_2 (which is presumably taking place at 21% O_2) effectively prevents photoinhibition under conditions of low external pCO_2. Furthermore, the presence of RuBP oxygenase activity in chemolithotrophic bacteria and photolithotrophic anaerobes[1, 70] and the absence of this protein and most of the other PCO cycle enzymes in the mesophyll cells of C_4 plants[42, 43, 94, 97] argue against photorespiratory metabolism being either a specific or the only adaptation for the dissipation of excess photochemical energy when the external CO_2 supply is limited. Indeed, chloroplasts have a variety of other mechanisms for protecting the photosynthetic apparatus against photooxidative damage, including direct photoreduction of molecular O_2,[85a] superoxide dismutase, ascorbate, reduced glutathione, α-tocopherol and the carotenoid pigments.[38a]

Since net photosynthesis in major C_3 agronomic crops is substantially increased at low pO_2 or elevated pCO_2, and since increases in photosynthetic productivity can be translated into increased crop yields,[39] the search for genetic or chemical control of photorespiration at the level of RuBP carboxylase/oxygenase must continue until mutants or active chemicals are identified or until an unequivocal role of photorespiration is established.

Acknowledgement

Some of our own research described herein was supported in part by grants from the National Science Foundation (PCM 78-06626) and the Competitive Research Grants Office of USDA/SEA (5901-0410-8-0119-0) to R.C.

REFERENCES

1. AKAZAWA, T. Ribulose-1,5-bisphosphate carboxylase. In M. GIBBS and E. LATZKO (Eds.) *Encyclopedia of Plant Physiology*, New Series, Vol. 6. Springer-Verlag, New York (1979) pp. 208–229.

2. ANDERSEN, K. Mutations altering the catalytic activity of a plant-type ribulose bisphosphate carboxylase/oxygenase in *Alcaligenes eutrophus*. *Biochim. biophys. Acta* **585** (1979) 1–11.

3. ANDREWS, T.J. and LORIMER, G.H. Photorespiration – still unavoidable? *FEBS Lett.* **90** (1978) 1–9.

4. APEL, P. and OHLE, H. CO_2-compensation point and leaf anatomy in species of the genus *Moricandia* DC. (Cruciferae). *Biochem. Physiol. Pflanz.* **174** (1979) 68–75.

5. APEL, P., TICHÁ, I. and PEISKER, M. CO_2 compensation concentrations in leaves of *Moricandia arvenis* (L.) DC. at different insertion levels and O_2 concentrations. *Biochem. Physiol. Pflanz.* **172** (1978) 547–552.

6. BADGER, M.R. and COLLATZ, G.J. Studies on the kinetic mechanism of ribulose-1,5-bisphosphate carboxylase and oxygenase reactions, with particular reference to the effect of temperature on kinetic parameters. *Carnegie Inst. Wash. Year Book* **76** (1977) 355–361.

7. BADGER, M.R., KAPLAN, A. and BERRY, J.A. The internal CO_2 pool of *Chlamydomonas reinhardtii:* response to external CO_2. *Carnegie Inst. Wash. Year Book* **76** (1977) 362–366.

8. BADGER, M.R., KAPLAN, A. and BERRY, J.A. A mechanism for concentrating CO_2 in *Chlamydomonas reinhardtii* and *Anabaena variablis* and its role in photosynthetic CO_2 fixation. *Carnegie Inst. Wash. Year Book* **77** (1978) 251–261.

9. BALDY, P., VERIN, C. and CAVALIÉ, G. 2 Phosphoglycolate phosphatase from french bean leaves: purification and some catalytic properties of two isozymes. Abs. Fourth Int. Congr. Photosynthesis Research, Reading (1977).

10. BAUMANN, G. and GÜNTHER, G. Effects of glyoxylate, glycine, and serine on the assimilation of $^{14}CO_2$ in mesophyll cells of *Chenopodium album. Biochem. Physiol. Pflanz.* **174** (1979) 160–168.

11. BAUWE, H. and APEL, P. Biochemical characterization of *Moricandia arvensis* (L.) DC., a species with features intermediate between C_3 and C_4 photosynthesis, in comparison with the C_3 species *Moricandia foetida* Bourg. *Biochem. Physiol. Pflanz.* **174** (1979) 251–254.

12. BERRY, J., BOYNTON, J., KAPLAN, A. and BADGER, M. Growth and photosynthesis of *Chlamydomonas reinhardtii* as a function of CO_2 concentration. *Carnegie Inst. Wash. Year Book* **75** (1976) 423–432.

13. BERRY, J.A., OSMOND, C.B. and LORIMER, G.H. Fixation of $^{18}O_2$ during photorespiration. Kinetic and steady-state studies of the photorespiratory carbon oxidation cycle with intact leaves and isolated chloroplasts of C_3 plants. *Pl. Physiol.* **62** (1978) 954–967.

14. BHAGWAT, A.S., RAMAKRISHNA, J. and SANE, P.V. Specific inhibition of oxygenase activity of ribulose-1,5-diphosphate carboxylase by hydroxylamine. *Biochem. biophys. Res. Commun.* **83** (1978) 954–962.

15. BIRMINGHAM, B.C. and COLMAN, B. Measurement of CO_2 compensation points of freshwater algae. *Pl. Physiol.* **64** (1979) 892–895.

16. BJORKMAN, O. Adaptive and genetic aspects of C_4 photosynthesis. In R.H. BURRIS and C.C. BLACK (Eds.) *CO_2 Metabolism and Plant Productivity.* University Park Press, Baltimore (1976) pp. 287–309.

16a. BOWES, G. and OGREN, W.L. O_2 inhibition and other properties of soybean ribulose 1,5-diphosphate carboxylase. *J. biol. Chem.* **247** (1972) 2171–2176.

17. BROWN, H.M., REJDA, J.M. and CHOLLET, R. Inhibition of ribulose 1,5-bisphosphate carboxylase-oxygenase activities by hydroxylamine. *Biochim. biophys. Acta* **614** (1980) (in press).

18. BROWN, R.H. Characteristics related to photosynthesis and photorespiration of *Panicum milioides.* In R.H. BURRIS and C.C. BLACK (Eds.) *CO_2 Metabolism and Plant Productivity.* University Park Press, Baltimore (1976) pp. 311–325.

19. BROWN, R.H. Photosynthesis of grass species differing in CO_2 fixation pathways. IV. Analysis of reduced oxygen responses in *Panicum milioides* and *Panicum schenckii. Pl. Physiol.* **65** (1980) 346–349.

20. CANVIN, D.T. Photorespiration: comparison between C_3 and C_4 plants. In M. GIBBS and E. LATZKO (Eds.) *Encyclopedia of Plant Physiology,* New Series, Vol. 6. Springer-Verlag, New York (1979) pp. 368–396.

21. CANVIN, D.T. and FOCK, H. Measurement of photorespiration. *Methods Enzymol.* **24B** (1972) 246–260.

22. CHENG, K.H. and COLMAN, B. Measurements of photorespiration in some microscopic algae. *Planta* **115** (1974) 207–212.

23. CHOLLET, R. C_4 control of photorespiration: studies with isolated mesophyll cells and bundle sheath strands. In R.H. BURRIS and C.C. BLACK (Eds.) *CO_2 Metabolism and Plant Productivity*. University Park Press, Baltimore (1976) pp. 327–341.

23a. CHOLLET, R. Effect of glycidate on glycolate formation and photosynthesis in isolated spinach chloroplasts. *Pl. Physiol.* **57** (1976) 237–240.

24. CHOLLET, R. The biochemistry of photorespiration. *Trends Biochem. Sci.* **2** (1977) 155–159.

25. CHOLLET, R. Evaluation of the light/dark ^{14}C assay of photorespiration. Tobacco leaf disk studies with glycidate and glyoxylate. *Pl. Physiol.* **61** (1978) 929–932.

26. CHOLLET, R. and ANDERSON, L.L. Regulation of ribulose 1,5-bisphosphate carboxylase–oxygenase activities by temperature pretreatment and chloroplast metabolites. *Archs. Biochem. Biophys.* **176** (1976) 344–351.

27. CHOLLET, R. and OGREN, W.L. Regulation of photorespiration in C_3 and C_4 species. *Bot. Rev.* **41** (1975) 137–179.

28. CHRISTELLER, J.T. and LAING, W.A. Effect of manganese ions and magnesium ions on the activity of soya-bean ribulose bisphosphate carboxylase/oxygenase. *Biochem. J.* **183** (1979) 747–750.

29. CHRISTELLER, J.T. and TOLBERT, N.E. Phosphoglycolate phosphatase. Purification and properties. *J. biol. Chem.* **253** (1978) 1780–1785.

30. COLMAN, B. and COLEMAN, J.R. Inhibition of photosynthetic CO_2 fixation in blue-green algae by malonate. *Pl. Sci. Lett.* **12** (1978) 101–105.

31. CORBETT, J.R. and WRIGHT, B.J. Inhibition of glycollate oxidase as a rational way of designing a herbicide. *Phytochemistry* **10** (1971) 2015–2024.

32. CORNIC, G. La photorespiration se déroulant dans un air sans CO_2 a-t-elle une fonction? *Can. J. Bot.* **56** (1978) 2128–2137.

33. CRESPO, H.M., FREAN, M., CRESWELL, C.F. and TEW, J. The occurrence of both C_3 and C_4 photosynthetic characteristics in a single *Zea mays* plant. *Planta* **147** (1979) 257–263.

34. CROOKSTON, R.K. Physiological characteristics and leaf anatomy of C_4 and C_3 species. Ph.D. Thesis, University of Minnesota (1972).

35. EHLERINGER, J. and BJÖRKMAN, O. Quantum yields for CO_2 uptake in C_3 and C_4 plants. Dependence on temperature, CO_2, and O_2 concentration. *Pl. Physiol.* **59** (1977) 86–90.

36. FINDENEGG, G.R. Correlations between accessibility of carbonic anhydrase for external substrate and regulation of photosynthetic use of CO_2 and HCO_3^- by *Scenedesmus obliquus*. *Z. PflPhysiol.* **79** (1976) 428–437.

37. GARRETT, M.K. Control of photorespiration at RuBP carboxylase/oxygenase level in ryegrass cultivars. *Nature, Lond.* **274** (1978) 913–915.

38. GÜNTHER, G., BAUMANN, G., KLOS, J. and BALFANZ, J. Effects of inhibitors and herbicides on net photosynthesis and photorespiration of sugar beet (*Beta vulgaris*) and white lambsquarter (*Chenopodium album*). *Biochem. Physiol. Pflanz.* **174** (1979) 616–628.

38a. HALLIWELL, B. The chloroplast at work. A review of modern developments in our understanding of chloroplast metabolism. *Prog. Biophys. Mol. Biol.* **33** (1978) 1–54.

39. HARDY, R.W.F., HAVELKA, U.D. and QUEBEDEAUX, B. The opportunity for and significance of alteration of ribulose 1,5-bisphosphate carboxylase activities in crop production. In H.W. SIEGELMAN and G. HIND (Eds.) *Photosynthetic Carbon Assimilation.* Plenum Press, New York (1978) pp. 165–178.

40. HATCH, A. and JENSEN, R.G. Activation of RuBP carboxylase by intermediates. *Pl. Physiol.* **63**, Suppl. (1979) 64.

41. HATCH, M.D. The C_4 pathway of photosynthesis. Evidence for an intermediate pool of carbon dioxide and the identity of the donor C_4-dicarboxylic acid. *Biochem. J.* **125** (1971) 425–432.

42. HATCH, M.D. C_4 pathway photosynthesis: mechanism and physiological function. *Trends Biochem. Sci.* **2** (1977) 199–202.

43. HATTERSLEY, P.W., WATSON, L. and OSMOND, C.B. *In situ* immunofluorescent labeling of ribulose-1,5-bisphosphate carboxylase in leaves of C_3 and C_4 plants. *Aust. J. Pl. Physiol.* **4** (1977) 523–539.

44. HEBER, U. and KRAUSE, G.H. What is the physiological role of photorespiration? *Trends Biochem. Sci.* **5** (1980) 32–34.

45. HOGETSU, D. and MIYACHI, S. Role of carbonic anhydrase in photosynthetic CO_2 fixation in *Chlorella. Pl. Cell Physiol., Tokyo* **20** (1979) 747–756.

46. HUNT, L.D. and OGREN, W.L. Glyoxylate stimulation of photosynthesis in isolated soybean leaf mesophyll cells. *Pl. Physiol.* **63**, Suppl. (1979) 38.

47. IMAI, K. and MURATA' Y. Effect of CO_2 concentration on growth and dry matter production of crop plants. VI. Effect of oxygen concentration on the CO_2-dry matter production relationship in some C_3- and C_4-crop species. *Jap. J. Crop. Sci.* **48** (1979) 58–65.

48. INGLE, R.K. and COLMAN, B. The relationship between carbonic anhydrase activity and glycolate excretion in the blue-green alga *Coccochloris peniocystis. Planta* **128** (1976) 217–223.

49. JEWESS, P.J., KERR, M.W. and WHITAKER, D.P. Inhibition of glycolate oxidase from pea leaves. *FEBS Lett.* **53** (1975) 292–296.

50. JONES, R.B. and WALEY, S.G. Spectrophotometric studies on the interaction between triose phosphate isomerase and inhibitors. *Biochem. J.* **179** (1979) 623–630.

51. KANAI, R. and KASHIWAGI, M. *Panicum milioides*, a Gramineae plant having Kranz leaf anatomy without C_4 photosynthesis. *Pl. Cell Physiol., Tokyo* **16** (1975) 669–679.

52. KECK, R.W. and OGREN, W.L. Differential oxygen response of photosynthesis in soybean and *Panicum milioides. Pl. Physiol.* **58** (1976) 552–555.

53. KENNEDY, R.A. Relationship between leaf development, carboxylase enzyme activities and photorespiration in the C_4-plant *Portulaca oleracea* L. *Planta* **128** (1976) 149–154.

54. KENNEDY, R.A. and LAETSCH, W.M. Plant species intermediate for C_3, C_4 photosynthesis. *Science, NY* **184** (1974) 1087–1089.

55. KESTLER, D.P., MAYNE, B.C., RAY, T.B., GOLDSTEIN, L.D., BROWN, R.H. and BLACK, C.C. Biochemical components of the photosynthetic CO_2 compensation point of higher plants. *Biochem. biophys. Res. Commun.* **66** (1975) 1439–1446.

56. KEYS, A.J., BIRD, I.F., CORNELIUS, M.J., LEA, P.J., WALLSGROVE, R.M. and MIFLIN, B.J. Photorespiratory nitrogen cycle. *Nature, Lond.* **275** (1978) 741–743.

57. KOIVUNIEMI, P., TOLBERT, N.E. and CARLSON, P.J. Ribulose-1,5-bisphosphate carboxylase/oxygenase and polyphenol oxidase in the tobacco mutant Su/su. *Pl. Physiol.* **61**, Suppl. (1978) 99.

58. KRAUSE, G.H., THORNE, S.W. and LORIMER, G.H. Glycolate synthesis by intact chloroplasts. Studies with inhibitors of photophosphorylation. *Archs. Biochem. Biophys.* **183** (1977) 471–479.

59. KRAUSE, G.H., KIRK, M., HEBER, U. and OSMOND, C.B. O_2-dependent inhibition of photosynthetic capacity in intact isolated chloroplasts and isolated cells from spinach leaves illuminated in the absence of CO_2. *Planta* **142** (1978) 229–233.

60. KRENZER, E.G., Jr., MOSS, D.N. and CROOKSTON, R.K. CO_2 compensation points of flowering plants. *Pl. Physiol.* **56** (1975) 194–206.

61. KU, S.B., EDWARDS, G.E. and KANAI, R. Distribution of enzymes related to C_3 and C_4 pathway of photosynthesis between mesophyll and bundle sheath cells of *Panicum hians* and *Panicum milioides. Pl. Cell Physiol., Tokyo* **17** (1976) 615–620.

62. KUMARASINGHE, K.S., KEYS, A.J. and WHITTINGHAM, C.P. Effects of certain inhibitors on photorespiration by wheat leaf segments. *J. exp. Bot.* **28** (1977) 1163–1168.

63. LAING, W.A., OGREN, W.L. and HAGEMAN, R.H. Regulation of soybean net photosynthetic CO_2 fixation by the interaction of CO_2, O_2, and ribulose 1,5-diphosphate carboxylase. *Pl. Physiol.* **54** (1974) 678–685.

64. LARSSON, C. Photosynthetic glycolate formation via phosphoglycolate in isolated spinach chloroplasts. In M. AVRON (Ed.) *Proc. Third Int. Congr. Photosynthesis.* Elsevier, Amsterdam (1974) pp. 1321–1328.

65. LLOYD, N.D.H. and CANVIN, D.T. Photosynthesis and photorespiration in sunflower selections. *Can. J. Bot.* **55** (1977) 3006–3012.

66. LLOYD, N.D.H., CANVIN, D.T. and CULVER, D.A. Photosynthesis and photorespiration in algae. *Pl. Physiol.* **59** (1977) 936–940.

67. LORIMER, G.H., OSMOND, C.B., AKAZAWA, T. and ASAMI, S. On the mechanism of glycolate synthesis by *Chromatium* and *Chlorella*. *Archs. Biochem. Biophys.* **185** (1978) 49–56.

68. LORIMER, G.H., WOO, K.C., BERRY, J.A. and OSMOND, C.B. The C_2 photorespiratory carbon oxidation cycle in leaves of higher plants: pathway and consequences. In D.O. HALL, J. COOMBS and T.W. GOODWIN (Eds.) *Photosynthesis '77: Proc. Fourth Int. Congr. Photosynthesis.* The Biochemical Society, London (1978) pp. 311–322.

69. McCASHIN, B.G. and CANVIN, D.T. Photosynthetic and photorespiratory characteristics of mutants of *Hordeum vulgare* L. *Pl. Physiol.* **64** (1979) 354–360.

70. McFADDEN, B.A. and PUROHIT, K. Chemosynthetic, photosynthetic, and cyanobacterial ribulose bisphosphate carboxylase. In H.W. SIEGELMAN and G. HIND (Eds.) *Photosynthetic Carbon Assimilation.* Plenum Press, New York (1978) pp. 179–207.

71. MILLER, A.G. and COLMAN, B. Evidence for HCO_3^- transport by the blue-green alga (cyanobacterium) *Coccochloris peniocystis*. *Pl. Physiol.* **65** (1980) 397–402.

72. MIYACHI, S. and OKABE, K.-I. Oxygen ehancement of photosynthesis in *Anacystis nidulans* cells. *Pl. Cell Physiol., Tokyo* **17** (1 976) 973–986.

73. MOORE, A.L., JACKSON, C., HALLIWELL, B., DENCH, J.E. and HALL, D.O. Intramitochondrial localisation of glycine decarboxylase in spinach leaves. *Biochem. biophys. Res. Commun.* **78** (1977) 483–491.

74. MORGAN, J.A. and BROWN, R.H. Photosynthesis in grass species differing in carbon dioxide fixation pathways. II. A search for species with intermediate gas exchange and anatomical characteristics. *Pl. Physiol.* **64** (1979) 257–262.

75. MORGAN, J.A., BROWN, R.H. and REGER, B.J. Photosynthesis in grass species differing in CO_2 fixation pathways. III. Oxygen response and enzyme activities of species in the *Laxa* group of *Panicum*. *Pl. Physiol.* **65** (1980) 156–159.

76. OGREN, W.L. Search for higher plants with modifications of the reductive pentose phosphate pathway of CO_2 assimilation. In R.H. BURRIS and C.C. BLACK (Eds.) *CO_2 Metabolism and Plant Productivity.* University Park Press, Baltimore (1976) pp. 19–29.

77. OGREN, W.L. Increasing carbon fixation by crop plants. In D.O. HALL, J. COOMBS and T.W. GOODWIN (Eds.) *Photosynthesis '77: Proc. Fourth Int. Congr. Photosynthesis.* The Biochemical Society, London (1978) pp. 721–733.

78. OKABE, K.-I., CODD, G.A. and STEWART, W.D.P. Hydroxylamine stimulates carboxylase activity and inhibits oxygenase activity of cyanobacterial RuBP carboxylase/oxygenase. *Nature, Lond.* **279** (1979) 525–527.

79. OKABE, K. and SCHMID, G.H. Properties of the tobacco aurea mutant Su/su var. aurea. On photorespiration and on the structure and function relationship in chloroplasts. In G. AKOYUNOGLOU and J.H. ARGYRONDI-AKOYUNOGLOU (Eds.) *Chloroplast Development.* Elsevier, Amsterdam (1978) pp. 501–506.

80. OKABE, K., SCHMID, G.H. and STRAUB, J. Genetic characterization and high efficiency photosynthesis of an aurea mutant of tobacco. *Pl. Physiol.* **60** (1977) 150–156.

81. OSMOND, C.B. and SMITH, F.A. Symplastic transport of metabolites during C_4-photosynthesis. In B.E.S. GUNNING and A.W. ROBARDS (Eds.) *Intercellular Communication in Plants: Studies on Plasmodesmata.* Springer-Verlag, New York (1976) pp. 229–241.

82. PELROY, R.A. and BASSHAM, J.A. Photosynthetic and dark carbon metabolism in unicellular blue-green algae. *Archs. Mikrobiol.* **86** (1972) 25–38.

83. POWLES, S.B. and OSMOND, C.B. Inhibition of the capacity and efficiency of photosynthesis in bean leaflets illuminated in a CO_2-free atmosphere at low oxygen: a possible role for photorespiration. *Aust. J. Pl. Physiol.* **5** (1978) 619–629.

84. POWLES, S.B., OSMOND, C.B. and THORNE, S.W. Photoinhinition of intact attached leaves of C_3 plants illuminated in the absence of both carbon dioxide and of photorespiration. *Pl. Physiol.* **64** (1979) 982–988.

85. QUEBEDEAUX, B. and CHOLLET, R. Comparative growth analyses of *Panicum* species with differing rates of photorespiration. *Pl. Physiol.* **59** (1977) 42–44.

85a. RADMER, R. Uptake of oxygen by light-generated reductant. *What's New in Pl. Physiol.* **10** (1979) 25–28.

86. RAGHAVENDRA, A.S., RAJENDRUDU, G. and DAS, V.S.R. Simultaneous occurrence of C_3 and C_4 photosynthesis in relation to leaf position in *Mollugo nudicaulis. Nature, Lond.* **273** (1978) 143–144.

87. RANDALL, D.D., NELSON, C.J. and ASAY, K.H. Ribulose bisphosphate carboxylase. Altered genetic expression in tall fescue. *Pl. Physiol.* **59** (1977) 38–41.

88. RANDALL, W.C., STREETER, K.B., CRESSON, E.L., SCHWAM, H., MICHELSON, S.R., ANDERSON, P.S. and CRAGOE, E.J., Jr. Quantitative structure-activity relationships involving the inhibition of glycolic acid oxidase by derivatives of glycolic and glyoxylic acids. *J. Med. Chem.* **22** (1979) 608–614.

89. RATHNAM, C.K.M. Metabolic regulation of carbon flux during C_4 photosynthesis. I. Evidence for parallel CO_2 fixation by mesophyll and bundle sheath cells *in situ. Z. PflPhysiol.* **87** (1978) 65–84.

89a. RATHNAM, C.K.M. C_4 photosynthesis: the path of carbon in bundle sheath cells. *Sci. Prog.* **65** (1978) 409–435.

90. RATHNAM, C.K.M. Metabolic regulation of carbon flux during C_4 photosynthesis. II. *In situ* evidence for refixation of photorespiratory CO_2 by C_4 phosphoenolpyruvate carboxylase. *Planta* **145** (1979) 13–23.

91. RATHNAM, C.K.M. and CHOLLET, R. CO_2 donation by malate and aspartate reduces photorespiration in *Panicum milioides*, a C_3–C_4 intermediate species. *Biochem. biophys. Res. Commun.* **85** (1978) 801–808.

92. RATHNAM, C.K.M. and CHOLLET, R. Phosphoenolpyruvate carboxylase reduces photorespiration in *Panicum milioides*, a C_3–C_4 intermediate species. *Archs. Biochem. Biophys.* **193** (1979) 346–354.

93. RATHNAM, C.K.M. and CHOLLET, R. Photosynthetic carbon metabolism in *Panicum milioides*, a C_3–C_4 intermediate species: evidence for a limited C_4 dicarboxylic acid pathway of photosynthesis. *Biochim. biophys. Acta* **548** (1979) 500–519.

94. RATHNAM, C.K.M. and CHOLLET, R. Photosynthetic carbon metabolism in C_4 plants and C_3–C_4 intermediate species. In L. REINHOLD, J.B. HARBORNE and T. SWAIN (Eds.) *Progress in Phytochemistry*, Vol. 6. Pergamon Press, Oxford (1980) pp. 1–48.

95. RATHNAM, C.K.M. and CHOLLET, R. Photosynthetic and photorespiratory carbon metabolism in mesophyll protoplasts and chloroplasts isolated from isogenic diploid and tetraploid cultivars of ryegrass (*Lolium perenne* L.). *Pl. Physiol.* **65** (1980) 489–494.

96. RAVEN, J.A. and GLIDEWELL, S.M. C_4 characteristics of photosynthesis in the C_3 alga *Hydrodictyon africanum. Plant Cell Environment* **1** (1978) 185–197.

97. RAY, T.B. and BLACK, C.C. The C_4 pathway and its regulation. In M. GIBBS and E. LATZKO (Eds.) *Encyclopedia of Plant Physiology*, New Series, Vol. 6. Springer-Verlag, New York (1979) pp. 77–101.

98. RHODES, P.R., KUNG, S.-D. and MARSHO, T.V. Relationship of ribulose-1,5-bisphosphate carboxylase–oxygenase specific activity to subunit composition. *Pl. Physiol.* **65** (1980) 69–73.

99. ROBISON, P.D., MARTIN, M.N. and TABITA, F.R. Differential effects of metal ions on *Rhodospirillum rubrum* ribulose-bisphosphate carboxylase/oxygenase and stoichiometric incorporation of HCO_3^- into a cobalt(III)–enzyme complex. *Biochemistry, NY* **18** (1979) 4453–4458.

100. SAYRE, R.T. and KENNEDY, R.A. Ecotypic differences in the C_3 and C_4 photosynthetic activity in *Mollugo verticillata*, a C_3–C_4 intermediate. *Planta* **134** (1977) 257–262.

101. SAYRE, R.T., KENNEDY, R.A. and PRINGNITZ, D.J. Photosynthetic enzyme activities and localization in *Mollugo verticillata* populations differing in the levels of C_3 and C_4 cycle operation. *Pl. Physiol.* **64** (1979) 293–299.

102. SCHNARRENBERGER, C. and FOCK, H. Interactions among organelles involves in photorespiration. In C.R. STOCKING and U. HEBER (Eds.) *Encylopedia of Plant Physiology*, New Series, Vol. 3. Springer-Verlag, New York (1976) pp. 185–234.

103. SERVAITES, J.C. and OGREN, W.L. Chemical inhibition of the glycolate pathway in soybean leaf cells. *Pl. Physiol.* **60** (1977) 461–466.

104. SERVAITES, J.C., SCHRADER, L.E. and EDWARDS, G.E. Glycolate synthesis in a C_3, C_4 and intermediate photosynthetic plant type. *Pl. Cell Physiol., Tokyo* **19** (1978) 1399–1405.

105. SETTER, T L., SCHRADER, L.E. and BINGHAM, E.T. Carbon dioxide exchange rates, transpiration, and leaf characters in genetically equivalent ploidy levels of alfalfa. *Crop. Sci.* **18** (1978) 327–332.

106. SOMERVILLE, C.R. and OGREN, W.L. A phosphoglycolate phosphatase-deficient mutant of *Arabidopsis. Nature, Lond.* **280** (1979) 833–836.

107. SOMERVILLE, C.R. and OGREN, W.L. Photorespiration mutants of *Arabidopsis thaliana* deficient in serine-glyoxylate amino-transferase activity. *Proc. natn. Acad. Sci. USA* **77**, No. 5 (1980) in press.

108. STEIGER, H.M., BECK, E. and BECK, R. Oxygen concentration in isolated chloroplasts during photosynthesis. *Pl. Physiol.* **60** (1977) 903–906.

109. TAKABE, T., OSMOND, C.B., SUMMONS, R.E. and AKAZAWA, T. Effect of oxygen on photosynthesis and biosynthesis of glycolate in photoheterotrophically grown cells of *Rhodospirillum rubrum. Pl. Cell Physiol., Tokyo* **20** (1979) 233–241.

110. TOLBERT, N.E. Glycolate metabolism by higher plants and algae. In M. GIBBS and E. LATZKO (Eds.) *Encyclopedia of Plant Physiology*, New Series, Vol. 6. Springer-Verlag, New York (1979) pp. 338–352.

111. WHITMAN, W.B., MARTIN, M.N. and TABITA, F.R. Activation and regula tion of ribulose bisphosphate carboxylase-oxygenase in the absence of small subunits. *J. biol. Chem.* **254** (1979) 10184–10189.

112. WILDNER, G.F. and HENKEL, J. The effect of divalent metal ions on the activity of Mg^{++} depleted ribulose-1,5-bisphosphate oxygenase. *Planta* **146** (1979) 223–228.

113. WILLIAMS, L.E. and KENNEDY, R.A. Relationship between early photosynthetic products, photorespiration, and stage of leaf development in *Zea mays. Z. PflPhysiol.* **81** (1977) 314–322.

114. YUN, S.-J., ISHII, R., HYEON, S.-B., SUZUKI, A., MURATA, Y. and TAMURA, S. Effects of some chemicals on photorespiration and photosynthesis in the excised rice leaves. *Agric. Biol. Chem., Tokyo* **43** (1979) 2207–2209.

115. ZELITCH, I. Photorespiration: studies with whole tissues. In M. GIBBS and E. LATZKO (Eds.) *Encyclopedia of Plant Physiology,* New Series, Vol. 6. Springer-Verlag, New York (1979) pp. 353–367.

116. ZELITCH, I. Photosynthesis and plant productivity. *Chemical & Engineering News* **57** (1979) 28–48.

117. ZELITCH, I. and DAY, P.R. Variation in photorespiration. The effect of genetic differences in photorespiration on net photosynthesis in tobacco. *Pl. Physiol.* **43** (1968) 1838–1844.

118. ZELITCH, I. and DAY, P.R. The effect on net photosynthesis of pedigree selection for low and high rates of photorespiration in tobacco. *Pl. Physiol.* **52** (1973) 33–37.

INDEX